Flow Cytometry Protocols

METHODS IN MOLECULAR BIOLOGY™

John M. Walker, Series Editor

91. **Flow Cytometry Protocols,** edited by *Mark J. Jaroszeski and Richard Heller, 1998*
90. **Drug–DNA Interaction Protocols,** edited by *Keith R. Fox, 1998*
89. **Retinoid Protocols,** edited by *Christopher P. F. Redfern, 1998*
88. **Protein Targeting Protocols,** edited by *Roger A. Clegg, 1998*
87. **Combinatorial Peptide Library Protocols,** edited by *Shmuel Cabilly, 1998*
86. **RNA Isolation and Characterization Protocols,** edited by *Ralph Rapley, 1998*
85. **Differential Display Methods and Protocols,** edited by *Peng Liang and Arthur B. Pardee, 1997*
84. **Transmembrane Signaling Protocols,** edited by *Dafna Bar-Sagi, 1998*
83. **Receptor Signal Transduction Protocols,** edited by *R. A. J. Challiss, 1997*
82. **Arabadopsis Protocols,** edited by *José M Martinez-Zapater and Julio Salinas, 1998*
81. **Plant Virology Protocols,** edited by *Gary D. Foster, 1998*
80. **Immunochemical Protocols (2nd. ed.),** edited by *John Pound, 1998*
79. **Polyamine Protocols,** edited by *David M. L. Morgan,* 1998
78. **Antibacterial Peptide Protocols,** edited by *William M. Shafer, 1997*
77. **Protein Synthesis:** *Methods and Protocols,* edited by *Robin Martin, 1998*
76. **Glycoanalysis Protocols,** edited by *Elizabeth F. Hounsel, 1998*
75. **Basic Cell Culture Protocols,** edited by *Jeffrey W. Pollard and John M. Walker, 1997*
74. **Ribozyme Protocols,** edited by *Philip C. Turner, 1997*
73. **Neuropeptide Protocols,** edited by *G. Brent Irvine and Carvell H. Williams, 1997*
72. **Neurotransmitter Methods,** edited by *Richard C. Rayne, 1997*
71. **PRINS and *In Situ* PCR Protocols,** edited by *John R. Gosden, 1997*
70. **Sequence Data Analysis Guidebook,** edited by *Simon R. Swindell, 1997*
69. **cDNA Library Protocols,** edited by *Ian G. Cowell and Caroline A. Austin, 1997*
68. **Gene Isolation and Mapping Protocols,** edited by *Jacqueline Boultwood, 1997*
67. **PCR Cloning Protocols:** *From Molecular Cloning to Genetic Engineering,* edited by *Bruce A. White, 1996*
66. **Epitope Mapping Protocols,** edited by *Glenn E. Morris, 1996*
65. **PCR Sequencing Protocols,** edited by *Ralph Rapley, 1996*
64. **Protein Sequencing Protocols,** edited by *Bryan J. Smith, 1996*
63. **Recombinant Proteins:** *Detection and Isolation Protocols,* edited by *Rocky S. Tuan, 1996*
62. **Recombinant Gene Expression Protocols,** edited by *Rocky S. Tuan, 1996*
61. **Protein and Peptide Analysis by Mass Spectrometry,** edited by *John R. Chapman, 1996*
60. **Protein NMR Protocols,** edited by *David G. Reid, 1996*
59. **Protein Purification Protocols,** edited by *Shawn Doonan, 1996*
58. **Basic DNA and RNA Protocols,** edited by *Adrian J. Harwood, 1996*
57. **In Vitro Mutagenesis Protocols,** edited by *Michael K. Trower, 1996*
56. **Crystallographic Methods and Protocols,** edited by *Christopher Jones, Barbara Mulloy, and Mark Sanderson, 1996*
55. **Plant Cell Electroporation and Electrofusion Protocols,** edited by *Jac A. Nickoloff, 1995*
54. **YAC Protocols,** edited by *David Markie, 1995*
53. **Yeast Protocols:** *Methods in Cell and Molecular Biology,* edited by *Ivor H. Evans, 1996*
52. **Capillary Electrophoresis:** *Principles, Instrumentation, and Applications,* edited by *Kevin D. Altria, 1996*
51. **Antibody Engineering Protocols,** edited by *Sudhir Paul, 1995*
50. **Species Diagnostics Protocols:** *PCR and Other Nucleic Acid Methods,* edited by *Justin P. Clapp, 1996*
49. **Plant Gene Transfer and Expression Protocols,** edited by *Heddwyn Jones, 1995*
48. **Animal Cell Electroporation and Electrofusion Protocols,** edited by *Jac A. Nickoloff, 1995*
47. **Electroporation Protocols for Microorganisms,** edited by *Jac A. Nickoloff, 1995*
46. **Diagnostic Bacteriology Protocols,** edited by *Jenny Howard and David M. Whitcombe, 1995*
45. **Monoclonal Antibody Protocols,** edited by *William C. Davis, 1995*
44. ***Agrobacterium* Protocols,** edited by *Kevan M. A. Gartland and Michael R. Davey, 1995*
43. **In Vitro Toxicity Testing Protocols,** edited by *Sheila O'Hare and Chris K. Atterwill, 1995*
42. **ELISA:** *Theory and Practice,* by *John R. Crowther, 1995*
41. **Signal Transduction Protocols,** edited by *David A. Kendall and Stephen J. Hill, 1995*
40. **Protein Stability and Folding:** *Theory and Practice,* edited by *Bret A. Shirley, 1995*
39. **Baculovirus Expression Protocols,** edited by *Christopher D. Richardson, 1995*
38. **Cryopreservation and Freeze-Drying Protocols,** edited by *John G. Day and Mark R. McLellan, 1995*
37. **In Vitro Transcription and Translation Protocols,** edited by *Martin J. Tymms, 1995*
36. **Peptide Analysis Protocols,** edited by *Ben M. Dunn and Michael W. Pennington, 1994*
35. **Peptide Synthesis Protocols,** edited by *Michael W. Pennington and Ben M. Dunn, 1994*
34. **Immunocytochemical Methods and Protocols,** edited by *Lorette C. Javois, 1994*
33. ***In Situ* Hybridization Protocols,** edited by *K. H. Andy Choo, 1994*
32. **Basic Protein and Peptide Protocols,** edited by *John M. Walker, 1994*

METHODS IN MOLECULAR BIOLOGY™

Flow Cytometry Protocols

Edited by

Mark J. Jaroszeski
Richard Heller
University of South Florida, Tampa, FL

Humana Press Totowa, New Jersey

999 Riverview Drive, Suite 208
Totowa, New Jersey 07512

This publication is printed on acid-free paper. ∞
ANSI Z39.48-1984 (American Standards Institute)
Permanence of Paper for Printed Library Materials.

Cover illustration:

Cover design by Jill Nogrady.

For additional copies, pricing for bulk purchases, and/or information about other Humana titles, contact Humana at the above address or at any of the following numbers: Tel.: 201-256-1699; Fax: 201-256-8341; E-mail: humana@mindspring.com; or visit our Website: http://humanapress.com

Printed in the United States of America. 10 9 8 7 6 5 4 3 2 1

Library of Congress Cataloging in Publication Data

Main entry under title:

Methods in molecular biology™.

Flow cytometry protocols/edited by Mark J. Jaroszeski, Richard Heller.
p. cm.—(Methods in molecular biology; 91)
Includes index.
ISBN 0-89603-354-6 (hardcover) (alk. paper); ISBN 0-89603-538-7 (combbound) (alk. paper)
1. Laboratory manuals. I. Jaroszeski, Mark J. II. Heller, Richard, 1954– . III. Series: Methods in molecular biology (Totowa, NJ); 91.
[DNLM: 1. Flow Cytometry—methods. W1 ME9616J v. 91 1998/QH 585.5.F56 F6447 1998]
QH585.5.F56F587 1998
611'.0181'0284—dc21
DNLM/DLC
for Library of Congress

97-37498
CIP

Preface

Flow cytometers are technically sophisticated instruments that utilize principles of physics, chemistry, engineering, digital electronics, optics, analog devices, laser technology, and biology to measure the light scattering and fluorescent characteristics of particles. These instruments have become invaluable tools in the area of cancer research, cell biology, and clinical diagnostics for analyzing a multitude of cellular and subcellular particles. Undoubtedly, flow cytometers have become widely used because they provide a very unique means of examining biological particles. Large quantities of cells/particles are individually analyzed, and it is typical for flow cytometers to quantitatively process thousands of individual particles in a matter of seconds. This a powerful analytic feat particularly if one relates it to the time required to examine several thousand individual cells using a microscope. This leaves little doubt regarding why the field of flow cytometry has advanced so rapidly since the first flow cytometers were commerically available in the 1970s. Today, flow cytometers are common in both large and small institutions.

Current flow cytometers have evolved with advances in computers and microelectronics into fast and reliable instruments. The science of flow cytometry has also benefitted from rapid growth in the fields of biochemistry and immunology over the past decade. A vast array of fluorochrome-conjugated antibodies and other reagents are now commercially available for probing structures and monitoring function in human cells as well as in many other species. The combination of excellent instrumentation and a vast array of probes has resulted in many new investigative protocols and has led to improvements in many existing methods.

Flow Cytometry Protocols has collected many improved and new applications that utilize this unique and powerful technology. The volume begins with a chapter that addresses the fundamentals of flow cytometry. Basic principles of instrumentation and data analysis are discussed in order to provide a foundation for beginning scientists who wish to use flow cytometry and the protocols presented throughout the text. The remaining twenty-one chapters present well-tested protocols. Five of these are dedicated to detection of terminal transferase ion leukemia, intracellular cytokines, intracellular/intranuclear antigens in leukemia and tumors, and cyclins. Six additional chapters describe protocols for performing functional studies. These include methods for assessing viability, transmembrane potential, oxidative metabolism, environmental particulate uptake, drug resistance, and cellular pharmacokinetics. Three protocols that describe methods for quantitating electroinserted proteins, electropermeabilization, and cell–cell electrofusion have also been included for those who utilize electricity to manipulate cells. Methods for examining DNA comprise a significant fraction of the uses for flow cytometry. This volume contains chapters on cell cycle analysis, DNA/RNA analysis of solid malignancies, solid-tumor DNA ploidy analysis, *in situ* hybridization, and apoptosis. The final two chapters present advanced protocols for chromosome sorting and 5/6 color analysis.

We would like to thank all of the contributors for their timely submission of high quality work. In addition, we would like to thank John Walker for his invaluable editorial advice and skill.

Mark J. Jaroszeski
Richard Heller

Contents

Contributors

ADORJAN ASZALOS • *Division of Applied Pharmacology Research, Office of Testing and Research, CDER, Laurel, MD*

JEANNE L. BECKER • *Department of Obstetrics and Gynecology, University of South Florida, Tampa, FL*

BRIAN E. CRUCIAN • *Laboratory Corporation of America, Tampa, FL*

ZBIGNIEW DARZYNKIEWICZ • *The Cancer Research Institute, New York Medical College, Elmsford, NY*

MARGARET S. DORDAL • *Abbott Laboratories, Abbott Park, IL*

ADEL K. EL-NAGGAR • *Department of Pathology, MD Anderson Cancer Center, University of Texas, Houston, TX*

KHALID EL OUAGARI • *IPBS CNRS, Toulouse Cedex, France*

RICHARD GILBERT • *Department of Chemical Engineering, College of Medicine, University of South Florida, Tampa, FL*

WOJCIECH GORCZYCA • *The Cancer Research Institute, New York Medical College, Elmsford, NY*

RICHARD HELLER • *Department of Surgery, College of Medicine, University of South Florida, Tampa, FL*

AMY IMRICH • *Physiology Program, Department of Environmental Health, Harvard School of Public Health, Boston, MA*

MARK J. JAROSZESKI • *Department of Surgery, College of Medicine, University of South Florida, Tampa, FL*

GLORIA JUAN • *The Cancer Research Institute, New York Medical College, Elmsford, NY*

LESTER KOBZIK • *Department of Environmental Health, Harvard School of Public Health Boston, MA*

MYRON R. MELAMED • *The Cancer Research Institute, New York Medical College, Elmsford, NY*

SIMON P. MONARD • *The Aaron Diamond AIDS Research Center for the City of New York, NY*

MICHAEL G. ORMEROD • *Reigate, UK*

ELISABETH PAIETTA • *Department of Oncology, Montefiore and Albert Einstein Cancer Center, Bronx, NY*

KENNETH J. PENNLINE • *Flow Cytometry Associates, Brentwood, TN*
GILBERT RADCLIFF • *Core Laboratory, Moffitt Research Center, H. Lee Moffitt Cancer Center and Research Institute, University of South Florida, Tampa, FL*
MARIE-PIERRE ROLS • *Department III: Glycoconjugues et Biomembranes, LPTF-CNRS, Toulouse Cedex, France*
KIM STECK • *Department of Pathology, MD Anderson Cancer Center, University of Texas, Houston, TX*
BRADLEY STRINGER • *Physiology Program, Department of Environmental Health, Harvard School of Public Health, Boston, MA*
MICHAEL K. TANNER • *Center for Vaccine Development, University of Maryland, Baltimore, MD*
JUSTIN TEISSIÉ • *IPBS CNRS, Toulouse Cedex, France*
JAMES L. WEAVER • *Division of Applied Pharmacology Research, Office of Testing and Research, CDER, Laurel, MD*
SAMUEL R. WELLHAUSEN • *Brown Cancer Center, Louisville, KY*
RAYMOND H. WIDEN • *Immunology Department, Tampa General Hospital, Tampa, FL*

1

Basics of Flow Cytometry

Gilbert Radcliff and Mark J. Jaroszeski

1. Introduction

Flow cytometry is a laser-based technology that is used to measure characteristics of biological particles. This technology is used to perform measurements on whole cells as well as prepared cellular constituents such as nuclei and organelles. Flow cytometers scan single particles or cells as they flow in a liquid medium past an excitation light source. The underlying principle of flow cytometry is that light is scattered and fluorescence is emitted as light from the excitation source strikes the moving particles. Light scattering and fluorescence is measured for each individual particle that passes the excitation source. Scattering and emission data can be used to examine a variety of biochemical, biophysical, and molecular aspects of particles. This unique and powerful technology is an important tool for many scientific disciplines because it allows characterization of cells or particles within a sample. Flow cytometry is particularly important for biological investigations because it allows qualitative and quantitative examination of whole cells and cellular constituents that have been labeled with a wide range of commercially available reagents, such as dyes and monoclonal antibodies.

Cells or particles are prepared as single-cell suspensions for flow cytometric analysis. This allows them to flow single file in a liquid stream past a laser beam. As the laser beam strikes the individual cells, two types of physical phenomena occur that yield information about the cells. First, light scattering occurs that is directly related to structural and morphological cell features. Second, fluorescence occurs if the cells are attached to a fluorescent probe. Fluorescent probes are typically monoclonal antibodies that have been conjugated to fluorochromes; they can also be fluorescent stains/reagents that are not conjugated to antibodies. Fluorescent probes are reacted with the cells or particles

From: *Methods in Molecular Biology, Vol. 91: Flow Cytometry Protocols*
Edited by: M. J. Jaroszeski and R. Heller © Humana Press Inc., Totowa, NJ

of interest before analysis; therefore, the amount of fluorescence emitted as a particle passes the light source is proportional to the amount of fluorescent probe bound to the cell or cellular constituent. The manner in which fluorescence is determined remains the same regardless of the probe. After acquisition of light scattering and fluorescence data for each particle, the resulting information can be analyzed utilizing a computer and specific software that are associated with the cytometer.

Flow cytometry has become a powerful tool for use in research as well as the clinical realm because cytometers have the capability to process thousands of individual particles in a matter of seconds. The unique advantage of flow cytometers relative to other detection instruments is that they provide a collection of individual measurements from large numbers of discrete particles rather than making a bulk measurement. This analysis strategy has made flow cytometry very popular and widely used. The applications of flow cytometry are diverse and include the interrogation of membrane, cytoplasmic, and nuclear antigens. Flow cytometry has been used to investigate whole cells and a number of cellular constituents, such as organelles, nuclei, DNA, RNA, chromosomes, cytokines, hormones, and protein content. Methods to perform a host of functional studies such as measurements of calcium flux, membrane potentials, cell proliferation rates, DNA synthesis, and DNA cell cycle analysis have also been developed for this technology. It appears that analysis of any cellular structure or function is possible using flow cytometry as long as an appropriate probe is available.

Flow cytometers function as particle analyzers in all of the applications mentioned above. There are two distinct types of flow cytometers that can be used to acquire data from particles. One type can perform acquisition of light scattering and fluorescence only. The other type is capable of acquiring scattering and fluorescence data but also has the powerful ability to sort particles. Both types function in a similar manner during acquisition. However, sorting instruments have the powerful ability to physically separate particles based on light scattering and/or fluorescent emission characteristics. Cytometers were originally designed to sort. The acronym FACS is often used as a synonym for flow cytometry and stands for fluorescent activated cell sorting. In recent years, particle analysis has been more widely used than sorting. Thus, cytometers that perform acquisition without sorting are the most common of the two types.

It should be noted that the theory and principles described hereafter are not intended to be manufacturer specific but can be applied to flow cytometers in general. Flow cytometry involves instrumentation that is complex and expensive. Usually large research facilities and hospitals have shared flow cytometers and trained personnel who are dedicated to operating them. Although these personnel perform sample acquisition or are available to assist in doing so, it is important that researchers and clinicians obtain basic knowledge of how flow cytometers work in order to intelligently design experiments and prepare samples. Researchers who wish to use flow cytometry, especially the beginner, also require a basic understanding of data interpretation. This basic flow cytometric knowledge is essential for performing experiments that will provide meaningful data. Understanding the basic principles of flow cytometry and data interpretation will facilitate the production of results that are not a consequence of inadvertently or unintentionally introduced artifacts.

This chapter should be viewed as a starting point for the individual unfamiliar with flow cytometry. The fundamental information presented in this chapter is intended to help beginning cytometer users, investigators, postdoctoral fellows, and technicians utilize flow cytometry in a manner that will yield high quality results. Instrument concepts will be stressed with an explanation of the theoretical basis behind them. Basic data presentation and interpretation methods that are used for analyzing flow cytometric data will also be detailed. In addition, this chapter will provide the beginner with a foundation that can be used to better understand and utilize the protocols presented throughout this volume.

2. History of Flow Cytometry

Throughout history, few other scientific techniques have involved the contributions of specialists from so many different backgrounds and disciplines as flow cytometry. A partial list of the various disciplines involved in the development of flow cytometry includes: biology, biotechnology, computer science, electrical engineering, laser technology, mathematics, medicine, molecular biology, organic chemistry, and physics. Flow cytometry experts are continually absorbing and combining knowledge from the aforementioned disciplines in an effort to advance the field.

The brief history of scientific developments listed below should enlighten the beginning user to what has transpired in the development of flow cytometry. Hopefully, a historical perspective will inspire an appreciation of the technology as it exists today:

1930 Caspersson and Thorell pioneered work in cytology automation.

1934 Moldaven attempted photoelectric counting of cells flowing through a capillary tube.

1940 Coons was credited with linking antibodies with fluorescent tags to mark specific cellular proteins.

1949 Coulter filed for a patent titled "Means for Counting Particles Suspended in a Fluid."

1950 Caspersson described microspectrophotometric measurement of cells in the UV and visible regions of the spectrum.

1950 Coons and Kaplan reported that fluorescein, conjugated as the isocyanate form, gave improved results over other dyes. Sometime thereafter, fluorescein became and has remained the fluorescent label of choice.

1967 Kamentsky and Melamed elaborated on Moldaven's method of forcing cells through a capillary tube and designed a sorting flow cell.

1969 Van Dilla, Fulwyler, and others at Los Alamos, NM (in what is now known as National Flow Cytometry Resource Labs) developed the first fluorescence detection cytometer that used the principles of hydrodynamic focusing, 90° optical configuration, and an argon ion laser excitation source.

1972 Herzenberg described an improved version of a cell sorter that could detect weak fluorescence of cells stained with fluorescein-labeled antibodies.

1975 Kohler and Milstein introduced monoclonal antibody technology which immediately provided the basis for highly specific immunological reagents for use in cell studies.

By the mid 1970s, the field of flow cytometry had matured to the point where commercial flow cytometers began to appear on the market. New focus was placed on fluorochrome development, methods of cell preparation, and enhanced electronic data handling capabilities. Scientists, commercial instrument manufacturers, and rapidly expanding biochemical industries perpetuated the development of flow cytometry throughout the 1980s and early 1990s.

3. Principles of Flow Cytometric Instrumentation

Flow cytometers can be described as four interrelated systems which are shown in **Fig. 1**. These four basic systems are common to all cytometers regardless of the instrument manufacturer and whether or not the cytometer is designed for analysis or sorting. The first is a fluidic system that transports particles from a sample through the instrument for analysis. The second is an illumination system that is used for particle interrogation. The third is an optical and electronics system for direction, collection, and translation, of scattered and fluorescent light signals that result when particles are illuminated. The fourth is a data storage and computer control system that interprets translated light and electrical signals and collates them into meaningful data for storage and subsequent analysis. Functional details of each system are described below.

3.1. Fluidic System

The fluidic system is the heart of a flow cytometer and is responsible for transporting cells or particles from a prepared sample through the instrument for data acquisition (**Fig. 1**). The primary component of this system is a flow chamber. The fluidic design of the instrument and the flow chamber determine how the light from the illumination source ultimately meets and interrogates

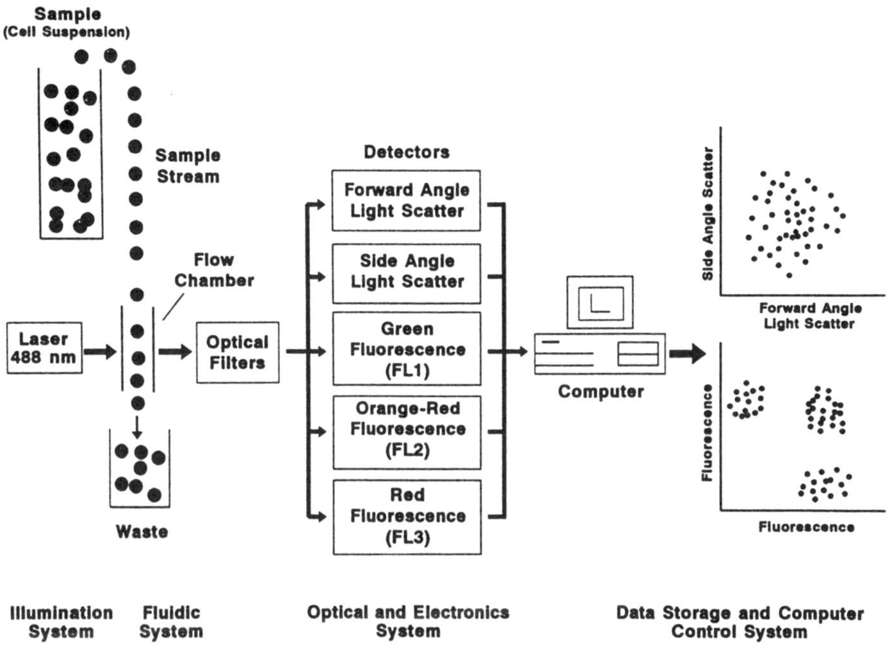

Fig. 1. A schematic of the primary components that comprise a flow cytometer. Dark arrows indicate the flow of particles and information. A fluidic system transports particles or cells from a prepared suspension past a focused laser beam that is generated by an illumination system. Particle interrogation takes place, one cell at a time, in a flow chamber. The resulting scattered light and fluorescence is gathered by an optical and electronics system that translates the light signals into information that is saved by the data storage and computer control system. After data from a sample has been stored, retrospective graphical data analysis can be performed with the aid of software.

particles. Typically, a diluent, such as phosphate-buffered saline, is directed by air pressure into the flow chamber. This fluid is referred to as sheath fluid and passes through the flow chamber after which it is intersected by the illumination source. The sample under analysis, in the form of a single particle suspension (*see* **Notes 1** and **2**), is directed into the sheath fluid stream prior to sample interrogation. The sample then travels by laminar flow through the chamber. The pressure of the sheath fluid against the suspended particles aligns the particles in a single-file fashion. This process is called hydrodynamic focusing and allows each cell to be interrogated by the illumination source individually while travelling within the sheath fluid stream.

Both types of cytometers, sorting and nonsorting, have fluidic systems that operate based on the same engineering principles. However, sorting instruments do not typically have flow chambers for interrogation. Instruments that have sorting capability are engineered in a manner that produces a hydrodynamically focused cell stream that passes through a nozzle. Intersection of the sample stream and laser occurs in air near the position where the stream exits the nozzle.

One problem that sometimes arises in fluidic systems during sample interrogation is called coincidence. All flow cytometry users should be aware of this potential problem that can occur in nonsorting systems that use flow chambers as well as in sorting instruments that use nozzles. A coincidence can occur under two types of conditions. If the distance between particles in a flow chamber is too small during interrogation because of high particle concentration (*see* **Note 3**), then the cytometer will be unable to resolve particles as individuals. A coincidence can also occur if two or more nonadherent particles exit a flow nozzle in such a manner that they are resolved as a single event in time. Irrespective of the cause, coincidence is a problem that defeats the one cell at a time analysis scheme of flow cytometry. Reducing the rate at which the sample passes through the cytometer is one means of avoiding coincidence (*see* **Note 4**).

3.2. Illumination System

Flow cytometers use laser beams that intercept a cell or particle that has been hydrodynamically focused by the fluidic system (**Fig. 1**). Light from the illumination source passes through a focusing apparatus before it intercepts the sample stream. This apparatus is a lens assembly that focuses the laser emission into a beam with an elliptical cross-section that ensures a constant amount of particle illumination despite any minor positional variations of particles within the sample stream. Light and fluorescence are generated when the focused laser beam strikes a particle within the sample stream. These light signals are then quantitated by the optical and electronics system to yield data that is interpretable by the user.

Lasers are the light sources of choice currently used in flow cytometric systems. Most flow cytometers utilize a single laser; however, some systems support the simultaneous use of two or more different lasers. The most commonly used laser is an argon ion laser that has been configured to emit light in the visible range of the spectrum. A 488-nm laser emission is used for most standard applications. The majority of fluorochromes that are available on the market today can be excited using this wavelength.

The reason lasers are used as the excitation source of choice in flow cytometers is attributed to coherence. A laser-generated beam diverges very little in terms of direction. Thus, laser beams remain compact and bright. In addition to directional coherence, laser-generated beams maintain very high spectral purity. Thus, lasers are excellent excitation sources because they provide a single wavelength beam that is also stable, bright, and narrow.

As previously stated, the majority of fluorochromes on the market today are capable of being excited by a wavelength of 488 nm. However, some experimental situations require use of a fluorochrome with an excitation wavelength other than 488 nm. For example, some fluorochromes are excited with UV light or by other wavelengths. Some types of lasers present in flow cytometers can be tuned to UV or other wavelengths. If the existing laser is not tunable, then another laser source that emits the desired wavelength is required. The principles of flow cytometry remain the same regardless of the illumination wavelength.

3.3. Optical and Electronics System

Light is scattered and emitted in all directions (360°) after the laser beam strikes an individual cell or particle that has been hydrodynamically focused. The optical and electronics system of a typical flow cytometer is responsible for collecting and quantitating at least five types of parameters from this scattered light and emitted fluorescence. Two of these parameters are light-scattering properties. Light that is scattered in the forward direction (in the same direction as the laser beam) is analyzed as one parameter, and light scattered at 90° relative to the incident beam is collected as a second parameter. This type of scheme for collecting forward and side-scattered light is referred to as optical orthogonal geometry. Most current cytometers in use today allow examination of three different types of fluorescent emission. These are acquired as the remaining three parameters that brings the total number collectable parameters to five (**Fig. 1**).

Forward-scattered light is a result of diffraction. Diffracted light provides basic morphological information such as relative cell size that is referred to as forward angle light scatter (FSC). Light that is scattered at 90° to the incident beam is the result of refracted and reflected light. This type of light scatter is referred to as side-angle light scatter (SSC). This parameter is an indicator of granularity within the cytoplasm of cells as well as surface/membrane irregularities or topographies.

Scattered light yields valuable information about the sample under examination. Correlating the measurements of FSC and SSC light signals allows for the discrimination of various cellular subpopulations in a heterogeneous sample and also allows identification of viable, less viable (i.e., cells tending toward death or apoptotic cells), and necrotic cells. FSC and SSC correlation also allows discrimination of cellular debris. Combined use of FSC and SSC signals improves the resolution of dissimilar populations within the same sample based on size, granularity, and cell surface topography. In addition, scattered light emission is typically monitored by the user in real time to assess instrument performance during acquisition. This is achieved by observation of computer graphics and/or oscilloscope screens. Real time monitoring is very important during sample acquisition because changes in light scattering patterns during acquisition allows observation of changes in cellular morphology. This yields important information regarding changes in cellular condition and can also give the cytometer user information regarding the fluidic condition of the instrument.

During cytometer operation, light scattered in the forward direction is first gathered by a collection lens and then directed to a photodiode. This lens collects light at approx 0.5–10° angles relative to the incident beam. The photodiode translates FSC light into electronic pulses that are proportional to the amount of forward light scattered by the cell or particle. Larger particles scatter more light in the forward direction than smaller particles. The electronic pulses for each particle in a sample are then amplified and converted to digital form for storage in a computer. Online or subsequent data analysis can be used to obtain a graphical display of the individual FSC measurements as well as mean and distributional FSC statistics from all or part of the analyzed sample.

SSC information is handled in a manner similar to FSC. A collection lens located at 90° to the intersection of the sample stream and laser collects the SSC signal. A fraction of this light signal is directed to a highly sensitive detector. This type of photodetector is called a photomultiplier tube (PMT). This form of highly sensitive detector is required because directed side-scatter accounts for approx 10% of the emitted light signal and is, therefore, not as bright as FSC light. PMTs detect and amplify weak signals. The amount of amplification can be adjusted by the operator in order to make the PMT more or less sensitive to the directed SSC light. Side-scatter light is ultimately converted to a voltage signal that is digitized and stored in a computer to yield SSC parameter information for each analyzed cell or particle. This information can be displayed and further analyzed in a manner identical to FSC data.

Light-scattering information, FSC and SSC, allows identification of various cell types based on their size and granularity/topography. Fluorescence results when fluorochrome-labeled particles or cells are illuminated by the laser beam and emit light with a specific spectral composition. This yields biochemical, biophysical, and molecular information about the cellular constituent to which the probe is attached. Use of fluorescence adds tremendous analytic dimension to the information that can be obtained from flow cytometric analysis because there are a vast number of probes that are commercially available for detecting surface and internal molecules in cells.

Most current laboratory bench-top flow cytometers are capable of detecting fluorescence from three different regions of the visible spectrum. Cytometers are optically configured to detect a narrow range of wavelengths in each region. This allows the use of up to three different fluorochromes in a single sample (*see* **Note 5**). Fluorescent emission is detected simultaneously along with FSC and SSC data; therefore, up to five parameters can be simultaneously measured for each analyzed sample. Correlation of any number of these fluorescent and light-scattering parameters is normally possible. This meets the analysis needs of most experimental applications.

Fluorescence is detected using networks of mirrors, optics, and beam splitters that direct the emitted fluorescent light toward highly specific optical filters. The filters collect light within the range of wavelengths associated with each of the three fluorescent channels. Filtered light is directed toward PMTs for conversion into electrical signals. The signals are then digitized, which results in a fluorescent intensity for each analyzed cell or particle.

Each of the three fluorescent channels is designed to detect a narrow range of wavelengths. Fluorescence generated from the green fluorochrome fluorescein isothiocyanate (FITC) is typically detected in a band of wavelengths that is designated as the FL1 parameter. Fluorescein isothiocyanate is the most commonly used fluorochrome in the field of flow cytometry. Similarly, orange-red light generated from the fluorochromes R-phycoerythrin (PE) and propidium iodide (PI) is typically detected in another range of wavelengths that is designated as the FL2 parameter. Red fluorescence is detected in a third wavelength range designated as FL3. Fluorochromes that emit in the FL3 channel are proprietary, and the names of these compounds differ depending on their manufacturer. Some examples of fluorochromes that can be detected in the FL3 channel are CyChrome (Pharmingen, La Jolla, CA); ECD (Coulter, Miami, FL); PerCP (Becton Dickinson, San Jose, CA); Quantum Red and Red-670 (Sigma, St. Louis, MO); and Tri-Color (Caltag, San Francisco, CA).

A simple form of flow cytometric analysis utilizes a single fluorochrome conjugated to an antibody to ascertain the absence or presence of an antigen. For this single color case, fluorescent cells are detected in one channel that corresponds to the primary wavelength emitted by the fluorochrome. A much more complex situation arises when analyzing cells that are labeled with two or more different fluorochromes (*see* **Note 6**). This added complexity is caused by overlap in the emission spectra of fluorochromes that are commonly used for flow cytometry. Fluorochromes do not emit a single wavelength of light. Usually, a particular fluorochrome will emit a spectrum of light that is strongest within a narrow band width that corresponds to the detection range of one fluorescent channel. However, fluorochromes also emit to a lesser degree in spectral regions outside of the wavelength range used for detection. If this weaker emission is within the range detected for another fluorescent channel, then cells labeled with the single fluorochrome will be detected in two channels.

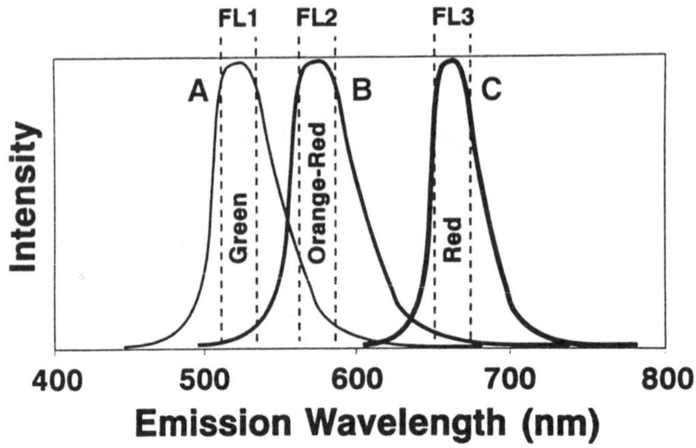

Fig. 2. Emission spectra from three hypothetical fluorochromes (A, B, and C) that illustrate spectral overlap. Vertical dashed lines indicate the range of wavelengths detected for each fluorescent channel (FL1, FL2, FL3). The fluorochromes that are used for flow cytometry have peak emissions that are centered within the wavelength range detected by one channel. The overlapping nature of emission spectra can result in detection of a single fluorochrome in two different channels.

A strong intensity will be detected in the proper channel, and a weak intensity will be detected in an inappropriate channel. **Figure 2** depicts this scenario. Spectral overlap is a problem when performing multicolor analysis because a cell that is labeled with a single fluorochrome may be detected by the optics of the cytometer as having fluorescence in two different channels.

The problems encountered when the emission spectra of two fluorochromes overlap can lead to false-positive results. For example, the emission from PE-labeled cells is normally detected as intense fluorescence in the orange-red (FL2) channel. Cells with a PE label may also be detected in the green (FL1) channel. Fluorescence in the green channel is typically reduced relative to the fluorescence in the proper orange-red channel. However, weak emission of PE-labeled cells within the wavelength range of the green channel can be detected by the cytometer. This fluorescence could be erroneously interpreted by the user as emission from a green fluorescing probe that was also present on the PE-labeled cells. The opposite case is also true. FITC is strongly detected in the green channel, but cells labeled with a FITC-conjugated antibody will typically fluoresce in the orange-red channel because of spectral overlap. Again, this can lead to false-positive results because the emission of FITC-labeled cells in the wavelength range detected as orange-red fluorescence could be misinterpreted.

Flow cytometers can be adjusted to electronically compensate for the complications that are associated with spectral overlap. Compensation subtracts

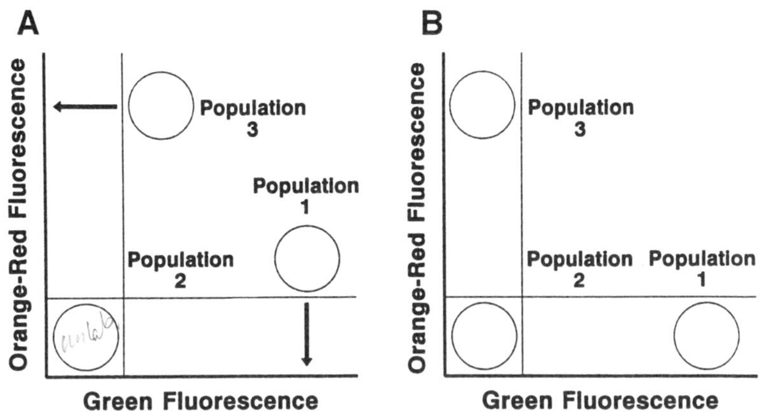

Fig. 3. Two-parameter fluorescent plots illustrating the effects of compensating for spectral overlap. Circles represent the position of analyzed cell populations. **(A)** An uncompensated situation shows Population 1 with a strong green fluorescence indicating, for example, positive labeling with FITC. Note that Population 1 also has a weaker orange-red fluorescence that is caused by overlap of the FITC emission spectrum into the wavelength range detected as orange-red by the cytometer. This weak fluorescence is greater than the fluorescence of unlabeled cells (background) shown as Population 2. Population 3 has a strong orange-red fluorescence indicating positive labeling for a PE. Spectral overlap can cause this population to have a green fluorescence that is weaker, but still above that of unlabeled cells. **(B)** Compensation circuitry within flow cytometers allows the user to overcome the problem of spectral overlap by electronically adjusting the instrument. Proper adjustment forces FITC and PE-positive populations to maintain high fluorescent magnitudes that correspond to the respective fluorochromes while decreading fluorescence caused by spectral overlap to that of unlabeled cells. Compensation adjustments are specific to fluorochromes used and can vary from experiment to experiment.

the overlapping signals from detection in an inappropriate fluorescent channel. The effects of proper compensation on the fluorescent intensities of analyzed cell populations are shown in **Fig. 3**. It is important to choose fluorochromes that have minimal spectral overlap when designing experiments. This will reduce the amount of compensation that is required.

3.4. Data Storage and Computer Control System

After light scattering and fluorescence is converted to electrical signals by the optical and electronics system, the information is converted into digital data that the computer can interpret (**Fig. 1**). The signals generated from cells or particles are referred to as events and are stored by the computer. Flow cytometry data files are known as list-mode files. A list-mode file contains unprocessed data of all the measured parameters along with coordinates for each event from the acquired sample. This type of file is stored on disk or other types of media during sample acquisition. The number of events acquired for each sample is always determined before analysis and is usually set using software designed to control cytometer operation. A conventional acquisition value is 10,000 events per sample. However, this value may vary and range upward of 100,000 events per sample depending on the experimental objective. For example, a large number of events might be acquired in a case in which rare subpopulations of cells are being sought for analysis (*see* **Note 7**).

In flow cytometry there are many situations in which one wishes to repeatedly view or print out variations of a data file. By acquiring list-mode data, retrospective data analysis can be performed. Therefore, saving list-mode files has become the method of choice for flow cytometric data collection. This mode of data storage is useful because no cytometric information with respect to the sample has been lost. Thus list-mode storage provides the most comprehensive information possible and should always be utilized when performing sample acquisition.

The computer is a very important part of flow cytometers because it is used to control most functions of the instrument. In order to obtain meaningful experimental information, it is imperative that the flow cytometer be appropriately configured prior to acquisition. Acquiring data is relatively easy. The difficult part is learning to configure the instrument correctly. It is highly probable that an inadequately trained user can obtain meaningless data without realizing it. For example, if light-scatter sensitivities are inappropriately set, specific cells or particles of interest could appear off scale and the information obtained would be noninformative. The beginning user should obtain adequate training from an expert or experienced user in the field (*see* **Note 8**). All flow cytometers analyze particles using the same principles; however, operation is manufacturer specific. Manufacturers offer educational courses specifically designed for the operation and applications of their respective instruments.

Although many of the specifics of operating the flow cytometer through the computer will be handled by a dedicated or experienced operator, the beginning user must be aware of several types of control samples that are critical. These controls allow proper adjustment of the flow cytometer so that experimental samples can be appropriately acquired. Data from these control samples serve as reference points for the information acquired from experimental samples. There are three basic types of control samples. Negative-control samples are used to adjust instrument parameters so that all data appears on scale. Positive controls are used to ensure that the antibodies used are capable of recognizing the antigen of interest. Compensation controls are employed when performing multifluorochrome analysis to adjust for spectral overlap.

Negative-control samples are used for two different purposes; most situations that use fluorochrome-labeled antibodies require two types of negative-control samples. The first type is simply a sample of cells that has not been reacted with a fluorochrome-labeled antibody. This sample is almost always acquired as the first sample in a set because it serves as a baseline reference point. FSC and SSC are usually adjusted so that the cells of interest appear on scale. In addition, the sensitivities of fluorescent channel PMTs are typically set so that these negative-control cells appear with intensities that are near zero but still on scale. In this regard, the nonfluorescing cells establish a reference point that can be used when describing the intensity of fluorochrome-labeled cells in subsequent experimental samples. This sample also allows the user to assess the natural or autofluorescence of the cells, and it gives the flow cytometer operator a valuable reference point that establishes that positively labeled cells from experimental samples will have higher intensities.

The second type of negative control is designed to investigate whether or not the cells of interest will nonspecifically bind the fluorochrome-labeled antibody. This type of sample is called an isotype control. Two types of labeling scenarios are commonly used. The first utilizes a single fluorochrome-conjugated antibody to identify an antigen. The correct isotype control is an antibody with exactly the same properties as the antibody used for experimental samples; however, the isotype control antibody has irrelevant specificity. Manufacturers list the appropriate isotype control antibody for each investigational antibody. The second labeling scenario uses an unconjugated primary antibody followed by a labeled secondary antibody. An appropriate isotype control would be prepared by simply adding the secondary antibody to the cells in the absence of the primary antibody. Fluorescent analysis of this second type of negative control sample allows the user to establish a nonspecific fluorescence intensity reference point that can be subtracted from the fluorescent values of experimental samples. This reference point can also be used to delineate a threshold fluorescence for judging positive/negative expression of the antigen of interest.

Positive controls are essential for establishing that the antibody used is capable of identifying the antigen of interest. This type of sample is typically prepared with a cell type that can be positively identified with the antibody. Cell lines that express the antigen of interest at high levels are good sources for positive-control cells. In addition, they also give the user and operator an approximation of the intensity that positive-expressing experimental cells will have.

Spectral overlap can lead to false-positive results, as discussed above, in samples that utilize multiple fluorochromes. Therefore, it is critical to prepare the proper control samples in order to facilitate compensation for this overlap. Control samples are processed along with the multifluorochrome-labeled experimental sample set. An identical preparation procedure is used except that only a single label is applied. Therefore, one control sample is required for each different fluorochrome. Compensation controls are analyzed before any experimental samples are acquired. Compensation adjustments are made, by computer control, on the flow cytometer while the cells in these control samples are under analysis so that subsequent samples will be correctly compensated for spectral overlap.

Fluorescent intensities of experimental samples are all relative to control samples. Therefore, it is important to prepare negative, positive, and compensation control samples. There can be considerable variation in the data obtained from day to day, when different instruments are used for analysis, or when different operators analyze samples. This can be true even when running the same type of samples. Consequently, it is critical that the correct control samples are prepared and analyzed with each sample set. This will ensure that the cytometer can be properly adjusted for easy acquisition of data from the experimental samples (*see* **Note 9**). Failure to prepare the correct control samples is a common mistake made by many beginning flow cytometer users. Often times, this mistake results in data that cannot be properly interpreted that ultimately translates to wasted time, energy, and reagents.

4. Data Analysis

Data analysis is a very critical part of any experiment that utilizes flow cytometry. The beginning user will probably have assistance from a dedicated flow cytometer operator when acquiring data; however, analysis of the acquired data is usually very specific to the experimental objectives (*see* **Note 7**). Therefore, the user is much more aware of what data will be required to achieve the experimental outcome. In order to conduct data analysis, the user must have a good working knowledge of what data analysis options are available, how to display data, and how to interpret data (*see* **Note 8**).

List-mode data is analyzed using a computer and software. The software is usually specific to flow cytometric data and is often part of the same computer system that is used to control the instrument during acquisition. Third-party companies also offer software for data analysis. These programs provide many ways to examine data; however, there are some very useful standard ways of presenting data that are common to all types of software. These are described below.

The most common display is a histogram. A typical histogram data plot is shown in **Fig. 4**. This type of plot is probably the easiest to interpret and understand because information from a single parameter is displayed. Histograms can be depicted using any parameter as long as the cytometer was configured to save the proper list-mode data for that particular parameter during acquisition. The figure is arranged with FSC on the X-axis and the relative number of

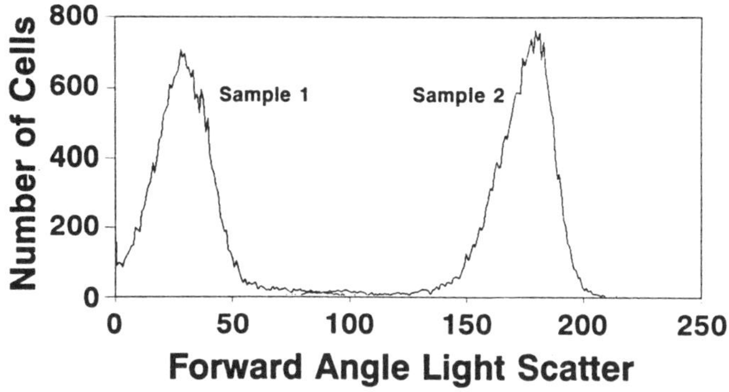

Fig. 4. A typical one-parameter histogram that shows data from two different samples that have been overlayed for comparison. The histogram illustrates that the cells from Sample 2 have a much higher forward angle light scatter than the cells from Sample 1.

cells are displayed on the Y-axis. The plot shows data from two different samples, 1 and 2, which have been overlayed for comparison.

Histograms are excellent tools for data analysis because they allow the user to visually see the distribution of a single measured parameter for the acquired events. A histogram format is commonly used to display results from samples that were treated using a variety or panel of antibodies conjugated to the same fluorochrome (*see* **Note 8**). It is then possible to compare these different samples by making individual histograms or by overlaying multiple samples on the same one parameter plot. Overlayed plots are an excellent means of qualitatively comparing fluorescence (or any other acquired parameter). Quantitative data can be obtained by graphically setting statistical markers based on control sample results. Mean and peak values on any type of histogram can be computed based on these markers. Percentages of positive-expressing events with parameter values above a threshold can also be determined by setting markers as an alternative format for interpretation.

It is also possible to display two parameters simultaneously such as FSC vs SSC or FL1 vs FL2. Any combination of acquired parameters can be used to depict a two-parameter data plot. For two-parameter plots, data from a population of individual particles can be displayed in the form of dots or as contours. Dot plots display data from each particle as a dot within both coordinate axes; one dot represents one acquired event. The positions of the dots reflect the relative intensities of the two measured parameters for that event. Contour density plots display the data from a population of cells as a series of concentric lines that correlate to different cell or particle densities within the axes. Contour plots are similar to topographical maps. The power of these two various types of data displays is that they allow an investigator to visualize two measured parameters on a single plot. Dot plots are probably the most common type of two-parameter plots, and they are also the easiest to understand. Contour displays require more experience to interpret.

Figure 5 shows three examples of two dimensional dot plots. All plots were derived from the same sample of cells that was treated with two different fluorescent probes. One probe utilized FITC (FL1) and the other contained PE (FL2). The plots illustrate a useful means of combining light scattering and fluorescence data for analysis.

Figure 5A is a two-dimensional dot plot of FSC vs SSC. The bulk of the cells appear as the most dense population of dots; each dot represents one acquired event. A gate, or region, has been drawn around the dense cell population of interest on the plot. Gates are a feature of analysis software that allow for definition of boundaries around populations of interest. Gating is a powerful analytic tool that is available on any type of two-dimensional plot. It is typically done by graphically drawing the region after a raw data plot has been constructed. Regions are most often drawn to isolate subsets of cells, as in the figure, for further analysis. Also, gating is used to exclude small cellular debris and/or large aggregates from subsequent analysis.

Figure 5B,C are both two-dimensional dot plots that were derived from the FSC vs SSC plot shown in **Fig. 5A**. Both fluorescent plots contain three distinct populations. **Figure 5B** shows the fluorescence of all events from the FSC vs SSC plot. **Figure 5C** is different in that it shows only those events within the gate drawn on the FSC vs SSC plot. Populations in the fluorescent plot that was made from gated cells (**Fig. 5C**) are much more resolved than those in the plot from the ungated sample (**Fig. 5B**). Increased resolution was the result of identifying the population of interest, gating, and then further analyzing those cells of interest. This type of procedure is a very common and extremely useful means for examining the characteristics of a population of interest.

The fluorescent plots in **Fig. 5** show three distinct cellular populations. These are a green population that is positive for FITC (FL1), an orange-red population that is positive for PE (FL2), and a third population that is positive for both FITC and PE (FL1 and FL2). Although fluorescence data could have been displayed and analyzed using separate single parameter histograms for FL1 and FL2 fluorescence, the two-parameter dot plot revealed much more information. The bivariate plot allowed identification of a dual fluorescing population and two mutually exclusive and distinct single-fluorescing populations. This information became evident on a two-parameter fluorescent dot plot that was obtained from a single-gated population on an FSC vs SSC plot.

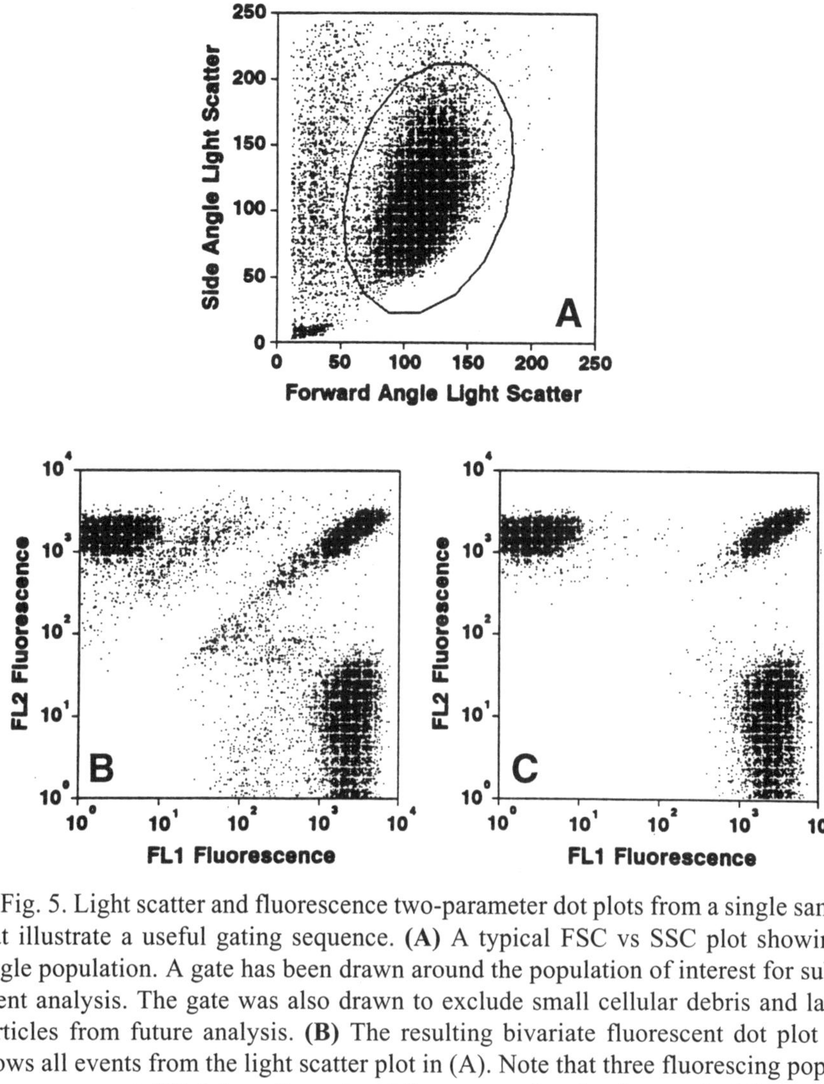

Fig. 5. Light scatter and fluorescence two-parameter dot plots from a single sample that illustrate a useful gating sequence. **(A)** A typical FSC vs SSC plot showing a single population. A gate has been drawn around the population of interest for subsequent analysis. The gate was also drawn to exclude small cellular debris and larger particles from future analysis. **(B)** The resulting bivariate fluorescent dot plot that shows all events from the light scatter plot in (A). Note that three fluorescing populations are present. **(C)** A two-dimensional fluorescent dot plot that resulted from showing only those cells that were within the gated region of the light scattering plot in (A). The three populations are more resolved as a result of gating.

Dot plots displaying both types of light scatter can provide important morphological characteristics such as cell size and granularity. They can also be used to identify viable cells and debris. This information is very useful for identifying a population of interest for subsequent analysis. Light-scattering properties (FSC or SSC), when combined with fluorescence data can also be an extremely valuable tool while undertaking analysis. These types of plots can assist the user in determining which acquired events elevate background fluorescence because of nonspecific binding of fluorochrome-labeled antibodies. Increased background fluorescence can also be because of a host of other reasons, such as entrainment of labeled antibody or probe in dead or dying cells as well as in cellular debris. This additional information assists identifying the population of interest so that events that contribute to elevated background fluorescence can be removed from further analysis by gating.

Figure 6 is a collection of data plots that illustrate how events that elevate background fluorescence can be identified and excluded from subsequent analysis. This is a common situation that arises during the analysis of cell samples that have been treated with fluorochrome-conjugated antibodies to ascertain the presence or absence of antigens. In these types of experiments, it is essential to first analyze an isotype control sample. Isotype control samples are used expressly for identifying the background fluorescence of cells/particles that is caused by nonspecific binding. This information serves as a reference point for comparing subsequent experimental samples. All six plots in the figure were derived from the same isotype control sample.

Figure 6A shows an ungated FSC vs SSC dot plot. An FL1 histogram, **Fig. 6B**, illustrates fluorescence that resulted from antibody treatment. The histogram has a common profile that has dual peaks. The first and largest peak corresponds to the majority of the cells in the sample. The second peak with increased fluorescence is most likely the result of nonspecific binding.

A useful method for determining the origin of secondary peaks in this type of control sample is to examine two types of plots. These are FSC vs FL1 and SSC vs FL1, which are given as **Figs. 6C** and **6D**, respectively. The FSC vs FL1 plot reveals a small population that has high fluorescence with lower FSC magnitude relative to the major population on the plot. The plot of SSC vs FL1 shows a population with higher SSC and increased fluorescence relative to the main population. Information from these two types of plots can be combined to identify those events that exhibit increased fluorescence caused by nonspecific binding. The plots show that events with low FSC and high SSC, relative to the major population, have increased fluorescence. This information can be used to draw a gate that excludes these types of unwanted events from the original FSC vs SSC plot, **Fig. 6E**. Gating results in a histogram of the control sample that does not have the artifactual secondary population as shown in **Fig. 6F**.

It is very important that the gate drawn from the isotype control sample, **Fig. 6E**, is used for analysis of all subsequent samples that will be related back to this control sample. It is also very important that gates are not applied to the population of interest using either of the light scatter vs fluorescence plots. Inadvertently drawing gates on these plots would only allow display of cells with fluorescence levels equivalent to this negative control. This would exclude any cells in subsequent samples that had fluorescence above the negative control. This would also completely exclude cells in experimental samples that exhibit fluorescence above the negative control. One should not hesitate to experiment with various combinations of light scatter and fluorescence plots in order to obtain the most highly resolved negative control population.

5. Summary

In summary, a beginner requires fundamental knowledge about flow cytometric instrumentation in order to effectively use this technology. It is important to remember that flow cytometers are very complex instruments that are composed of four closely related systems. The fluidic system transports particles from a suspension through the cytometer for interrogation by an illumination system. The resulting light scattering and fluorescence is collected, filtered, and converted into electrical signals by the optical and electronics system. The data storage and computer control system saves acquired data and is also the user interface for controlling most instrument functions. These four systems provide a very unique and powerful analytical tool for researchers and clinicians. This is because they analyze the properties of individual particles, and thousands of particles can be analyzed in a matter of seconds. Thus, data for a flow cytometric sample are a collection of many measurements instead of a single bulk measurement.

Basic knowledge of instrumentation is a tremendous aid to designing experiments that can be successfully analyzed using flow cytometry. For example, it is important to know the emission wavelength of the laser in the instrument that will be used for analysis. This wavelength is critical knowledge for selecting probes. It is also important to understand that a different range of wavelengths is detected for each fluorescent channel. This will aid selection of probes that are compatible with the flow cytometer. Understanding the complication that emission spectra overlap contributes to detection can be used to guide fluorochrome selections for multicolor analysis. All of these experiment design considerations that rely on knowledge of how flow cytometers work are a very practical and effective means of avoiding wasted time, energy, and costly reagents.

Data analysis is a paramount issue in flow cytometry. Analysis includes interpreting as well as presenting data that has been stored in list-mode files. Data analysis is very graphically oriented. There are a number of types of graphic representation that are available to visually aid data analysis. Two standard types of displays are used. These data plots are one-parameter histograms and bivariate plots. A user must be familiar with these two fundamental types of display in order to effectively analyze data.

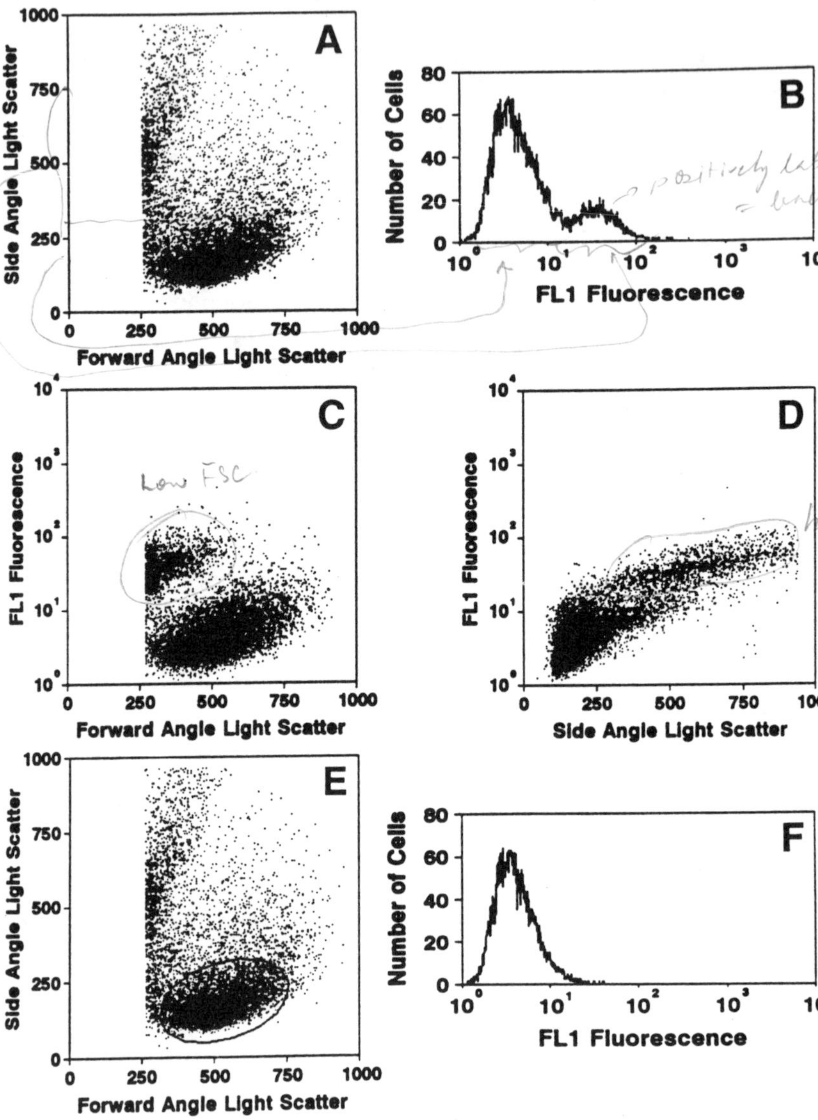

Fig. 6. An example of how gating can be utilized to determine the source of nonspecific binding and background fluorescence. All plots were generated from the same isotype control sample for a single antibody-conjugated fluorochrome. **(A)** shows the FSC vs SSC data. Examination of a green fluorescence (FL1) histogram **(B)** shows a common pattern that results isotype controls samples. Note that the histogram has two peaks. The smallest peak has increased fluorescence. This peak represents cells that were positively labeled using the antibody. In an ideal situation, no cells in this type of

Histograms are the most simple modes of data representation. Histograms allow visualization of a single acquired parameter. Mean fluorescence and distributional statistics can be obtained based on markers that the user can graphically set on the plot. Percentages of positively expressing particles relative to a control sample can also obtained in a similar manner. In addition, multiple histograms can be overlayed on one another to depict qualitative and quantitative differences in two or more samples.

Two-parameter data plots are somewhat more complicated than histograms; however, they can yield more information. Two-parameter dot plots of FSC vs SSC allow visualization of both light-scattering parameters that are important for identifying populations of interest. Bivariate fluorescent plots allow discrimination of dual-labeled populations that might remain hidden if histograms were used to display fluorescent data. Two-parameter plots that combine one light-scattering parameter and a fluorescent parameter are useful for analyzing control samples to elucidate the origin of nonspecific binding.

Data analysis is very graphically oriented. Experience and pattern recognition become important when using two-parameter data plots for qualitative as well as quantitative analysis. The technique of gating or drawing regions on dual parameter light-scatter plots allows one to exclude information and examine the population of interest by disallowing particles that might confound or interfere with analysis. This is one of the fundamental uses for gating. In addition, more elaborate gating scenarios can also be used eliminate particles that are the result of nonspecific binding.

6. Notes

1. Cells or particles are typically prepared as a suspension in a buffered saline solution. However, cells suspended in a liquid growth media can be used if appropriate precautionary measures are used between experimental acquisitions. Since most growth media is supported by some form of protein, buildup in the sample lines can lead to artifactual "carry over" effects. For example, running alcohol to clean sample lines after such an experiment will fix proteins in the sample lines and can lead to undesired effects and artifacts that will appear the next time that the flow cytometer is used. Drawing a 10–30% bleach solution through the fluidic system followed by sterile deionized water appears to be the best measure of protection to avoid carry-over effects while maintaining a clean fluidic system.

---

sample should be positive. Therefore, the cells within the secondary peak represent background fluorescence or cells that have nonspecifically bound to the antibody. Examination of light scatter vs FL1 fluorescence in **(C,D)** reveal that cells with increased fluorescence have low FSC and high SSC. This information can be used to draw a gate using FSC vs SSC information **(E)** that excludes low FSC and high SSC events. Examination of the gated cells on a FL1 histogram **(F)** shows that the secondary peak has been removed.

2. The fluidic system on some instruments can produce aerosols; therefore, it is important to identify any biohazardous materials and take the necessary precautions.
3. Cell concentration can easily be adjusted prior to running cells through the flow cytometer by counting them using a hemacytometer. These counts should be conducted using the completely processed cells; cell counts prior to multiple manipulations such as centrifugation or washing will not accurately predict cell concentrations after cell preparation has been completed. This is due to losses that typically occur during cell transfer and decantation. Including trypan blue as a vital dye to determine cell viability just prior to acquiring samples will ensure time saving and efficient use of resources. Unfortunately, this is not always done even though it is easy to do and requires minimal time relative to the hours or days that are spent preparing an entire experiment. There are instances of course when the ideal number of cells required is not always available. The only option at this juncture is to use the available cells to obtain results even if they are only qualitative in nature.
4. Higher sample flow rates during acquisition can result in lower data resolution. When high resolution is required, as in DNA cell-cycle analysis or rare-event analysis, slower sample rates will result in higher resolution.
5. It is critical to ascertain that all monoclonal antibodies, probes, stains, and other reagents are compatible with the flow cytometer. In addition, it is also important to select fluorochromes that can be detected using the optical configuration of the specific flow cytometer that will be used for analysis. Consultation with the instrument manufacturer or personnel that normally operate the flow cytometer are the most time efficient means of determining compatibility.
6. It is imperative that the investigator clearly define the objective of the experiment. It is important to decide which parameters will be used for acquisition, which appropriate control samples will be prepared, and what type of data analysis will be performed. This will help ensure that the defined objective will be met.
7. If the samples are to be acquired by a dedicated operator, it would be prudent to discuss the objective of the experiment. This is especially important for beginning flow cytometry users. This discussion is typically not a critical review of the experiment but an excellent means for ensuring that appropriate controls are prepared so that the operator can properly configure the instrument to meet the experimental objective.
8. Information pertaining to the various types of special treatments the cells may have been exposed to are an invaluable source of information to a flow cytometer operator. Some treatments may alter fluorescent and light-scatter properties. For example, fixation can alter fluorescent and/or morphological cellular patterns. Make the cytometer operator aware of any type of special treatment. This will enable the operator to properly set instrument parameters, acquire, and/or analyze samples correctly.
9. Organization is a key factor for efficiently adjusting the flow cytometer using control samples and then acquiring data from experimental samples. It is very useful to have a protocol for all control and experimental samples. This protocol should also identify the reagents that were used to prepare each sample. In addi-

tion, all sample tubes (including control samples) should be labeled for easy identification. Well-labeled tubes and a sample list save time and eliminate confusion. It is prudent to schedule sample acquisition time. Since most flow cytometers are shared equipment, scheduling will avoid conflicts with other investigators.

Flow Cytometry Information Resources

1. The International Society for Analytical Cytology (ISAC), a world-wide professional organization publishes the journals *Cytometry* (published monthly) and *Cytometry: Communications in Clinical Cytometry* (published quarterly). These journals publish review articles as well as research reports relating to flow cytometry and related areas. ISAC also runs international meetings. Membership in ISAC includes subscription to the aforementioned journals, which are the premier journals in the field of cytometry.
2. A large percentage of papers in the American Association of Immunologists' *Journal Of Immunology* also report extensive flow cytometric data.
3. The ISAC World Wide Web Home Page (address; http://nucleus.immunol.washington.edu/ISAC.html). This page includes updated information of ISAC Congresses and other related meetings, additional links to other Internet resources in cytometry, an updated section flow cytometry related software, job vacancies and wanted section, and Electronic Congress Hall. Online discussion areas where members of the cytometry community can participate in on-going forums and/or create new topics are also included.
4. A cytometry mailing list/bulletin board service where open, on-going discussions of flow cytometry issues are shared (address: <cytometry@flowcyt.cyto.purdue.edu>). Purdue University has a web site that includes contact information on societies related to cytometry and companies that sell cytometry-related products. Almost every cytometry-related web site in the world is also listed (address: http://www.cyto.purdue.edu).
5. Flow cytometry user's meetings are held in numerous geographical (scientific/academic) communities around the world where cytometrists share information by providing round table discussions, open forums, manufacturer-sponsored presentations, and a variety of notable guest speakers. These user meetings are informal and typically occur within an institution and/or among several institutions. Flow cytometry users in a particular geographical location are aware of these informal types of meetings and are very receptive to fostering the flow-cytometry community in an effort to further this field of technology.

References

1. Longobardi-Given, A. (1992) *Flow Cytometry, First Principles.* Wiley-Liss, New York.
2. Melamed, M., Lindro, T., and Mendelsohn, M., eds. (1990) *Flow Cytometry and Cell Sorting, 2nd Ed.* Wiley-Liss, New York.
3. Parks, D. and Herzenberg, L. (1989) Flow cytometry and fluorescence-activated cell sorting, in *Fundamental Immunology* (Paul, W., ed.), Raven, New York.

4. Robinson, J. P., ed. (1993) *Handbook of Flow Cytometry Methods.* Wiley-Liss, New York.
5. Rose, N., DeMacrio, E., Fahey, J., Friedman, H., and Penn, G. (1992) *Manual of Clinical Laboratory Immunology* American Society for Microbiology, Washington, DC, pp. 156–200.
6. Shapiro, H. (1994) *Practical Flow Cytometry,* 3rd ed. Liss, New York.
7. Owens, M. A. and Loken, M. R. (1995) *Flow Cytometric Principles for Clinical Laboratory Practice.* Wiley-Liss, New York.
8. Radbruch, A., ed. (1992) *Flow Cytometry and Cell Sorting.* Springer-Verlag, New York.
9. Ormerod, M. G., ed. (1994) *Flow Cytometry: A Practical Approach,* 2nd ed. IRL Press, Oxford, UK.

2

Detection of Terminal Transferase in Leukemia

Elisabeth Paietta

1. Introduction

The mere presence of terminal deoxynucleotidyl transferase (TdT), a DNA polymerase, in leukemic cells provides no help in assigning these blast cells to a particular cell lineage *(1)*. Differential levels of TdT gene transcription, however, result in diagnostically significant expression patterns of the enzyme with lower biochemical activity and weaker staining intensity by antibody recognition in myeloid as compared to lymphoid leukemia *(2–4)*. One major advantage of measuring TdT by flow cytometry lies in its ability to objectively reflect staining intensities, a challenging task otherwise when one evaluates antibody staining under the microscope using the standard slide technique, thereby alleviating the need for cumbersome and expensive biochemical enzyme assays. The weak fluorescence staining of TdT-expressing myeloid leukemia cells, however, until recently has caused significant technical problems in the flow cytometric TdT detection, whereas several approaches have proven successful in the flow cytometric evaluation of TdT in the intensely staining lymphoid cells *(3)*. Using optimal experimental conditions, the combined analysis of nuclear TdT and surface antigens in all types of leukemia now allows for the detection of minimal residual disease at levels as low as 0.02–0.5% of abnormal cells.

Although in normal hematopoiesis TdT is detected predominantly in cortical thymocytes, with few (<5%) bone marrow cells (originally termed "prothymocytes"), and none of peripheral blood cells expressing appreciable TdT activity *(5)*, TdT has been convincingly demonstrated in lineage–antigen-negative, $CD34^+$-normal bone marrow progenitor cells *(6)*, identifying this enzyme as a lineage-uncommitted hematopoietic marker. The occurrence of TdT in lymphoid malignancies is uncontested, with highest levels of the

From: *Methods in Molecular Biology, Vol. 91: Flow Cytometry Protocols*
Edited by: M. J. Jaroszeski and R. Heller © Humana Press Inc., Totowa, NJ

enzyme being found in blast cells from all but the most mature forms of acute lymphocytic leukemia (ALL), in lymphoid blast crisis of chronic myelogenous leukemia (CML), and in T-cell lymphoblastic lymphoma *(5)*. Controversy exists, however, regarding the frequency of TdT expression in myeloid leukemias *(2)*. Between 5% and 75% of acute myeloid leukemia (AML) cases have been reported to express TdT, whereby methodological differences in TdT detection and the unfortunate use of arbitrary cut-off levels to define TdT positivity are the major culprits for those discrepancies. The technical challenge in determining TdT expression in AML cells may further explain the conflicting information regarding a prognostic significance of TdT expression in this disease. Given the inherent limitations of the slide technique and the added costs and tediousness of biochemical-enzyme determinations, ultimate answers to the biologic significance of TdT in myeloid leukemia will depend on the standardization of flow cytometric assays with optimal levels of sensitivity and the option to double-label for leukemia-associated surface antigens, thereby confirming the lineage affiliation of TdT-expressing blast cells.

A number of protocols have been described for fixation and permeabilization of cells aiming at achieving a satisfactory TdT evaluation by flow cytometry. Not surprisingly, most of these protocols have been judged by their performance on ALL lymphoblasts and, even if included, flow cytometric results in AML cells have been rarely compared with those obtained by the slide technique, the established reference method. Aside from variations in fixation/permeabilization procedures, differences in the sensitivity of flow cytometric TdT detection, particularly when measured in AML cells, are attributable mostly to the type (monoclonal vs polyclonal) and condition (conjugated vs unconjugated) anti-TdT antibody, and the inclusion of blocking reagents to reduce nonspecific background fluorescent staining. Although labeled as unsuitable for staining TdT in AML cells, some of the earlier protocols (e.g., those developed prior to the existence of monoclonal anti-TdT antibodies) may well be useful provided that newer, better anti-TdT antibodies are used, unless proven otherwise, as in the case of the ORTHO Permeafix™ procedure *(7)*. In the following, special emphasis is given to important technical advantages, such as preservation of scatter characteristics or the possibility of double-labeling for surface antigens, and the suitability of each protocol for TdT staining in myeloid leukemia, as far as data are available.

1.1. The Beginnings of Flow Cytometric TdT Staining

The nuclear localization of TdT required that cell fixation in single-cell suspension be established before TdT detection by flow cytometry could first be attempted by McCaffrey et al. *(8)*. Using a modification of this fixation method (10% formalin fixation and 0.05% Tween-20 permeabilization followed by methanol or acetone), Hirata and Okamoto ***(9)*** demonstrated TdT flow cytometrically with excellent comparison to the manual slide technique. Polyclonal rabbit anti-TdT antibody was used in indirect immunofluorescent staining. The authors noted that acetone resulted in cell aggregation and subsequently intolerably high nonspecific staining and that their method did not allow for double-staining with surface antigens. Furthermore, they commented on the markedly reduced fluorescence intensity of TdT staining in AML cells.

Loftin et al. ***(10)*** took a different approach by allowing cells to swell in 0.1 *M* KCl followed by fixation in cold methanol before addition of rabbit anti-TdT antiserum. Problems encountered were nonspecific uptake of primary antibody, more pronounced in myeloid than lymphoid leukemia cells, and the inability to discriminate between weakly TdT staining and negative cells.

Double-labeling for TdT and surface markers for flow cytometry was successfully performed by Slaper-Cortenbach et al. ***(11)*** in buffered formalin acetone-fixed cells. Despite alterations in cell size, right-angle light scatter properties appeared to be preserved in the fixed cells. Fixation conditions of 8s, however, had to be precisely adhered to in order to maintain TdT fluorescence intensity. Although flow cytometric results in ALL cells compared favorably to TdT staining in cytospin preparations, TdT could not be measured in the 12 cases of AML tested.

Permeabilization of cells with saponin (0.25%) after paraformaldehyde fixation was used by Bardales et al. ***(12)*** in conjunction with indirect immunofluorescence staining. Interestingly, despite the majority of patients tested being ALLs, there was no relationship between the amount of TdT activity measured by biochemical assay and the number of TdT-positive cells detected by flow cytometry. Whereas such discrepancy is not unexpected in AML, given the smaller quantities of TdT protein expressed in TdT^{+}AML than in ALL at equal numbers of immunologically TdT^{+} blast cells ***(2)***, excellent correlations between level of enzyme activity and TdT positivity by flow cytometry have been reported in ALL ***(3)***.

1.2. Methods Developed in the 1990s

The big breakthrough in making flow cytometric TdT detection a valuable diagnostic tool came with the development of directly fluorescein (FITC)-conjugated monoclonal anti-TdT antibodies (Supertechs, Bethesda, MD). It opened the door to two- (and later three-) color immunofluorescence staining for nuclear TdT and surface antigens, previously shown to be of imminent significance in leukemia cell characterization by the slide technique ***(13)***. Aside from providing insight in double-marker expression as a leukemic cell feature, two-color immunofluorescence is essential in all TdT-staining methods in which scatter properties of the cells are distorted after fixation/permeabilization.

This was recognized by Gore et al. *(14)* whose TdT staining assay, although exhibiting exquisite sensitivity (level of detection of 0.035% ALL blast cells in mixing experiments), failed to retain cell characteristics in terms of size and granulation. Nonspecific FITC-binding by granulocytes made double-labeling for surface antigens essential to confirm the lineage of TdT-expressing cells. Their method involved fixing the cells in 1% paraformaldehyde in phosphate-buffered saline (PBS), followed by permeabilization in 0.1% Triton X-100. Human AB serum was included at 2% in all antibody incubation steps to block nonspecific binding even when directly conjugated anti-TdT antibody was used. No data are provided on the success of this method in AML cells by these authors. However, the published experience with this method in AML is dismal unless an aldehyde-blocking step is included into the fixation regimen *(15)*.

The paraformaldehyde/methanol fixation protocol described by Drach et al. *(16)* allowed for the simultaneous detection of TdT and surface antigens and was able to detect as few as 0.02% of TdT^+ blast cells in mixing experiments with normal peripheral blood lymphocytes. Unfortunately, no direct comparison is reported between flow cytometry and results by microscopic slide evaluation. Such data would have been particularly valuable given the unexpected background staining, predominantly of granulocytic and monocytic cells, reported with this method when using monoclonal anti-TdT antibody and the rather low incidence of TdT expression in AML (15.6%).

The first commercially available fixation/permeabilization solution successfully used for TdT staining in flow cytometry was Beckton-Dickinson's (Mountainview, CA) diethylene glycol-based FACS red cell-lysing solution *(17)*. Because of its triple properties as formaldehyde-containing cell fixative, permeabilizer, and red cell-lysing reagent, it facilitated TdT staining in unseparated, whole blood or bone-marrow samples. Successful permeabilization was confirmed by comparing the results with FACS-lysing solution with those obtained after cell-membrane permeabilization with octyl glucoside, following a published method *(18)*. Use of the FACS-lysing reagent represents a distinct simplification of flow cytometric TdT detection. It preserves cell-scatter characteristics, and is suitable for double- or triple-color analysis *(19)*. The performance of this method in detecting TdT in AML cells is questionable. Although Syrjälä et al. *(17)* make no mention of problems with nonspecific background fluorescence, the weak fluorescence intensity shown in their paper for an example of TdT staining in ALL blast cells raises serious doubts that TdT can be accurately measured in AML cells with their well-documented low TdT staining intensity.

Another commercial cell fixative, Ortho Permeafix™ (Ortho Diagnostic Systems, Raritan, NJ) perfectly maintains cell structure and morphology, allows simultaneous detection of TdT and surface antigens in unseparated peripheral blood, and offers as an additional advantage that long-term fixation does not impair immunostaining ***(20)***. Although yielding satisfactory numerical results in ALL blast cells, fluorescence intensity of TdT staining after Ortho Permeafix is weak even in ALL when compared to other fixation methods and unacceptably low in TdT^+ AML ***(7)***.

A method that showed consistent and reproducible staining of TdT^+ cells in AML, superimposable to results obtained by the standard slide technique, involves paraformaldehyde fixation followed by blocking of free aldehyde groups with excess glycine prior to Triton X-100 cell permeabilization ***(15)***. The same approach successfully reduced nonspecific background staining of formalin-fixed cells in the demonstration of intracellular B-cell antigens by flow cytometry ***(21)***. The major disadvantage of this method lies in a marked distortion of scatter characteristics, occasionally excessive cell loss, and the addition of an additional step to the already time-consuming procedure.

Processing of cells with OPTI-lyse (Immunotech, Westbrook, ME), another formaldehyde-based red blood cell-lysing reagent, has been proposed as an alternative method ***(22)***. Although claiming to do so, this procedure does not offer any advantage over other published flow cytometric TdT-detection methods and, most notably, presents no data on TdT detection in AML cells.

The most reliable results of TdT measurements in AML cells by flow cytometry have come from work with the Fix & Perm reagents produced by An Der Grub Bio Research (Austria) and distributed in the United States by Caltag (San Francisco, CA) ***(3)***. While preserving cell size and structure, this method reliably detects TdT in AML cases with sensitivity levels completely comparable to those achieved by the slide technique. Since this method also works very well for the detection of intracellular myeloperoxidase in combination with intracytoplasmic CD22, CD3, or lactoferrin (***23,*** unpublished results), it can serve multiple purposes in a routine leukemia diagnostic immunophenotyping laboratory.

2. Materials

Reagents and solutions used in satisfactory procedures of flow TdT determination are presented (*see* **Note 1**).

2.1. Cell Permeabilization and Fixation

1. 1X PBS: 120 m*M* NaCl, 2.7 m*M* KCl, 10 m*M* phosphate buffer, pH 7.4 at room temperature (Commercially available from Sigma, St. Louis, MO).
2. 1% Paraformaldehyde in PBS: 25 parts EM grade 4% paraformaldehyde are mixed with 10 parts of 10X PBS and 65 parts distilled water. To prepare 4% paraformaldehyde, dissolve 4 g paraformaldehyde in 100 mL distilled water under a chemical fume hood in a warm water bath while adjusting to pH 7.0 with NaOH (*see* **Note 2**).

3. 0.1% Triton X-100: weigh 0.1 g of Triton X-100 (use a dropper for this viscous solution) into 100 mL distilled water; stir until the Triton is dissolved.
4. Aldehyde blocking buffer: 3.75 g glycine, 10 g sucrose, in 500 mL of 1X PBS.
5. FACS-lysing solution: the 10X solution is commercially available from Beckton-Dickinson; dilute 1:10 in distilled water before use.
6. PBS/BSA/azide: dissolve 2–5 g (according to your own preference) of bovine serum albumine (BSA) and 0.1 g of sodium azide in 100 mL of 1X PBS.
7. Immunofluorescence assay medium (IFA) (*see* **Note 3**): 10 m*M* HEPES, pH 7.4, 150 m*M* NaCl, 4% calf serum (heat inactivated at 65°C for 30 min). Prepare 1 *M* HEPES solution, pH 7.4 (260.3 g/L of distilled water) and a 1.5 *M* NaCl solution (87.7 g/L of distilled water). For 100 mL of IFA, mix 1 mL of 1 *M* HEPES, pH 7.4, 10 mL of 1.5 *M* NaCl and 4 mL calf serum, and add 85 mL of distilled water.
8. Fix & Perm: the solutions are commercially available from Caltag.

2.2. Antibody Sources

Either a mixture of FITC-conjugated mouse monoclonal anti-human TdT immunoglobulins or a single FITC-conjugated monoclonal anti-TdT antibody is recommended. From the information available, good experiences have been reported with the antibodies distributed by Supertechs, (Bethesda, MD); Dako, (Carpinteria, CA), or Immunotech. It is important to use FITC-conjugated mouse monoclonal immunoglobulins with irrelevant specificity as negative controls. If unconjugated anti-TdT antibody is used, counterstaining with FITC-conjugated secondary immunoglobulin is performed following standard procedures. It is recommended to test for antibody specificity and suitability in your own test system using known TdT-positive and -negative control cells (*see* **Note 4**).

3. Methods

This section summarizes the various protocols described for flow cytometric TdT staining and focuses on discussing their technical and diagnostically relevant advantages and disadvantages. Methodological details for proven satisfactory procedures in both myeloid and lymphoid leukemia are presented.

3.1. The FACS-Lysing Solution (BD) Procedure (17,19)

Although not proven to be reliable in TdT staining of myeloid leukemia cells, this method is discussed because of its cost effectiveness and because it is a procedure routinely used for red cell lysis prior to acquisition of samples on the flow cytometer. It can be applied for whole blood or bone marrow as well as for mononuclear cells isolated by ficoll density gradient centrifugation.

1. Adjusted cell concentration to between 5.0 and 10.0 × 10^6/mL of IFA.
2. Combined staining for surface antigens will be discussed in the next section.

3. Incubate cells with antibody to the surface antigen of choice and, subsequently, add 2 mL of 1:10 diluted FACS-lysing solution to the cell suspension under vortexing for 10 min at room temperature.
4. Centrifuge the cell suspensions for 5 min at 300*g*.
5. Aspirate the supernatant and gently resuspend the cell pellet in 2–5 mL of PBS combined with 2–5% BSA and sodium azide.
6. Centrifuge the cells again for 5 min at 300*g*, aspirate the supernatant, and gently vortex the cell pellets.
7. Add the manufacturer's recommended amount of FITC-conjugated anti-TdT antibody to the test sample (*see* **Note 2**). Add an isotype-specific FITC-labeled irrelevant mouse IgG antibody to the negative control cells.
8. Incubate the cells for 30 min at 4°C in the dark.
9. Add 2 mL of PBS/BSA/azide solution, centrifuge the cells for 5 min at 300*g*, and aspirate the supernatant.
10. Repeat the wash step once.
11. Add 0.5 mL of 0.5% formaldehyde to fix the cells.
12. Store samples in the dark at 4°C until acquisition on the flow cytometer.

3.2. The Aldehyde Blocking Procedure (15)

Fixation of cells with formalin or paraformaldehyde, which are aldehyde fixatives, for subsequent immunostaining increases nonspecific binding of antibodies when compared to fresh cells because of the creation of free aldehyde groups. Blocking of these reactive sites with excess glycine can markedly reduce nonspecific binding and thus allow for better separation of weakly stained cells from background staining, an important issue particularly in AML cells (*see* **Note 1**).

1. Resuspend the mononuclear cells in immunofluorescence assay medium (IFA) to a concentration of 10^7/mL, and react 50 µL of cell suspension with the selected phycoerythrin (PE)-conjugated monoclonal anti-surface antigen antibody.
2. Wash the cells twice in IFA by centrifugation at 300*g* for 5 min at room temperature.
3. After the second wash, fix the pelleted cells in 2 mL of 1% paraformaldehyde/PBS for 15 min at room temperature.
4. Add 1 mL of aldehyde blocking solution to the pellet of fixed cells. Incubate for 30 min at room temperature with one buffer change at 15 min.
5. Subsequently, pellet the cells, aspirate the supernatant, and add 2 mL of 0.1% Triton X-100 in IFA for 3 min at room temperature.
6. Spin the permeabilized cells at 500*g* for 10 min at 4°C and aspirate the supernatant.
7. Resuspend the cells in 100 µL of IFA with 5% human AB serum, add FITC-conjugated monoclonal anti-TdT antibody or FITC-conjugated control immunoglobulin to the test or control tube, respectively, at the manufacturer's recommended concentration, for 1 h at 4°C in the dark (*see* **Note 2**).
8. Wash the stained cells twice with 2-mL aliquots of 0.1% Triton X-100 in IFA. Then, resuspend the washed cells in PBS and acquire them on the flow cytometer.

3.3. The Fix & Perm Cell Preparation (3)

This method is recommended for the routine detection of TdT in ALL and AML cells. The method preserves structural cell characteristics very well, can be used with whole blood or bone marrow as well as mononuclear cell isolates, allows double-staining for surface antigens, is quick, works equally well in lymphoid and myeloid blast cells, and TdT staining (with appropriate controls) may simply be added on to the simultaneous determination of myeloperoxideas with cytoplasmic CD3, CD22, or lactoferrin by the same fixation procedure *(23)*. The concomitant evaluation of these intracellular lymphoid and myeloid antigens with TdT provides a helpful tool in the analysis of TdT positive cells.

1. Following staining for PE- and/or PerCP-conjugated surface antigens, place approx 10^6 cells (In 50 μL) in IFA in a 5-mL tube.
2. Add 100 μL of Caltag reagent A (FIXation medium, at room temperature) without vortexing for a 15 min incubation at room temperature.
3. Add 5 mL of PBS and centrifuge at 300*g* for 5 min.
4. Aspirate the supernatant then add 100 μL of Caltag reagent B (PERMeabilization medium, at room temperature) plus the manufacturer recommended amount of FITC-conjugated monoclonal anti-TdT antibody or control antibody to the cell pellet (*see* **Note 2**).
5. Gently vortex for 1–2 s.
6. Incubate for 15 min at room temperature in the dark.
7. Add 5 mL of PBS to the cell suspension(s) and centrifuge for 5 min at 300*g* at room temperature.
8. Remove the supernatant and resuspend the cells in PBS for acquisition on the flow cytometer. For storage, cells should be kept at 4°C in the dark.

3.4. Analysis of TdT in Combination with Surface Antigens

To identify blast cells and confirm the cell lineage of TdT-positive cells, multicolor staining for nuclear TdT and surface antigens should be performed. For double-staining of TdT and one surface antigen, 50–100 μL of adjusted cells are incubated for 15–30 min at room temperature in the dark with the appropriate amount of PE-conjugated antibody to the surface antigen of choice, e.g., CD19, CD10, or CD5 in a case of B- or T-ALL, respectively, CD33, CD13 in a case of AML, or CD34, HLA-DR as general markers of immaturity. To facilitate recognition of the blast cell population, triple-staining with PerCP-conjugated CD45 antibody can be performed or, alternatively, to save costs, an extra aliquot of cells may be stained with FITC- or PE-conjugated CD45 antibody (usually used in a laboratory for setting of the leukocyte gate) and subjected to cell fixation and permeabilization.

In the analysis, the gate is set around the mononuclear cell population. The intensity of CD45 staining in a CD45 vs right-angle light scatter display can help distinguishing blast cells with dull CD45 expression from brightly staining T-lymphocytes contaminating the mononuclear cell population so that the gate of analysis can be adjusted appropriately. Blast cells are identified by the simultaneous expression of FITC-labeled TdT and the PE-labeled surface antigen of choice previously shown to be expressed by the majority of blast cells in a given leukemia population.

3.5. Detection of Minimal Residual Disease (MRD)

When using TdT detection as a means of identifying immunologic minimal residual disease (MRD), all procedural considerations discussed above apply as well. **Figure 1** shows an example for detecting MRD in a patient who initially presented with TdT$^+$, HLA-DR$^+$ undifferentiated AML (contourplot A). After completion of induction chemotherapy and achievement of a clinical and hematologic complete remission, <1% HLA-DR$^+$, TdT$^+$ cells are still detected in the peripheral blood of this patient (**Fig. 1**, contourplot B), in which normally no TdT$^+$ should be present. If seen in the bone marrow, these TdT$^+$, HLA-DR$^+$ cells could not have been definitely identified as residual blast cells simply based on antigen profile since normal TdT$^+$ cells in the bone marrow are HLA-DR$^+$ as well. As a general rule and whenever possible, antigen combinations should be chosen in the detection of residual leukemia that distinguish between TdT$^+$ normal precursor cells detectable by routine immunophenotyping in normal bone marrow and residual TdT$^+$ leukemic cells. An example would be double-labeling for TdT and CD11b in cases of immature monocytic leukemia since mature monocytes, although being CD11b positive, lack TdT as do mature CD11b positive myeloid cell; or double-labeling for TdT and CD15, a rather mature myeloid antigen, provided the initial blast cell population was CD15-positive; even double-labeling for TdT and CD33 can be useful since the frequency of CD33$^+$/TdT$^+$ normal precursor cells in the bone marrow is extremely low. In cases of ALL after treatment, double-labeling for TdT and CD19 or CD10 in B-cell ALL or with CD2 in T-cell ALL has a good chance of detecting residual disease, even though normal precursor cells expressing these antigen combinations do exist, albeit they are very rare *(1)*.

4. Notes

1. As outlined above, most of the published protocols for flow cytometric TdT determination will work in cases of ALL in which strong TdT-staining intensity is the rule. One must remember, though, that certain ALL blast cells that express myeloid antigens tend to demonstrate weaker TdT staining than myeloid antigen-negative lymphoblasts *(2)* and, weaker staining has also been seen in cases of

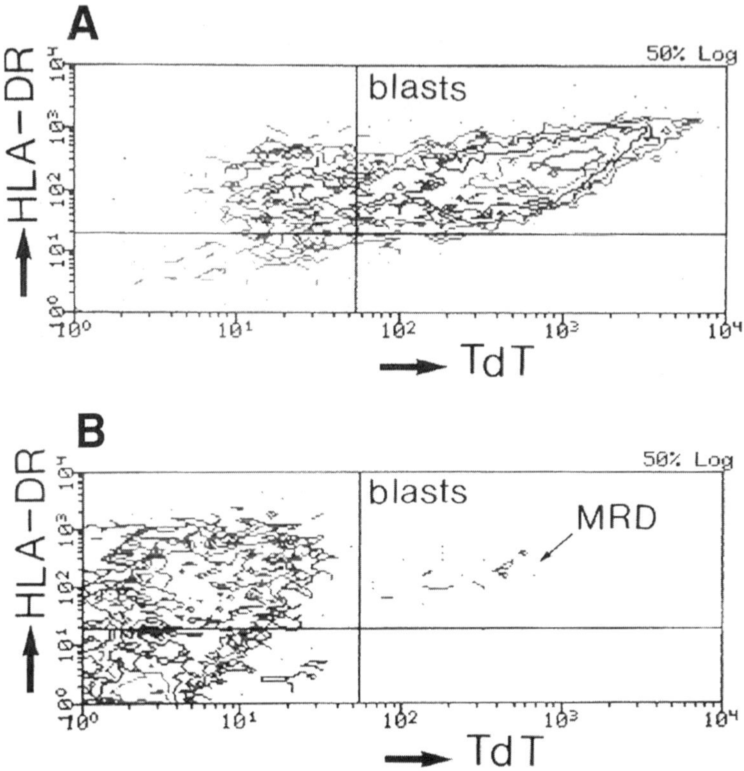

Fig. 1. Contourplot of peripheral blood mononuclear cells in a patient with TdT^+, $HLA\text{-}DR^+$ AML at the time of presentation **(A)** and at the time of clinical and hematologic remission **(B)**. In the remission sample, 2% peripheral blood of mononuclear cells still express the TdT/HLA-DR-positive immunophenotype typical of this patient's original blast cells.

pre-T ALL *(23)*. Therefore, irrespective of the predominant immunophenotype of a given leukemia cell population, optimal conditions for TdT staining should be aimed for, suitable to reliably discern the particularly weakly TdT-staining cells in AML. To date, of the protocols presented in the previous section, only two procedures are proven to yield satisfactory results in AML, the aldehyde-buffer blocking of formaldehyde/Triton X-100 fixed and permeabilized cells *(15)*, and the procedure involving the commercially available (Caltag Labs) Fix & Perm reagents *(3)*. In addition to these procedures, the FACS lysing solution method is also discussed in detail because it involves a reagent routinely used for red-cell lysis in immunodiagnostic laboratories and because the data available on its performance in myeloid leukemia are not clear. If the FACS-lysing solution procedure is used in diagnostic TdT determinations, it is recommended that a comparative study with the slide technique be initiated in a group of patients with

AML (at least 5–10 TdT-positive patients by microscopic evaluation) before this method is fully incorporated into the leukemia diagnostic assay panel.

2. Whatever approach is taken in terms of choice of fixation/permeabilization conditions, it is highly advisable to use directly conjugated monoclonal anti-TdT antibodies since they facilitate double- and triple-labeling for TdT and surface antigens and because they result in considerably lower background staining than when secondary antibodies to unconjugated primary monoclonal anti-TdT antibody or polyclonal antisera are employed.
3. The immunofluorescent assay (IFA) buffer should contain between 2 and 5% of human AB serum. Some investigators may opt to also include fetal bovine serum into the buffer system. An example for an immunofluorescent medium would be 10 m*M* HEPES, pH 7.4, 150 m*M* NaCl, 4% fetal bovine serum to which 5% human AB serum is added during all antibody incubation steps ***(15)***. The general abbreviation of IFA is used to refer to individually chosen immunofluorescence assay media.
4. Control staining of known TdT-negative and TdT-positive cells is advisable. TdT-positive cell lines, such as the MOLT3 T-cell lymphoblastoid cell line or the Nalm-6 pre-B leukemia cell line, and TdT-negative cell lines, such as Daudi cells, a Burkitt's lymphoma cell line, or HL-60 cells, a myeloid cell line, are convenient sources for control cells. One must keep in mind however, that depending on culture conditions and growth status of cell lines, their level of TdT expression may vary greatly on a daily basis. TdT test control cells have recently become available from Supertechs. This control cell suspension contains a 1:1 mixture of TdT-positive and TdT-negative lymphoblastoid human cells and, if stored at 4°C, has a guaranteed life-span of 90 d.

References

1. Paietta, E. (1995) Immunobiology of acute leukemia, in *Neoplastic Diseases of the Blood, 3rd Ed.* (Wiernik, P. H., Canellos, G., Dutcher, J. P., and Kyle, R., eds.) Churchill Livingstone, New York, pp. 211–247.
2. Paietta, E., Racevskis, J., Bennett, J. M., and Wiernik, P. H. (1993) Differential expression of terminal transferase (TdT) in acute lymphocytic leukemia expressing myeloid antigens and TdT positive acute myeloid leukemia as compared to myeloid antigen negative acute lymphocytic leukemia. *Br. J. Haematol.* **84,** 416–422.
3. Meenan, B., Heavey, C., Lichtenstein, A., Andersen, J., and Paietta, E. (1996) Terminal transferase expression in the differential diagnosis of acute leukemias. *Leukemia Lymphoma* **22,** 265–269.
4. Farahat, N., Lens, D., Morilla, R., Matutes, E., and Catovsky, D. (1995) Differential TdT expression in acute leukemia by flow cytometry: a quantitative study. *Leukemia* **9,** 583–587.
5. Bollum, F. J. (1979) Terminal deoxynucleotidyl transferase as a hematopoietic cell marker. *Blood* **54,** 1203–1215.
6. Gore, S. D., Kastan, M. B., and Civic, C. I. (1991) Normal human bone marrow precursors that express terminal deoxynucleotidyl transferase include T-cell precursors and possible lymphoid stem cells. *Blood* **77,** 1681–1690.

7. Murray, M., Heavey, C., and Paietta, E. (1995) ORTHO Permeafix™ fixation is not suitable for the flow cytometric detection of nuclear terminal transferase in acute myeloid leukemia. *Leukemia* **9,** 226, 227.
8. McCaffrey, R., Lillquist, A., Sallan, S., Cohen, E., and Osband, M. (1981) Clinical utility of leukemia cell terminal transferase measurements. *Cancer Res.* **41,** 4814–4820.
9. Hirata, M. and Okamoto, Y. (1987) Enumeration of terminal deoxynucleotidyl transferase positive cells in leukemia/lymphoma by flow cytometry. *Leukemia Res.* **11,** 509–518.
10. Loftin, K. C., Reuben, J. M., Dalton, W., Hersh, E. M., and Sujansky, D. (1986) Terminal transferase in leukemias by flow cytometry. *Diag. Immunol.* **4,** 165–169.
11. Slaper-Cortenbach, I. C. M., Admiraal, L. G., Kerr, J. M., van Leeuwen, E. F., von dem Borne, A. E. G. Jr., and Tetteroo, P. A. T. (1988) Flow-cytometric detection of terminal deoxynucleotidyl transferase and other intracellular antigens in combination with membrane antigens in acute lymphatic leukemias. *Blood* **72,** 1639–1644.
12. Bardales, R. H., Carrato, A., Fleischer, M., Schwartz, M. K., and Koziner, B. (1989) Detection of terminal deoxynucleotidyl transferase (TdT) by flow cytometry in leukemic disorders. *J. Histochem. Cytochem.* **37,** 509–513.
13. Bettelheim, P., Paietta, E., Majdic, O., Gadner, H., Schwarzmeier, J., and Knapp, W. (1982) Expression of a myeloid marker on TdT-positive acute lymphocytic leukemic cells: evidence by double-fluorescence staining. *Blood* **60,** 1392–1396.
14. Gore, S. D., Kastan, M. B., Goodman, S. N., and Civin, C. I. (1990) Detection of minimal residual T-cell acute lymphoblastic leukemia by flow cytometry. *J. Immunol. Methods* **132,** 275–286.
15. Paietta, E., Meenan, B., Heavey, C., and Thomas, D. (1994) Detection of terminal transferase in acute myeloid leukemia by flow cytometry. *Cytometry* **16,** 256–261.
16. Drach, J., Gattringer, C., and Huber, H. (1991) Combined flow cytometric assessment of cell surface antigens and nuclear TdT for the detection of minimal residual disease in acute leukemia. *Br. J. Haematol.* **77,** 37–42.
17. Syrjälä, M. T., Tiirikainen, M., Jansson, S.-E., and Krusius, T. (1993) Flow cytometric analysis of terminal deoxynucleotidyl transferase. *Hematopathology* **99,** 298–303.
18. Hallden, G., Andersson, U., Hed, J., and Johansson, S. G. O. (1989) A new membrane permeabilization method for the detection of intracellular antigens by flow cytometry. *J. Immunol. Methods* **124,** 103–109.
19. Horvatinovich, J. M., Sparks, S. D., and Borowitz, M. J. (1994) Detection of terminal deoxynucleotidyl transferase by flow cytometry: a three color method. *Cytometry* **18,** 228–230.
20. Pizzolo, G., Vincenzi, C., Nadali, G., Veneri, D., Vinante, F., Chilosi, M., Basso, G., Connelly, M. C., and Janossy, G. (1994) Detection of membrane and intracellular antigens by flow cytometry following ORTHO Permeafix™ fixation. *Leukemia* **8,** 672–676.
21. Caldwell, C. W. (1994) Preservation of B-cell associated surface antigens by chemical fixation. *Cytometry* **16,** 243–249.
22. Serke, S. (1995) Detection of terminal deoxynucleotidyl transferase by permeabilization of cells using a standard red blood cell lyse reagent. *Cytometry* **19,** 189, 190.
23. Knapp, W., Strobl, H., and Majdic, O. (1994) Flow cytometric analysis of cell-surface and intracellular antigens in leukemia diagnosis. *Cytometry* **18,** 187–198.

3

The Use of Flow Cytometry to Detect Intracellular Cytokine Production in Individual Cells

Brian E. Crucian and Raymond H. Widen

1. Introduction

The current methods commonly employed to detect cytokine production have several drawbacks. Bioassays are not necessarily cytokine-specific in that they measure functional properties. The production of supernatant cytokine protein can be readily measured by enzyme-linked immunosorbent assay (ELISA) methods, but unless a highly purified cell population was cultured, this method does not identify the population of cells responsible for the cytokine production. In addition, the results of ELISA assays reflect the net outcome of produced, absorbed and degraded cytokine and do not distinguish between biologically active and inactive substances. The detection of cytokine RNA (*in-situ* hybridization and reverse transcriptase-polymerase chain reaction) adequately detect gene expression, but this does not guarantee the translation of the message into cytokine protein. Thus, methods were developed to detect cytokine production at the individual cell level. These methods usually also possessed the ability to positively identify the cell population of interest. The various strategies that have been utilized have been reviewed by Lewis ***(1)***.

The intracellular detection of cytokine protein by flow cytometry, which can be used in conjunction with surface-marker analysis (by using two or more color analysis) serves nicely to positively identify cytokine-producing cells, even when analyzing a mixed population of cells. The use of this method for cytokine analysis was described in detail by Sander et al. as well as by Jung et al. in the early 1990s ***(2–4)*** and more recently by Prussin et al. ***(5)***, and has been used with increasing frequency in the recent literature and applied to a variety of experimental situations ***(6–10)***. Briefly, these papers

From: *Methods in Molecular Biology, Vol. 91: Flow Cytometry Protocols*
Edited by: M. J. Jaroszeski and R. Heller © Humana Press Inc., Totowa, NJ

described a method by which the cell membranes would first be fixed to prevent the leakage of the intracellular contents during permeabilization. The cells would then be permeabilized to expose the intracellular contents to detection antibodies, and then surface markers could also be stained to identify the cells of interest.

Permeabilization of cell membranes is a well established technique that has been often used to investigate intracellular processes ***(11–14)***. There are slight variations in reagents and technique used between the various published methods. A unique strength of these methods is the ability to detect the production of multiple cytokines simultaneously at the single cell level.

Potential shortcomings of these techniques are the realization that the intracellular presence of a cytokine need not necessarily be equated with secreted cytokine and the fact that the method is far more qualitative than quantitative. A relative measure of quantitation between various cell types can be achieved by using the relative fluorescence intensities. It is also of note that the possibility exists that this method will detect the presence of absorbed cytokine; however, the literature indicates that so far this has not been a major limitation ***(2)***. Definitive studies correlating the synthesis and intracellular storage of cytokines with their subsequent release have yet to be performed.

2. Materials

2.1. Activation of Cells During Culture

1. Complete medium requirements will vary from cell type to cell type. Use the designated complete medium supplemented as indicated for the cell type of interest. For the culture of human peripheral blood mononuclear cells (PBMCs), we have used RPMI medium 1640 containing 10% fetal bovine serum, 25 m*M* HEPES buffer, 1×10^5 µg/mL penicillin and streptomycin, 25 µg/mL fungizone, and 10 µg/mL gentamicin.
2. To activate the cells to secrete cytokines during culture, any combination of the following mitogens added to the medium may be used: 5 µg/mL phytohemagglutinin (PHA), 10 ng/mL phorbol myristate acetate (PMA), 1 µg/mL ionomycin, and 5 µg/mL of LPS. We have had success culturing human PBMCs in medium containing 5 µg/mL PHA, 10 ng/mL PMA and 1 µg/mL ionomycin for either 5 or 24 h for T-cell analysis or medium with 5 µg/mL of lippopolysaccharide (LPS) for 24 h for monocyte analysis. The conditions selected will depend on the cell population to be assayed and the expression kinetics for the cytokine of interest.
3. In most cases an inhibitor of extracellular protein transport must be added to the cell cultures for the final 5–6 h of culture to shut down extracellular transport of cytokines and to allow for intracellular accumulation to reach detectable levels ***(2,15)***.

2.2. Intracellular Staining of Cytokines

1. 1X Dulbecco's phosphate-buffered saline (PBS), Ca^{2+} Mg^{2+} free.
2. Paraformaldehyde fixation buffer: For a 4.0% paraformaldehyde fixation buffer dissolve 4 g of paraformaldehyde powder in 100 mL PBS, heating at 56°C for 30–60 min to facilitate dissolving of the powder. Because of the toxicity of paraformaldehyde, perform all weighing and heating steps in a fume hood and wear appropriate personal protective equipment.
3. 1X Permeabilization buffer (PB): generally consists of blocking agents combined with saponin in PBS. We have had success using a buffer consisting of 5.0% nonfat dry milk (NFDM) and 0.5% saponin in PBS.
4. Detection antibodies: The detection antibodies will vary depending on the cytokine to be detected and antibody panel configuration to be used. In general, directly labeled antibodies are best as they eliminate the need for second step reagents and reduce the possibility of nonspecific binding.
5. Second step conjugates: If it was necessary to use an unlabeled antibody for cytokine assessment, then a detection conjugate must be used. Fluorochrome-labeled anti-isotype antibodies serve this purpose and labeled strepavadin may be used with biotinylated primary antibodies.
6. Surface marker antibodies: Standard fluorochrome-labeled antibodies to surface markers can be employed following intracellular cytokine staining to identify cells of interest provided they will use a fluorescence channel not used by the cytokine detection antibodies.

3. Methods

3.1. Isolation and Culture of Cells for the Detection of Cytokine

Given that cytokine protein will be detected by the binding of photoactive fluorochromes that must be analyzed by an instrument, it follows that if more cytokine protein is present then detection will be easier. For this reason the activation of cells in vitro is desirable, and serves to upregulate the production of cytokine protein. In addition, the abrogation of the extracellular protein transport allows intracellular accumulation that greatly enhances the sensitivity of this method. We have found that the addition of the carboxylic ionophore monensin to our cell culture medium during the final 5 h of culture greatly enhanced the signal-to-noise ratio.

1. Cells used for analysis must be isolated by conventional means, either by density gradient centrifugation, cell sorting or filtration, depending on the experimental design. The culture of mixed cell types is acceptable, as the cytokine-producing cells can be positively identified during analysis by the staining of a surface marker, if one is uniquely expressed on the cells of interest.
2. Place cells in culture medium containing the appropriate mitogenic stimulus. The mitogen selected and the time of culture will vary depending on the cell type and cytokines selected for analysis.

3. Add monensin to the culture medium for the final 5 h of culture at a final concentration of 3 μ*M* to allow intracellular accumulation of cytokine (*see* **Note 1**).
4. Wash cells in PBS and subject to the cell staining procedure.

3.2. The Staining of Intracellular Cytokines in Activated Cells

In general, it is important to remember that a nearly endless combination of labeled or unlabeled primary antibodies, a variety of second-step conjugates, and surface marker antibodies exists. Care must be exercised to select a combination in which there will be no crossbinding, which can lead to false-positive results. Therefore, the specific application of the procedure must be highly individualized to fit the requirements of the investigator. The procedures that follow are a generalized guide that assume the investigator wishes to examine the expression of one or more cytokines in conjunction with subsequent surface marker analysis.

Multicolor flow cytometry analysis is most easily performed using a combination of directly labeled monoclonal or polyclonal antibodies; however, there may be instances in which an antibody reagent is available only in an unlabeled format. We have included two methods below to accommodate studies using only directly labeled antibodies and experiments in which indirect-labeling procedures are necessary. Obviously, if looking at only a single marker (single color), the problems associated with secondary detection antibodies possibly reacting with more than one primary antibody will not be faced. In addition, for laboratories which will use this procedure on a limited basis, it may be desirable to purchase a commercial kit containing premade reagents (*see* **Note 2**).

The use of the proper controls is essential in experiments utilizing the intracellular detection of cytokines by flow cytometry. The appropriate positive and negative controls most commonly used are discussed in **Note 3**.

3.2.1. Staining Protocol Using Directly Labeled Antibodies for Multicolor Analysis

If the experiment involves surface-marker staining, this should be accomplished first. If all markers are intracellular, proceed to **step 3** below.

1. After completion of the cell culture activation protocol, wash approx $0.5–1.0 \times 10^6$ cells per marker combination to be assayed in 3 mL PBS. Resuspend the cell pellet in 100 μL PBS and add the appropriate monoclonal antibodies(s) (MAbs) to stain the surface marker(s) of interest.
2. Incubate the cells/antibody combination for 15–30 min at 4°C in the dark.
3. Wash the cells once with 3 mL PBS and resuspend the pellet in 200 μL fixation buffer (*see* **Note 4**). Incubate the mixture for 10 min at room temperature in the dark.

4. Wash the cells once with 3 mL PBS and resuspend the pellet in 200 µL of PB (*see* **Notes 5** and **6**) along with 0.5–2.0 µg/mL of the intracellular cytokine detection antibodies.
5. Incubate the cell/antibody combination for 30 min at room temperature in the dark.
6. Wash the cells three times in PB to remove any excess unbound antibody and resuspend the cell pellet in 0.5–1.0 mL PBS and analyze on a flow cytometer (*see* **Note 7**).

3.2.2. Staining Protocol Using Unlabeled Primary Antibody(s) for Multicolor Analysis

Whenever one of the reagents used in a multicolor flow cytometry experiment is unlabeled, stain with that reagent first, regardless of the cellular location of the target.

3.2.2.1. Unlabeled Primary Antibody Directed Against Surface Marker

1. After completion of the cell culture activation protocol, wash approx 0.5–1.0 × 10^6 cells per marker combination to be assayed in 3 mL PBS and resuspend the cell pellet in 200 µL PBS.
2. Add 2–5 µg of the unlabeled antibodies to surface marker to the mixture and incubate 15–30 min at 4°C in the dark.
3. Wash the cells once in 3 mL of PBS and resuspend the cell pellet in 200 µL PBS.
4. Add the manufacturers suggested amount (usually 2–5 µg) of the fluorochrome-labeled second-step conjugate (*see* **Note 8**), which will bind specifically to the primary antibody.
5. Incubate the cell/antibody mixture for 15–30 min at 4°C in the dark.
6. Wash the cells once in 3 mL PBS and resuspend the cell pellet in 200 µL PBS.
7. Add at least a 10-fold excess of an unlabeled antibody of the same species and isotype of the primary unlabeled antibody (*see* **Note 9**) and incubate the cell/antibody mixture for 30 min at 4°C in the dark.
8. Wash the cells once with 3 mL PBS and resuspend the pellet in 200 µL fixation buffer (*see* **Note 4**). Incubate the mixture for 10 min at room temperature in the dark.
9. Wash the cells once with 3 ml PBS and resuspend the pellet in 200 µL of PB (*see* **Note 5**) along with .5–2.0 µg/mL of the fluorochrome-labeled intracellular cytokine detection antibodies.
10. Incubate the cell/antibody combination for 30 min at room temperature in the dark.
11. Wash the cells three times in PB to remove any excess unbound antibody and resuspend the cell pellet in 0.5–1.0 mL of PBS and analyze on a flow cytometer (*see* **Note 7**).

3.2.2.2. Unlabeled Primary Antibody Directed Against Intracellular Cytokine(s)

1. After completion of the cell culture activation protocol, wash approx 0.5–1.0 × 10^6 cells per marker combination to be assayed in 3 mL PBS and resuspend the cell pellet in 200 µL fixation buffer (*see* **Note 4**). Incubate the mixture for 10 min at room temperature in the dark.

2. Wash the cells once with 3 mL PBS and resuspend the pellet in 200 μL of PB (*see* **Note 5**) along with .5–2.0 μg/mL of the unlabeled intracellular cytokine detection antibodies.
3. Incubate the cell/antibody combination for 30 min at room temperature in the dark.
4. Wash the cells once in 3 mL PB and resuspend the cell pellet in 200 μL PB.
5. Add at least a 10-fold excess of an unlabeled antibody of the same species and isotype of the primary unlabeled antibody (*see* **Note 9**) and incubate the cell/antibody mixture for 30 min at 4°C in the dark.
6. Wash the cells twice in PB and resuspend the pellet in 200 μL PB. At this point, other cytokines can be stained using directly labeled anticytokine antibodies. If other cytokines are to be stained in this manner, proceed to **step 7**. If no other cytokines are to be stained and you wish to now stain surface markers, proceed to **step 10**. If no staining of surface markers is required, proceed to **step 13**.
7. Wash the cells once with 3 mL PBS and resuspend the pellet in 200 μL PB along with 0.5–2.0 μg/mL of the directly labeled intracellular cytokine detection Antibody(s).
8. Incubate the cell/antibody combination for 30 min at room temperature in the dark.
9. Wash the cells once in 3 mL PB and resuspend the cell pellet in 200 μL PBS.
10. Wash the cells once in 3 mL PBS and resuspend the cell pellet in 200 μL PBS. Resuspend the cell pellet in 100 μL PBS, and add the appropriate directly labeled MAb(s) to stain the surface marker(s) of interest.
11. Incubate the cells/antibody combination for 15–30 min at room temperature in the dark.
12. Wash the cells three times in PBS to remove any excess unbound antibody and resuspend the cell pellet in 0.5–1.0 mL PBS and analyze on a flow cytometer (*see* **Note 7**).

3.3. Flow Cytometric Analysis

There are at least two gating strategies that may be applied for flow cytometric analysis of intracellular cytokines. They are conventional FWD vs SSC (light scatter) gating and immunoscatter gating, in which one of the fluorescence parameters is plotted against the SSC signal. Examples of both types of analysis strategy are demonstrated in **Fig. 1**. Light-scatter gating works well with short-term activated cells in which the light-scatter properties have not been altered significantly (for example, PBMCs cultured for 4 h with PMA + ionomycin). For studies involving longer-term culture of cells with mitogenic stimuli (for example, PBMCs cultured with PHA for 24–48 h), the light-scatter properties are altered, making identification of gating clusters more difficult. Immunoscatter gating is particularly useful in these situations since one may pull the target cluster out of the mixed population of cells by identifying a specific target marker (such as CD3) and plotting it against SSC. The CD3-gated events may then be analyzed for expression of intracellular cytokines. It is the ability to specifically identify the cytokine-

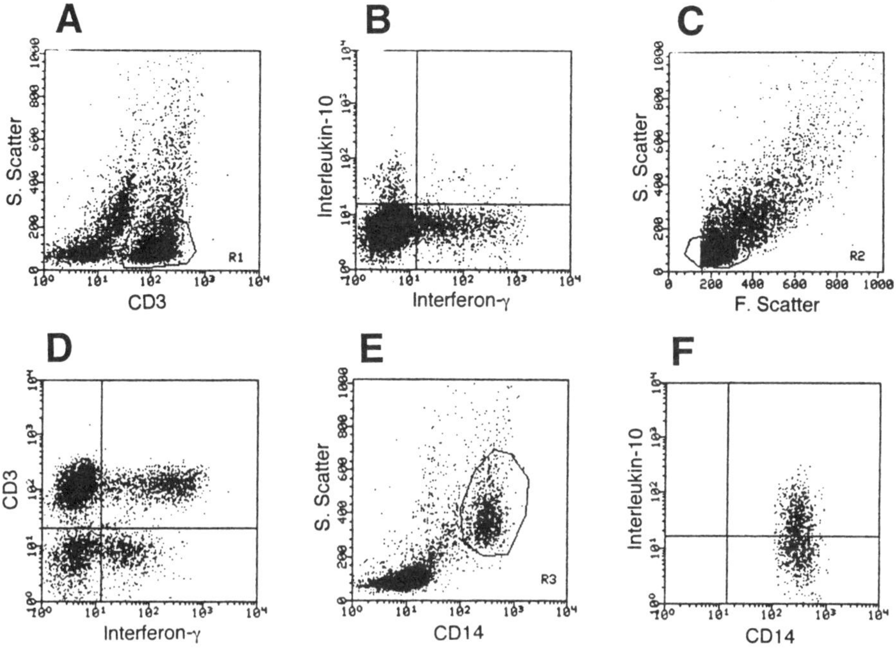

Fig 1. Representative scatter plots demonstrating the analysis of intracellular cytokines by flow cytometry in human PBMCs activated during culture in vitro. **(A)** T-cell gate; **(B)** T-cell intracellular IFN-γ and IL-10 following PBMC activation with PMA, PHA, and ionomycin for 24 h; **(C)** lymphocyte gate; **(D)** IFN- γ detection in T-cells vs non T-cells differentiated by expression of CD3; **(E)** monocyte gate; **(F)** monocyte IL-10 production from PBMCs activated with LPS for 24 h.

producing cells that makes the flow cytometric analysis of intracellular cytokines such a powerful tool for studying immune regulation.

4. Notes

1. In most instances the activation of cytokine producing cells in vitro is not enough to boost cytokine production to a detectable level. This is because cytokine protein is secreted from the cell as it is made. In contrast, surface markers are readily detected without manipulation since they are anchored in the cell membrane. This deficit can be overcome by the addition of an inhibitor of peptide transport to the cell culture for approximately the final 5 h of culture prior to analysis. These agents arrest the extracellular transport of protein, leading to an accumulation of intracellular protein without altering the *de novo* synthesis. The carboxylic ionophore monensin ***(16)*** is the most common agent used ***(2,8,10)***. The use of brefeldin-A has also been reported ***(6)***.

2. Commercial kits consisting of ready-made fixation and permeabilization mediums and buffers are available from several sources (PharMingen, Becton Dickinson, Caltag, Ortho). Although these kits generally work very well, they may be suited more for use in a clinical setting where a limited number of samples may be run. In a research setting, where large numbers of samples may need to be analyzed, it may be more cost effective to utilize in-house procedures, such as the one described here.
3. The use of appropriate controls is essential when conducting cytokine analysis by this method. Standard positive controls for the expression of cytokines usually include a mitogen-stimulated cell type that has been previously demonstrated to result in cytokine detection by this method. When purchasing directly labeled antibodies to cytokines from a vendor, the appropriate cell mitogen system is usually outlined for achieving detection and comparison to the experimental system. In addition, flooding the staining reaction with an excess of unlabeled recombinant cytokine has been reported to serve as an excellent (although potentially expensive) control for the function of the detection antibodies. Unstimulated cells, cytokine gene-deletion mutant cell lines, preincubation with an excess of unlabeled anti-cytokine antibody, and activated cells from which the monensin step is eliminated may all serve as negative controls for this assay system.
4. The ideal fixative agent must preserve cell morphology, surface marker antigenicity, intracellular antigenicity with minimal cell aggregation, and cell loss. A solution of 4.0% paraformaldehyde is by far the most widely used fixing agent. It is important to test the effects of paraformaldehyde treatment on the antigens of interest prior to analysis to be sure antigenicity is not compromised.
5. Permeabilization of the cells to be analyzed is required to give the battery of detection antibodies access to the intracellular antigens. Saponin added to the PB at a concentration of 0.1–0.5% seems to be the most widely used permeabilization agent in the literature. Saponins are plant glycosides with a high affinity for cholesterol that intercalate into the cell membranes and form pores of approx 8 nm or larger ***(3)***. Permeabilization with saponin is reversible and it is important that saponin be constantly present during all steps, including washes, until the intracellular staining is complete.
6. It has been our experience that the permeabilization of cells, followed by subsequent staining with antibodies, results in an unusually high background level of fluorescence. Although the precise reasons for this phenomenon are as yet unclear, the addition of agents that block nonspecific binding to the permeabilization buffer seem to be common. We have had success by adding 5.0% NFDM to our permeabilization buffer. Other laboratories have reported the use of 0.1% bovine serum albumin (BSA) ***(7)*** and heat-inactivated serum ***(3,4)***.
7. The activation of a mixed population of cells during culture tends to alter the typical forward vs side-scatter, light-scatter properties. During the activation of mononuclear cells derived from peripheral blood, for example, culture in the presence of PHA, PMA, and ionomycin results in enlargening of the cells and a dramatic increase in the forward scatter. For this reason, great care must be exer-

cised in the drawing of the analysis gate, particularly during the analysis of mixed cell populations. It is sometimes beneficial to gate the population of interest using immunoscatter (the plot of forward or side scatter vs a surface marker identifying the population). This technique will essentially pull the population out of a mix of cells with similar scatter properties and allow a clean analysis. In addition, another approach to a more precise analysis is to use the typical forward vs side-scatter gating, analyzing the gated events as a lineage marker (i.e., CD3) vs the cytokine of interest. Examples of these gating techniques are presented in **Fig. 1**.

8. The use of a conjugate to detect nonfluorochrome-labeled primary antibodies will depend on the type of primary antibody used. To detect unlabeled primary antibodies, a labeled polyclonal antibody to the primary antibody isotype (ie., a antimouse IgG FITC-labeled) works well. If the primary antibody is biotin labeled, a fluorochrome-labeled-streptavidin second step reagent is adequate, although background from the detection of endogenous biotin must be considered. In addition to the use of blocking agents to reduce nonspecific binding, the use of fluorescence-labeled goat antimouse F(ab')2 immunoglobulin G (heavy and light chains) as a conjugate tended to reduce the nonspecific binding levels as compared to labeled-whole antibody conjugates. Presumably the elimination of the Fc portion of the molecule is responsible for this improvement.
9. Following indirect primary antibody staining and prior to staining with an additional antibody (same isotype) it is necessary to block the unbound sites on the primary conjugate. This step will prevent the conjugate from binding the secondary antibody resulting in a false-positive reading. Flooding the reaction system with an excess of an irrelevant unlabeled isotype-matched antibody accomplishes this goal. Several manufacturers have unlabeled isotype controls available. We have had success using 1.0–5.0% of heat-inactivated mouse serum.

References

1. Lewis, C. E. (1991) Detecting cytokine production at the single cell level. *Cytokine* **3,** 184–188.
2. Jung, T., Schauer, U., Heusser, C., Neumann, C., and Rieger, C. (1993) Detection of intracellular cytokines by flow cytometry. *J. Immunol. Meth.* **159,** 197–207.
3. Sander, B., Andersson, J., and Andersson, U. (1991) Assessment of cytokines by the immunofluorescence and the paraformaldehyde-saponin procedure. *Immunol. Rev.* **119,** 65–93.
4. Sander, B., Hoiden, I., Andersson, U., Moller, E., and Abrams, J. S. (1993) Similar frequency and kinetics of cytokine producing cells in murine peripheral blood and spleen. *J. Immunol. Methods* **166,** 201–214.
5. Prussin, C. and Metcalf, D. D. (1995) Detection of intracytoplasmic cytokine using flow cytometry and directly conjugated anti-cytokine antibodies. *J. Immunol. Methods* **188,** 117–128.
6. Picker, L. J., Singh, M. K., Zdraveski, Z., Treer, J. R., Waldrop, S. L., Bergstresser, P. R., and Maino, V. C. (1995) Direct demonstration of cytokine

synthesis heterogeneity among human memory/effector T cells by flow cytometry. *Blood* **86,** 1408–1419.

7. Assenmacher, M., Schmitz, J., and Radbruch, A. (1994) Flow cytometric determination of cytokines in activated murine T helper lymphocytes: expression of interleukin-10 in interferon-γ and in interleukin-4–expression cells. *Eur. J. Immunol.* **24,** 1097–1101.
8. Elson, L., Nutman, T., Metcalfe, D., and Prussin, C. (1995) Flow cytometric analysis for cytokine production identifies T helper 1, T helper 2, and T helper 0 cells within the human CD4+CD27- lymphocyte subpopulation. *J. Immunol.* **154,** 4294–4301.
9. Kreft, B., Singer, G. G., Diaz-Gallo, C., and Rubin Kelly, V. (1992) Detection of intracellular interleukin-10 by flow cytometry. *J. Immunol. Methods* **156,** 125–128.
10. Crucian, B., Dunne, P., Friedman, H., Ragsdale, R., Pross, S., and Widen, R. (1996) Detection of altered Th1 and Th2 cytokine production by peripheral blood cells in multiple sclerosis using intracellular cytokine detection by flow cytometry in conjunction with surface marker analysis. *Clin. Diag. Lab. Immunol.* **3,** 411–416.
11. Andersson, J., Abrams, J., Bjork, L., Funa, K., Litton, M., Agren, K., and Andersson, U. (1994) Concomitant in vivo production of 19 different cytokines in human tonsils. *Immunol.* **83,** 16–24.
12. Fiskum, G., Craig, S. W., Decker, G. L., and Lehninger, A. L. (1980) The cytoskeleton of digitonin-treated rat hepatocytes. *Proc. Natl. Acad. Sci. USA* **77,** 3430–3434.
13. Anderson, P., Blue, M. L., O'Brien, C., and Schlossman, S. F. (1989) Monoclonal antibodies reactive with the T cell receptor chain: production and characterization using a new method. *J. Immunol.* **143,** 1899–1904.
14. Aargon, J. J., Feliu, J. E., Frenkel, R. A., and Sols, A. (1980) Permeabilization of animal cells for kinetic studies of intracellular enzymes: in situ behavior of the glycoslytic enzymes of erytherocytes. *Proc. Natl. Acad. Sci. USA* **77,** 6324–6328.
15. Henter, J. I., Soder, O., and Andersson, U. (1988) Identification of individual tumor necrosis factor/cachectin producing cells after lippopolysaccharide induction *Eur. J. Immunol.* **18,** 983–988.
16. Tartakoff, A. (1983) Perturbation of vesucular traffic with the carboxylic ionophore monensin. *Cell* **32,** 1026–1028.

4

Detection of Intracellular/Intranuclear Antigens

Applications in Leukemia/Lymphoma Analysis

Raymond H. Widen

1. Introduction

The increased utilization of flow cytometry for the study of cell-associated antigens has paralleled the continuing development of new monoclonal antibodies (MAbs) specific for cell-associated antigens. Availability of these reagents, combined with the introduction of newer fluorochromes such as phycoerythrin (PE) and peridinin chlorophyll (PERCp) or tandem conjugates such as PE-Cy5 that may be used in conjunction with fluorescein isothiocyanate (FITC) allow simultaneous measurement of coexpression of three or four different antigens. Flow cytometry instrumentation also has evolved to the point that three- and four-color analysis is routinely available in simple benchtop instruments, not just on high powered sorting instruments.

Although clinical flow cytometry often is associated simply with determining CD4 counts in HIV-infected individuals, multicolor flow cytometry has become an important tool for phenotyping leukemia or lymphoma samples in many laboratories. Indeed, the broad array of markers available to identify cell types and subtypes allows for precise identification of specific lineages, along with providing a means to detect abberant expression of lineage markers on cells that would not normally express such antigens (e.g., CD19 on acute myelogenous leukemia cells). Generally speaking, the vast majority of markers applied to leukemia/lymphoma typing to date have been cell surface proteins. Indeed, although specific markers for a variety of intracellular antigens are available from a variety of commercial sources, they have been underutilized because of some of the difficulties encountered in using flow cytometry to

From: *Methods in Molecular Biology, Vol. 91: Flow Cytometry Protocols*
Edited by: M. J. Jaroszeski and R. Heller © Humana Press Inc., Totowa, NJ

study intracellular markers. The principle challenge in performing analysis of cytoplasmic or nuclear antigens is the need to permeabilize the cells to allow access of the antibodies to the intracellular constituents. The ideal permeabilization system should:

1. Allow for retention of the cells' light-scattering characteristics.
2. Preserve the antigenicity of the target marker.
3. Prevent the leakage of the marker protein out of the permeabilized cell.
4. Allow the simultaneous detection of surface and intracellular markers.

A number of protocols using various methods for cell permeabilization have been described for the flow-cytometric measurement of intracellular antigens. Some examples of permeabilizing agents that have been utilized include alcohols and a variety of detergents or detergent-like materials ***(1–5)***. Although all of these methods were adequate for allowing the detection of intracellular antigens, they all had the critical problem of altering the normal light scattering characteristics of the cells, thereby compromising the ability to differentiate populations of cells. The loss of scatter properties may be of little importance when working with purified cell lines, however if one is working with a mixed population of cells from blood, bone marrow, or tissues it is of great importance to retain the ability to separate cells by their scatter characteristics to help define which cells express the antigen(s) of interest. Recently, a number of manuscripts describing in-house developed procedures and several commercially produced kit methods have been published that address the problem of obtaining adequate permeabilization to allow staining of intracellular antigens while preserving acceptable light-scattering characteristics to allow for differentiation of mixed populations ***(6–11)***. We describe herein the application of one of these commercial methods to detecting intracellular markers that provide useful information in the characterization of leukemia/lymphoma cells. The apparent common factor in all of these methods is the stabilization of the cell membrane by fixation with formaldehyde prior to the addition of the permeabilization agent (proprietary for the commercial products), thereby preserving the normal scatter characteristics and helping to reduce the leakage of markers from the permeabilized cells. An important consideration with this procedure is that paraformaldehyde fixation tends to lead to reduced resolution in DNA content analysis, so it is recommended that alternative procedures ***(12–14)*** be considered for studies in which accurate S phase or DNA indices need to be examined.

Table 1 provides vendor information for four of the commercially available fixation and permeabilization kits. **Table 2** summarizes some of the markers that we have had success with or have been published that pertain to the topic of leukemia/lymphoma phenotyping.

Table 1
Commercially Available Kits for Fixation/Permeabilization for Intracellular Antigen Detection by Flow Cytometry

Product	Vendor	Address
FACS lysing and FACS permeabilizing solution	Becton Dickinson Immunocytometry Systems	2350 Qume Drive San Jose, CA 95132 800–223–8226
Fix and Perm	Caltag Laboratories	36 Rozzi Place S. San Francisco, CA 94080 800–874–4007
Permeafix	Ortho Diagnostic Systems	Raritan, NJ 08869 800–322–6374
Cytoperm/Cytofix	Pharminigen	San Diego, CA 94121 800–848–6227

Table 2
Intracellular/Intranuclear Markers Applied to Leukemia/Lymphoma Phenotyping

1. Myeloperoxidase/lactoferrin: myeloid lineage[a]
2. TdT: immature lymphoid and a subset of myeloid leukemia[a]
3. Immunoglobulin heavy and light chains: B-cell, plasma cell[a]
4. Cytoplasmic CD3: immature T-cell[a]
5. Lysozyme: myeloid[b]
6. Cytoplasmic CD22: B lineage (immature cells lacking surface CD22)[b]
7. Cytoplasmic CD13: myeloid lineage[b]

[a]Test has been utilized in our laboratory with described procedure.
[b]Test performed using current procedure, *see* **ref.** *7*.

2. Materials

1. Dulbecco's phosphate-buffered saline (PBS): Dissolve powder in 1 L distilled or deionized reagent water, adjusting to pH 7.4 if needed. Filter sterilize using 0.22-μ filter and store at 4°C until use.
2. 1% Paraformaldehyde in PBS: Dissolve 5 g of paraformaldehyde (wear mask, gown, and gloves when handling and weigh under fume hood) in 500 mL PBS. Heat (under a fume hood) to 56°C for 60 min, mixing occasionally. Check and adjust to pH 7.4 if necessary. Store at 2–8° C for no more than 30 d.
3. Fix and Perm Permeabilization Kit (Caltag, South San Francisco, CA).
4. Monoclonal antibodies (MAbs) (various vendors).

3. Methods

3.1. Surface Marker Staining (if Needed)

See **Note 1** for recommendations concerning use of unlabeled primary antibodies in multicolor experiments. **Note 2** covers the appropriate use of isotype control antibodies or preimmune polyclonal control sera in the staining process as a means to set thresholds for positively stained cells. **Note 3** provides suggestions pertaining to color compensation when dealing with multicolor experiments. **Note 4** provides suggestions for addressing problems of excessive autofluorescence or high background staining.

1. Place 100 μL of the cell suspension containing 1–2 × 10^5 cells in a 12 × 75-mm tube along with the appropriate concentration of labeled test or control antibody.
2. Incubate in the dark at 4°C for 15–30 min.
3. Wash the cells with PBS by centrifugation at approx 300–500*g* for 5 min.
4. The pelleted cells may be handled following **step 1, Subheading 3.2.**

3.2. Caltag Method for Fixation/Permeabilization

1. Place 50 μL of the fresh cell suspension (no surface marker stain), or resuspend the surface marker stained cells, in 100 μL of Reagent A (fixation medium) in a 5-mL tube.
2. Incubate 15 min at room temperature (in the dark if surface marker staining was performed previously).
3. Add 3–5 mL of PBS and centrifuge for 5 min at 300–500*g*.
4. Remove supernatant carefully and completely.
5. Resuspend pellet in 100 μL Reagent B (permeabilization medium) along with the appropriate volume of the labeled antibody to the appropriate tube. Vortex gently for 1–2 s to suspend the pellet.
5. Incubate for 15 min at room temperature in the dark.
6. Wash the cells with PBS as described in **step 3** above.
7. Remove the supernatant and resuspend the pellet in 0.5–1 ml PBS containing 1% paraformaldehyde and store at 2–8°C for up to 24 h. For optimal results, cells should be analyzed within 24 h of staining.
8. Since the scatter and fluorescent intensity/compensation settings generally are similar to nonpermeabilized cells, start off collecting events using settings you have optimized previously for the cell type you are testing. See **Note 3** for further comments on compensation settings and gating strategies.
9. Examples of studies utilizing intracellular antigen detection in multicolor experiments are depicted in **Figs. 1** and **2**.

4. Notes

1. Cell-associated proteins may be detected with either monoclonal or polyclonal antibodies; generally speaking the use of MAbs is less problematic due to less likelihood of nonspecific reactivity and background staining. When monoclonal

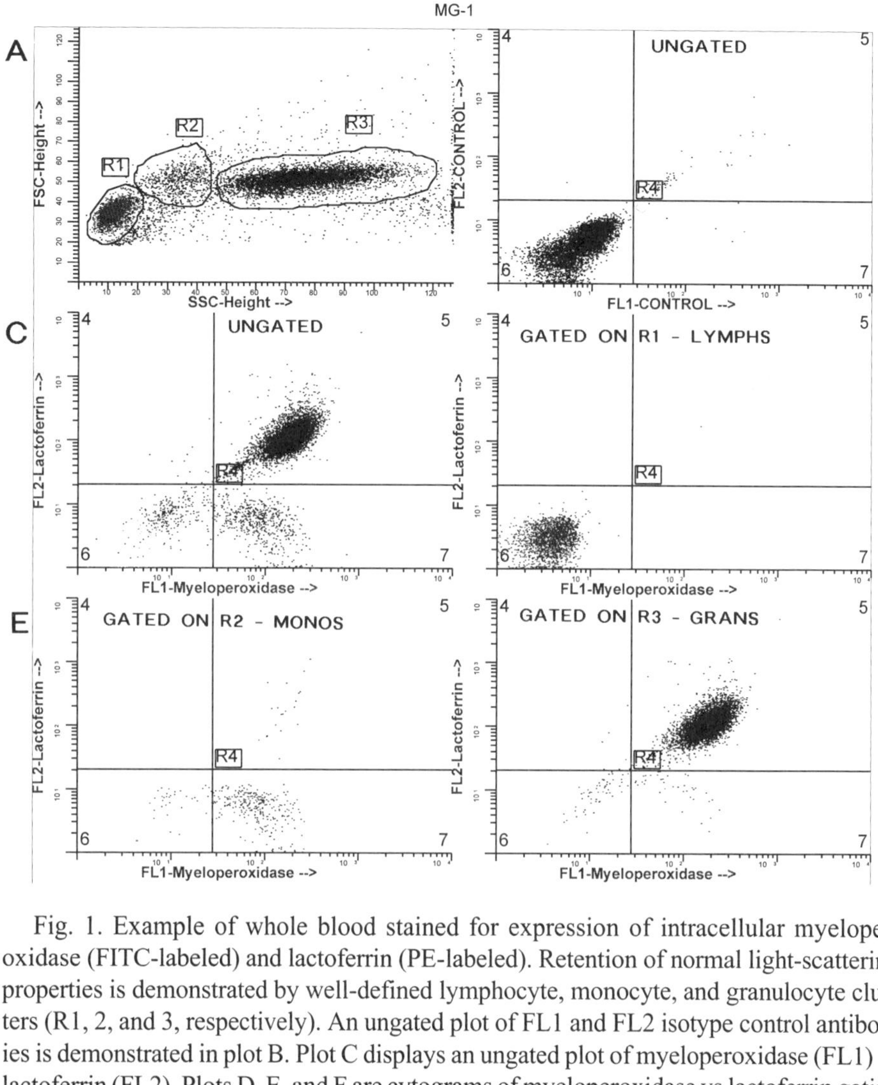

Fig. 1. Example of whole blood stained for expression of intracellular myeloperoxidase (FITC-labeled) and lactoferrin (PE-labeled). Retention of normal light-scattering properties is demonstrated by well-defined lymphocyte, monocyte, and granulocyte clusters (R1, 2, and 3, respectively). An ungated plot of FL1 and FL2 isotype control antibodies is demonstrated in plot B. Plot C displays an ungated plot of myeloperoxidase (FL1) vs lactoferrin (FL2). Plots D, E, and F are cytograms of myeloperoxidase vs lactoferrin gating on lymphocytes, monocytes, and granulocytes, respectively. As expected, lymphocytes are negative for both markers, monocytes are positive for myeloperoxidase and negative for lactoferrin, and granulocytes are positive for both cytoplasmic markers.

reagents are not available, affinity purified preparations of polyclonal antibodies and F(ab)'2 preparations often provide cleaner assays than do whole polyclonal antisera preparations. Additionally, the use of directly fluorochrome-conjugated antibodies greatly simplifies performance of multicolor flow cytometric analyses since using directly conjugated reagents allows for staining multiple markers simultaneously.

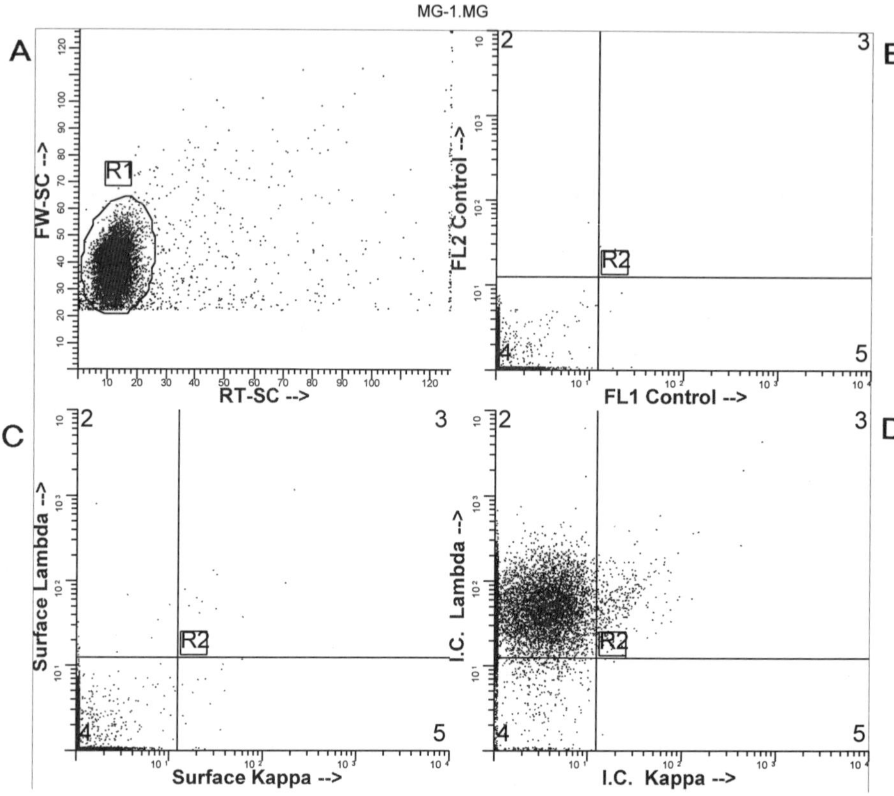

Fig. 2. Application of intracytoplasmic light-chain analysis in a B-cell lymphoma that had undetectable surface immunoglobulin. **(A)** Forward vs side-scatter plot showing a predominance of small lymphocytes. **(B)** Plot of isotype control monoclonal antibodies for the intracellular assay. **(C)** Plot of analysis of surface light chains using κ (FL1) vs λ (FL2). **(D)** Plot of cytoplasmic κ (FL1) vs λ (FL2).

The use of unlabeled primary antibodies often is necessary however, and although it poses greater challenges, the suggestions that follow often will be successful. Obviously, if one wishes to perform single-color analyses, the task involves simply performing the first incubation of the permeabilized cells with the primary (unlabeled) antibody, followed by a single wash in buffer and staining with a fluorochrome-labeled antibody specific for the species and appropriate isotype of the primary antibody (i.e., goat antimouse IgG if it was a mouse IgG primary antibody). The processing would proceed as described in **Subheading 3.** from here on. On the other hand, if multicolor experiments are desired, the first step is to stain with the unlabeled primary antibody reagent first (regardless of the cellular location of the target marker). After incubation, washing and staining with the fluorochrome-labeled secondary antibody, excess normal serum of the same species as the primary antibody is added (for approx 10–15 min) to block any free binding sites on the labeled detecting antibody. After an additional wash step, staining with the second or third color reagents may begin with the target being either intracellular or on the cell's surface. The staining process now proceeds as described in **Subheading 3.** An alternative approach that has been described for analysis of two markers using two unlabeled primary antibodies involves the use of secondary antibodies specific for subclasses of the primary antigen-specific immunoglobulins. For example, if the primary monoclonal antibodies are of IgG1 and IgG2a types, one may perform follow-up staining with anti-IgG1 and anti-IgG2a-specific antibodies labeled with different fluorochromes. Similarly, if one of the primary antibodies is an IgM class and the other is IgG, then detection may be accomplished with specific anti-IgM and IgG labeled with different fluorochromes. The key is the narrow specificity of the second antibody and the design of appropriate controls to verify no crossreactivity of the reagents with the alternate targets (i.e., the anti IgG2a indeed does not react with the IgG1 primary antibody and vice versa). Another factor that must be considered when using indirect staining procedures to detect intracellular antigens followed with a direct label for a surface marker is the stability of the surface marker to the fixation/permeabilization procedure.

2. The use of fluorochrome-labeled control antibodies of the same isotype, or labeled preimmune serum for polyclonal reagents is important for flow cytometric analysis of intracellular as well as surface marker studies. The control reagents should be used at the same concentration as the specific antibodies. Such controls are particularly important when studying antigens that are present at low densities, resulting in fluorescence distributions that trail off the level of the negative control—reagent/antigen combinations that give clearly resolved positive and negative patterns are not as critical.
3. Compensation for spectral overlap of the different fluorochromes used in multicolor flow cytometry is addressed in Chapters 1 and 2. In our experience, using the methods described in this chapter, compensation settings that are effective for surface marker studies work fairly well for intracellular antigen studies. The most common need for adjustment occurs when staining antigens with high levels of expression with excess overlap of FITC (FL1) into the PE (FL2) channel and occasionally similar problems with excess FL2 signal in the FL3 channel using a dye emitting in the 670-nm range. Gating strategies may include standard FWD vs SSC gating since scatter characteristics are preserved. Alternatively, gating may be performed based on "immunoscatter" ***(15,16)*** in which one of the fluorescence parameters is plotted against one of the scatter parameters. We have successfully utilized fluorescence vs SSC with parameters such as CD19 FL3 vs SSC for intracellular light-chain or TdT analysis and CD38 vs SSC to gate plasma cells for light- or heavy-chain studies.
4. One problem that may occur with flow cytometric immunofluorescence analysis of either surface or intracellular markers is the potential for high background or nonspecific binding of the antibody to the cells. If autofluorescence is detected,

use of fluorochromes, such as PE and PE-Cy5 or PercP, that emit in the orange and red wavelengths may prove useful. If nonspecific binding of the antibody appears to be occuring, the use of isotype or preimmune serum controls for monoclonal and polyclonal antibodies, respectively, provides some degree of control for this problem. Additionally, you may need to further titrate the antibody since excess antibody may contribute to this problem. When indirect immunofluorescence assays are utilized, controls for nonspecific binding of both components is necessary. If nonspecific binding remains a problem, a simple solution that often works to reduce nonspecific binding without significantly reducing specific antigen antibody interactions is to include 0.5–5% concentrations of an alternative protein, such as bovine serum albumin, fetal or newborn bovine serum, rehydrated sterile nonfat dry milk, normal autologous or isologous serum/plasma, or normal sera from the same host used to generate the antibodies used in the test. These agents function by competing with the relatively low concentrations of test antibodies in binding to the cells. An approach for documenting antigen specificity is the use of blocking procedures in which excess free antigen is added to specifically and competitively block the interaction of the antibody with the cell associated target; further controls using excess unrelated free antigen will not result in competitive inhibition. If excess free antigen is not readily available, an alternative approach is to block labeled/specific antibody with unlabeled but not with antibodies specific for unrelated antigens.

References

1. Levitt, D. and King, M. (1987) Methanol fixation permits flow cytometric analysis of immunofluorescent stained intracellular antigens. *J. Immunol. Methods* **96,** 233–237.
2. Schroff, R. W., Bucana, C. D., Klein, R. A., Farrell, M. M., and Morgan, A. C. (1984) Detection of intracytoplasmic antigens by flow cytometry *J. Immunol. Methods,* **70,** 167–177.
3. Landay, A., Jennings, C., Forman, M., and Raynor, R. (1993) Whole blood method for simultaneous detection of surface and cytoplasmic antigens by flow cytometry. *Cytometry* **14,** 433–440.
4. Drach, J., Gattringer, C., and Huber, H. (1991) Combined flow cytometric assessment of cell surface antigens and nuclear TdT for the detection of minimal residual disease in acute leukemia. *Brit. J. Hematol.* **77,** 37–42.
5. Gore, S. D., Kastan, M. B., Goodman, S. N., and Civin, C. (1990) Detection of minimal residual T cell acute lymphoblastic leukemia by flow cytometry. *J. Immunol. Methods* **132,** 275–286.
6. Syrjala, M. T., Tiirkainen, M., Jansson, S. E., and Krusius, T. (1993) Flow cytometric analysis of terminal deoxynucleotidy transferase: a simplified method. *Am. J. Clin. Pathol.* **99,** 298–303.
7. Knapp, W., Strobl, H., and Majdic, O. (1994) Flow cytometric analysis of cell surface and intracellular antigens in leukemia diagnosis. *Cytometry (Commun. Clin. Cytometry)* **18,** 187–198.

8. Tiirikainen, M. (1995) Evaluation of red blood cell lysing solutions for the detection of intracellular antigens by flow cytometry. *Cytometry* **20,** 341–348.
9. Francis, C. and Connelly, M. C. (1996) Rapid single step method for flow cytometric detection of surface and intracellular antigens using whole blood. *Cytometry* **25,** 58–70.
10. Hallden, G., Andersson, U., Hed, J., and Johansson, S. G. O. (1989) A new membrane permeabilization method for the detection of intracellular antigens by flow cytometry. *J. Immunol. Methods* **124,** 103–109.
11. Sander, B., Andersson, J., and Andersson, U. (1991) Assessment of intracellular cytokines and the paraformaldehyde saponin procedure. *Immunol. Rev.* **119,** 65–93.
12. Begg, A. C. and Hofland, I. (1991) Cell kinetic analysis of mixed populations of cells using three color fluorescence flow cytometry. *Cytometry* **12,** 445–454.
13. Schmid, I., Uittenbogaart, C. H., and Giorgi, J. V. (1991) A gentle fixation and permeabilization method for combined cell surface and intracellular staining with improved precision in DNA quantification. *Cytometry* **12,** 279–285.
14. Schutte, B., Tinnemans, M. M. F. J., Pijpers, G. F. P., Lenders, M.-H. J. H., and Ramakers, F. C. S. (1995) Three parameter flow cytometric analysis for simultaneous detection of cytokeratin, proliferation associated antigens and DNA content. *Cytometry* **21,** 177–186.
15. Stelzer, G. T., Shultz, K., and Loken, M. R. (1993) CD45 gating for routine flow cytometric analysis of human bone marrow specimens. *Ann. NY Acad. Sci.* **677,** 265–279.
16. Mercolino, T. J., Connelly, M. C., Meyer, E. J., Knigfht, M. D., Parker, J. W., Stelzer, G. T., and DeChirico, G. (1995). Immunologic differentiation of absolute lymphocyte count with an integrated flow cytometric system: a new concept for absolute T cell determinations. *Cytometry (Commun. Clin. Cytometry)* **22,** 48–59.

5

Detection of Intracellular/Intranuclear Antigens in Tumor Cells

Raymond H. Widen and Jeanne L. Becker

1. Introduction

The application of flow cytometry for the study of cell surface antigens has evolved significantly in the last 20 yr. One of the factors involved is the continuing development of a vast array of monoclonal antibodies specific for cell-associated antigens. Another important factor is the development of newer fluorochromes, such as phycoerythrin (PE) and peridinin chlorophyll (PERCp), or tandem conjugates, such as PE-Cy5, that complement fluorescein isothiocyanate (FITC) to allow simultaneous measurement of coexpression of three different antigens. With the availability of dyes that allow for three- and now four-color analyses, along with the demand from investigators for this ability, the major manufacturers of flow cytometry instrumentation have produced instruments that readily measure three fluorescence colors along with forward and side scatter. Simple bench-top flow cytometers capable of four-color measurements have made their way to the marketplace.

Despite the availability of the instrumentation and although many monoclonal and polyclonal antibodies specific for cytoplasmic or nuclear antigens have been available for several years, their use in flow cytometric analyses have been somewhat limited because of technical and methodologic problems related to the need to permeabilize the cytoplasmic membrane to allow access to the intracellular antigens.

A number of protocols using various methods for cell permeabilization have been described for the flow cytometric measurement of intracellular and/or intranuclear antigens. Examples of permeabilizing agents that have been utilized include acetone, alcohols, and detergents or detergent-like materials *(1–11)*. These methods allowed the detection of intracellular antigens but they altered the light-scattering characteristics of the cells. Several newer methods

From: *Methods in Molecular Biology, Vol. 91: Flow Cytometry Protocols*
Edited by: M. J. Jaroszeski and R. Heller © Humana Press Inc., Totowa, NJ

achieve adequate permeabilization for staining of intracellular antigens and preserve the normal light scattering characteristics of the cells to allow for differentiation of mixed populations ***(12–17)*** . We have utilized a pretreatment of cells with 1% paraformaldehyde in PBS for 18–24 h prior to permeabilization with 0.1% saponin; this procedure allowed for essentially unchanged light-scatter characteristics and successful permeabilization. Sander et al. ***(12)*** described a method of brief fixation (15–30 min) with 4% paraformaldehyde followed by permeabilizing with saponin that also yielded excellent results for both light scatter and intracellular antigen detection. Several commercial kits for fixation and permeabilization are now available ***(14–16)*** that accomplish the same goal of providing access to intracellular antigens for staining and retention of normal light scatter characteristics. A good review of some of the general strategies for cell fixation and permeabilization may be found in the article coauthored by Bauer and Jacobberger ***(18)***. The critical factor in all of these methods is the stabilization of the cell membrane by fixation with formaldehyde prior to the addition of the permeabilization agent. The stabilization of the membrane preserves the normal scatter characteristics of the cells and helps to prevent leakage of the intracellular constituents during the staining process.

These improved fixation/permeabilization methods have created the opportunity to simultaneously measure surface, and cytoplasmic and intranuclear antigens in mixed populations of cells, allowing for selective gating to analyze coexpression of a variety of markers. These methods will have numerous important uses in areas such as expression of tumor promotors and/or suppressors, cyclins, and other cell cycle-associated molecules, cytokines, and other regulatory proteins as well as other markers of cell function yet to be described. In this chapter we will describe applications of the paraformaldehyde/saponin method in analysis of intracellular proteins in cultured tumor cells.

2. Materials

1. Dulbecco's phosphate-buffered saline (PBS): Dissolve powder in 1 L of distilled or deionized reagent water, add 0.5 g sodium azide and adjust to pH 7.4 if needed. Filter sterilize using 0.22 µm filter and store at 4°C until use.
2. 1% Paraformaldehyde in PBS: Dissolve 5 g paraformaldehyde (wear mask and gloves and weigh under fume hood) in 500 mL PBS. Heat (under a fume hood) to 56°C for 60 min, mixing occasionally. Check and adjust to pH 7.4 if necessary. Store at 2–8°C for no more than 30 d.
3. 4% Paraformaldehyde in PBS: Dissolve 4 g paraformaldehyde in 100 mL PBS, using the methods and precautions described for **step 2** above. Store at 2–8°C for no more than 30 d.
4. Staining buffer for the paraformaldehyde/saponin procedure: Aseptically add 5 mL sterile FBS to 500 mL PBS. Store at 2–8°C for 6 min.

5. Permeabilization buffer for the paraformaldehyde/saponin procedure: Dissolve 0.1 g saponin in 100 mL of the staining buffer (mixed in **step 4** above) and store at 2–8°C for 1 min.

3. Methods

The methods described below assume staining with directly conjugated antibodies in multicolor experiments. *See* **Note 1** for suggestions for use of unlabeled primary antibodies in multicolor protocols. **Notes 2** and **3** discuss the use of isotype or preimmune serum control antibodies and color compensation respectively. **Note 4** deals with possible techniques to reduce background or autofluorescence levels. If your application involves measurement of DNA content and intracellular antigens, refer to **Note 5** for modifications of the paraformaldehyde/saponin procedure or for references to other methods that may prove useful.

3.1. Surface Marker Staining (if Required)

1. Prepare the cell suspension from fresh tissues or cell culture using the methods normally followed in your laboratory.
2. Place 100 μL of the cell suspension containing 1–2 × 10^5 cells in a 12 × 75-mm tube along with the appropriate concentration of labeled test or control antibody.
3. Incubate in the dark at 4°C for 15–30 min.
4. Wash the cells by centrifugation at approx 300–500*g* for 5 min.
5. Suspend the cells in 3 mL PBS and wash by centrifugation (300–500*g* for 5 min).
6. Suspend the cell pellet in the appropriate fixative according to the method you will use.

3.2. Paraformaldehyde/Saponin Procedure for Staining Intracellular Antigens

1. Place 100 μL fresh cell suspension in 12 x 75-mm tubes for each sample and antibody to be tested and add 0.5 mL PBS containing 4% paraformaldehyde or 0.5 mL PBS containing 1% paraformaldehyde and incubate at 4°C for 15 min or 18–24 h, respectively. If cells were prestained for surface marker, simply suspend the final wash pellet in 0.5 mL paraformaldehyde solution.
2. After the appropriate time for fixation, wash the cells once by centrifugation (300–500*g*) with 4 mL PBS. Decant the supernatant.
3. Resuspend the cell pellet in 3 mL permeabilization buffer and spin at 300–500*g* for 5 min.
4. Resuspend the cell pellet in 100 μL of permeabilization buffer along with the appropriate volume of the test or control antibody to the appropriate tubes.
5. Incubate 30 min at 4°C in the dark.
6. Wash with 3 mL permeabilization buffer by centrifugation at 300–500*g* for 5 min.
7. Resuspend pellet in PBS or PBS with 1% paraformaldehyde if samples will not be analyzed within a few hours. All samples should be analyzed within 24 h for optimal results.

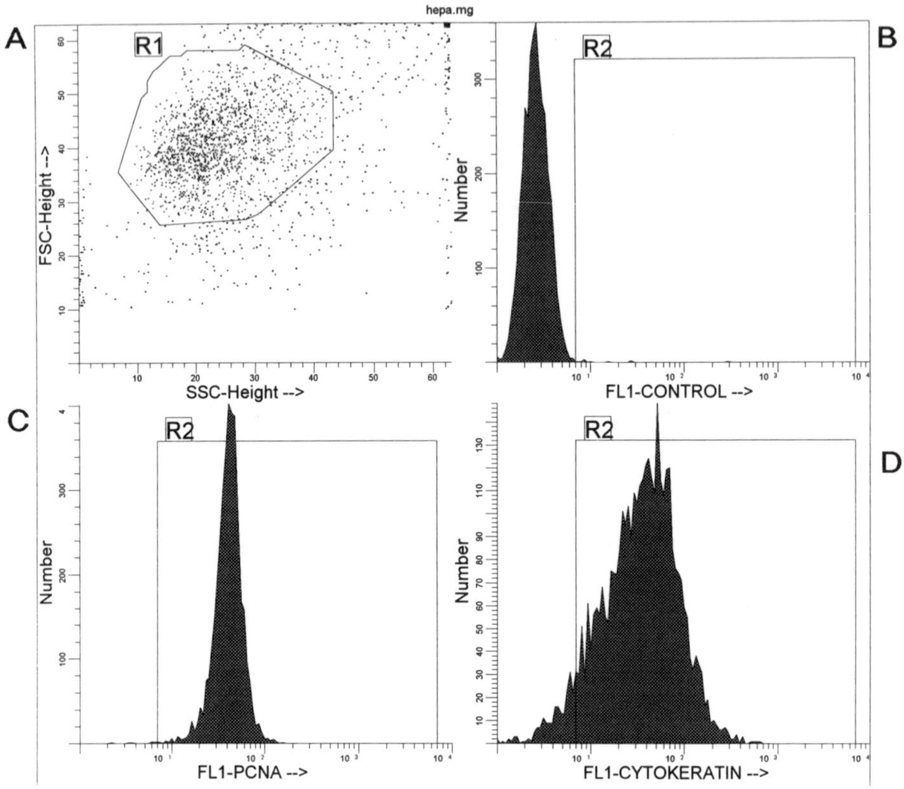

Fig. 1. Analysis of cytokeratin and proliferating cell nuclear antigen (PCNA) expression in Hep 2 carcinoma cells treated with the paraformaldehyde/saponin procedure. **(A)** Cytogram of forward vs side scatter. **(B)** Single-color histogram of FL1 isotype control antibody staining. **(C)** Single-color histogram of PCNA demonstrating accessibility to intranuclear antigens. **(D)** Single-color histogram of cytoplasmic cytokeratin expression.

8. Since the scatter and fluorescent intensity/compensation settings generally are similar to nonpermeabilized cells, start off collecting events using settings you have optimized previously for the cell type you are testing. Fine tune the gain settings for scatter and fluorescence channels to optimize the instrument for your experiment. Results of studies using cultured Hep 2 cells and a mixed muellerian ovarian tumor line *(19)* are depicted in **Figs. 1** and **2**. *See* **Note 2** for a discussion of controls that may be applicable to specific study parameters. *See* **Note 3** for suggestions relating to color compensation in multicolor experiments. **Note 4** provides recommendations for addressing problems with autofluorescence or high background staining. Finally, **Note 5** discusses a modification to the

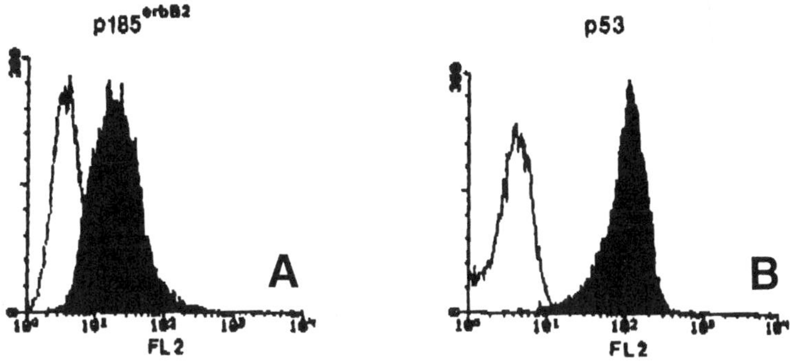

Fig. 2. Flow cytometric examination of regulatory protein expression in a mixed mullerian ovarian cancer cell line (LN1). The data in plots A and B demonstrate the applicability of the paraformaldehyde-saponin procedure to detection of erbB2 (Her2/neu) and p53 (solid plot), respectively. Isotype control antibody staining is shown in the open plot.

paraformaldehyde/saponin procedure that may provide adequate staining of DNA content with propidium iodide in conjunction with intracellular antigen detection (*see* **Fig. 3**).

4. Notes

1. The immunofluorescent detection of cell-associated antigens is most easily accomplished using antibodies that are directly conjugated to the fluorochrome. These may be in the form of monoclonal antibodies (MAbs) or polyclonal antisera; the former generally offer lower likelihood of nonspecific or cross reactions than do the polyclonal reagents. However, several vendors offer affinity purified polyclonal reagents, with the additional improvement of being available in F(ab)'2 preparations that lack the Fc portion that may contribute to nonspecific binding through cellular Fc receptors. The use of directly fluorochrome-conjugated antibodies greatly simplifies performance of multicolor flow cytometric analyses to detect two or more markers simultaneously. The use of unlabeled primary antibodies often is necessary however, presenting greater complexity in experimental design and controls. If the particular experiment calls for only a single color of immunofluorescence (only one marker) the performance of indirect assays is less problematic. Simply perform the first incubation of the permeabilized cells with the primary (unlabeled) antibody, and after a single wash step in permeabilization buffer, add the fluorochrome-conjugated secondary antibody and proceed with the rest of the described procedure. In contrast, if the experiment involves multiple markers (multiple antibodies therefore), the first step is to stain with the unlabeled primary antibody reagent. After incubation, washing, and staining with the fluorochrome-labeled secondary antibody, excess normal serum of the same

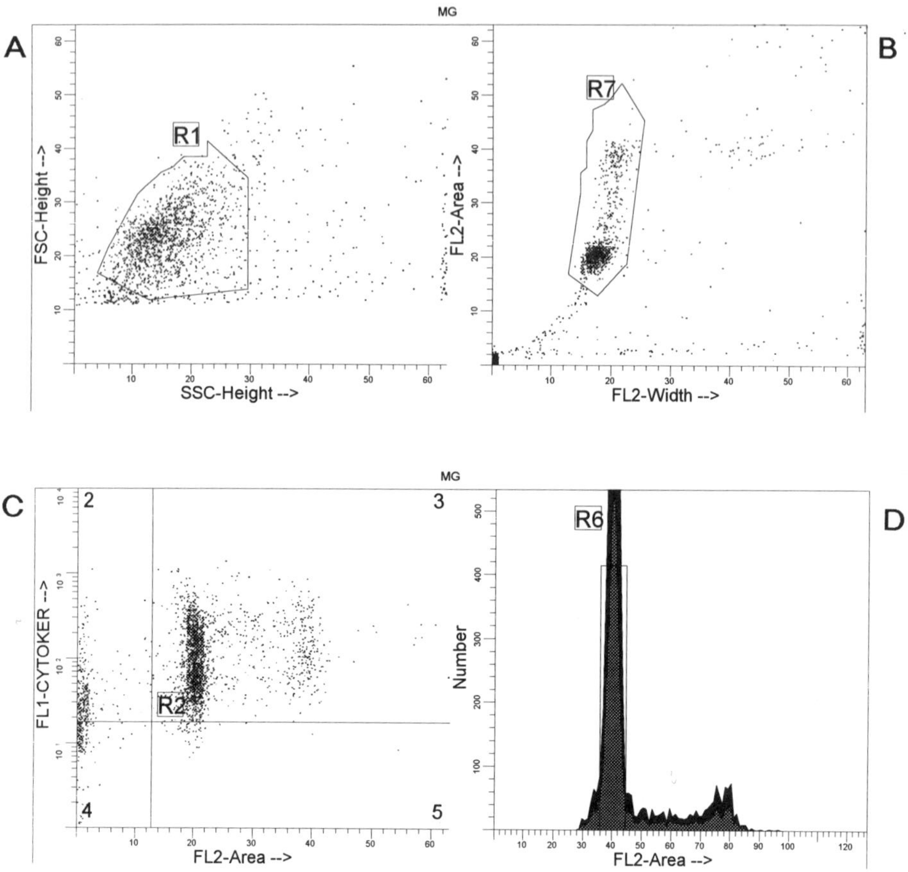

Fig. 3. Demonstration of two parameter analysis of DNA content (propidium iodide, PI) (FL2) vs cytokeratin (FL1) in Hep 2 cells heated 1 h at 70°C to restore DNA-staining characteristics after paraformaldehyde fixation. Plot A demonstrates retention of light-scatter characteristics after paraformaldehyde fixation, heat treatment, and permeabilization with saponin. Plot B is a cytogram of FL2 width vs FL2 area to demonstrate minimal doublet formation. Plot C is a cytogram of cytoplasmic cytokeratin (FL1) vs PI (FL2) to demonstrate retention of the ability to detect intracellular antigens and DNA staining characteristics using this treatment. Plot D is a single-color histogram of FL2 area gated off the FL2 width vs FL2 area plot to demonstrate acceptable DNA staining, with G0/G1 peak CV of 4.2 after the heat treatment (CV 9.9 in cells processed without the heat treatment).

species as the primary antibody must be added for approx 15 min to block any free binding sites on the labeled detecting antibody. Another wash step is performed after which you may begin staining with the second or third color reagents. The staining process now proceeds as described in the methods. An alternative approach may be applied for analysis of two markers using two unlabeled primary antibodies that have different isotypes. For example, if the primary MAbs are of IgG1 and IgG2a types, one may perform follow up staining with FITC-labeled anti-IgG1 and PE-labeled anti-IgG2a. It is critical that the detecting reagents have documented narrow specificity for only the particular isotypes.

2. Isotype control or preimmune polyclonal antibody preparations should be used for intracellular antigen tests, unless the target is present at levels to provide obvious differentiation between positive and negative cells. Even under the latter conditions, it is advisable to perform pilot studies including these nonspecific-binding controls to document test validity.
3. Compensation for spectral overlap of the different fluorochromes used in multicolor flow cytometry is addressed in Chapters 1 and 2. In general, compensation settings that work for surface marker combinations apply quite well to intracellular antigen studies. It is advisable to perform pilot studies with individually labeled samples to verify efficacy of compensation in specific applications. For example, run a separate tube with cells stained only with FITC and another tube with cells stained only with PE. Display the live data in a two parameter plot of FL1 vs FL2. In the case of the FITC-labeled cells, all of the positive events should be observed in the quadrant associated with FL1+/FL2- events if the compensation is adequate. Similarly, the tube with cells labeled only with the FL2 reagent should have positive cells appearing only in the quadrant with FL1-/FL2+ events. The intensity of fluorescence in the irrelevant channel (i.e., FL2 for the FL1 reagent) should be the same as that in unlabeled cells or in the double-negative quadrant (FL1-/FL2-).
4. Autofluorescence and high background may present problems in any flow cytometry study. For cells with high levels of autofluorescence, try to use fluorochromes such as PE and PE-Cy5 or PERCp that emit in the orange and red wavelengths since most cellular autofluorescence is highest in the FITC range. Nonspecific binding may be addressed by using isotype or preimmune serum controls. Another approach that often works to reduce nonspecific binding without significantly reducing specific antigen–antibody interactions is to include 0.5–5% concentrations of an alternative protein (albumin, bovine serum, and so on). These agents compete with the lower concentrations of test antibodies in binding to the cellular component that may nonspecifically interact with the test antibody. Antigen specificity may be documented by using blocking with excess soluble antigen or by using blocking with an unlabeled form of the same antibody used in the test. In the latter situation, demonstration of blocking with unlabeled specific antibody but not by unlabeled irrelevant antibody supports the argument that you are measuring an antigen-specific reaction in your test.
5. Formaldehyde fixation of cells leads to reduced binding of dyes, such as propidium iodide, to cellular DNA, thereby reducing intensity of fluorescence and raising the measured coefficient of variation. As a result, measurement of ploidy and S phase is compromised. Use of the paraformaldehyde/saponin procedure as described is not recommended if accurate ploidy or S phase measurements based on DNA-binding dyes are important. Instead, we recommend trying

the following modifications to the paraformaldehyde/saponin procedure or an alternative procedure for fixation/permeabilization *(19–21)*. The modification to the paraformaldehyde/saponin procedure is based on a recent observation by Overton and McCoy *(22)* that heating formalin-fixed cells at 75°C for 1 h restores DNA staining to prefixation levels in terms of both CV of the G0/G1 peak and staining intensity. We have found that adding this heating step after fixation with either 1% or 4% paraformaldehyde and a single wash in PBS and resuspension in PBS achieves good DNA staining while maintaining light scatter and intracellular antigen stainability. After the 1-h heat treatment, the cells may be washed once with PBS and once with permeabilization buffer and then processed from **step 3** on in the standard paraformaldehyde/saponin procedure. Results of a study using Hep 2 cells in a two color cytokeratin vs DNA (propidium iodide) protocol are shown in **Fig. 3**.

References

1. Levitt, D. and King, M. (1987) Methanol fixation permits flow cytometric analysis of immunofluorescent stained intracellular antigens. *J. Immunol. Methods* **96,** 233–237.
2. Schroff, R. W., Bucana, C. D., Klein, R. A., Farrell, M. M., and Morgan, A. C. (1984) Detection of intracytoplasmic antigens by flow cytometry *J. Immunol. Methods* **70,** 167–177.
3. Landay, A., Jennings, C., Forman, M., and Raynor, R. (1993) Whole blood method for simultaneous detection of surface and cytoplasmic antigens by flow cytometry. *Cytometry* **14,** 433–440.
4. Drach, J., Gattringer, C., and Huber, H. (1991) Combined flow cytometric assessment of cell surface antigens and nuclear TdT for the detection of minimal residual disease in acute leukemia. *Brit. J. Hematol.* **77,** 37–42.
5. Gore, S. D., Kastan, M. B., Goodman, S. N., and Civin, C. (1990) Detection of minimal residual T cell acute lymphoblastic leukemia by flow cytometry. *J. Immunol. Methods* **132,** 275–286.
6. Pollice, A. A., McCoy, J. P., Shackney, S. E., Smith, C. A., Agarwal, J., Burholt, D. R., Janocko, L. E., Hornicek, F. J., Singh, S. G., and Hartsock, R. J. (1992) Sequential paraformaldehyde and methanol fixation for simultaneous flow cytometric analysis of DNA, cell surface proteins and intracellular proteins. *Cytometry* **13,** 432–444.
7. Hirata, M. and Okamoto, Y. (1987) Enumeration of terminal deoxynucleotide transferase positive cells in leukemia/lymphoma by flow cytometry. *Leukemia Res.* **11,** 509–518.
8. Clevenger, C. V., Bauer, K. D., and Epstein, A. L. (1985) A method for simultaneous nuclear immunofluorescence and DNA content measurement using monoclonal antibodies and flow cytometry. *Cytometry* **6,** 208–214.
9. Slaper-Cortenbach, I. Cm, Admiraal, L. G., Kerr, J. M., van Leeuwen, E. F., von dem Borne, A. E. G. Kr., and Tetteroo, P. A. T. (1988) Flow cytometric detection of terminal deoxynucleotidyl transferase and other intracellular antigens in combination with membrane antigens in acute lymphatic leukemias. *Blood* **72,** 1639–1644.

10. Dent, G. A., Leglise, C., Przwansky, K. B., and Ross, D. W. (1989) Simultaneous paired analysis by flow cytometry of surface markers, cytoplasmic antigens or oncogene expression with DNA content. *Cytometry* **10,** 192–198.
11. Jacobberger, J. W., Fogelman, D., and Lehman, J. M. (1986) Analysis of intracellular antigens by flow cytometry. *Cytometry* **7,** 356–364.
12. Sander, B., Andersson, J., and Andersson, U. (1991) Assessment of intracellular cytokines and the paraformaldehyde saponin procedure. *Immunol. Rev.* **119,** 65–93.
13. Syrjala, M. T., Tiirkainen, M., Jansson, S. E., and Krusius, T. (1993) Flow cytometric analysis of terminal deoxynucleotidyl transferase: a simplified method. *Am. J. Clin. Pathol.* **99,** 298–303.
14. Knapp, W., Strobl, H., and Majdic, O. (1994) Flow cytometric analysis of cell surface and intracellular antigens in leukemia diagnosis. *Cytometry (Commun. Clin. Cytometry)* **18,** 187–198.
15. Tiirikainen, M. (1995) Evaluation of red blood cell lysing solutions for the detection of intracellular antigens by flow cytometry. *Cytometry* **20,** 341–348.
16. Francis, C. and Connelly, M.C. (1996) Rapid single step method for flow cytometric detection of surface and intracellular antigens using whole blood. *Cytometry* **25,** 58–70.
17. Hallden, G., Andersson, U., Hed, J., and Johansson, S. G. O. (1989) A new membrane permeabilization method for the detection of intracellular antigens by flow cytometry. *J. Immunol. Methods* **124,** 103–109.
18. Bauer, K. D. and Jacobberger, J. W. (1984) Analysis of intracellular proteins. *Methods Cell. Biol.* **41,** 352–373.
19. Becker, J. L., Prewett, T. L., Spaulding, G. F., and Goodwin, T. J. (1993) Three-dimensional growth and differentiation of ovarian tumor cell line in high aspect rotating-wall vessel: morphologic and embryologic considerations. *J. Cell. Biochem.* **51,** 283–289.
20. Begg, A. C. and Hofland, I. (1991) Cell kinetic analysis of mixed populations of cells using three color fluorescence flow cytometry. *Cytometry* **12,** 445–454.
21. Schmid, I., Uittenbogaart, C. H., and Giorgi, J. V. (1991) A gentle fixation and permeabilization method for combined cell surface and intracellular staining with improved precision in DNA quantification. *Cytometry* **12,** 279–285.
22. Schutte, B., Tinnemans, M. M. F. J., Pijpers, G. F. P., Lenders, M.-H. J. H., and Ramakers, F. C. S. (1995) Three parameter flow cytometric analysis for simultaneous detection of cytokeratin, proliferation associated antigens and DNA content. *Cytometry* **21,** 177–186.
23. Overton, W. R. and McCoy, J. P. (1994) Reversing the effect of formalin on the binding of propidium iodide to DNA. *Cytometry* **16,** 351–356.

6

Detection of Cyclins in Individual Cells by Flow and Laser Scanning Cytometry

Gloria Juan and Zbigniew Darzynkiewicz

1. Introduction

Progression of cells through successive phases and checkpoints of the cell cycle is maintained by sequential phosphorylation of different sets of nuclear and cytoplasmic proteins by cyclin-dependent kinases (CDKs) *(1–12)*. By activating their partner CDKs and targeting them to the respective protein substrates cyclins play a key regulatory role in this process. Cyclins B1, A, E, and D are expressed discontinuously during the cycle. The synthesis and degradation of these cyclins occurs at well-defined time points of the cell cycle (**Table 1**).

Cyclin B1 activates CDC2 whose kinase activity is essential for cell transition from G2 to M *(6,11)*. The onset of cyclin B1 accumulation is seen at the time of cell exit from S. Maximal levels of cyclin B1 exhibit cells entering mitosis. This cyclin is degraded rapidly during the transition to anaphase *(13)*. Cyclin A associates with either CDC2 or CDK2; the kinase activity of the complex drives the cell through S and G2 phases of the cycle *(14–16)*. Cellular accumulation of cyclin A starts early in S and its maximal expression is seen at the end of G2. This protein is rapidly degraded in prometaphase and the metaphase cells are essentially cyclin A negative. The kinase partner of cyclin E is CDK2 and this holoenzyme is essential for cell transition from G1 to S phase. Cyclin E starts to accumulate in the cell in mid-G1 and is maximally expressed at the time of cell entrance to S. Its continuous breakdown takes place as the cell progresses through S *(17–19)*. Expression of the different members of D family of cyclins (D1, D2, and D3) is tissue and cell-type specific. These cyclins are maximally expressed in response to mitogenic stimulation of G0 cells or by mitogens and growth factors. During exponential phase of cell growth their level appears to decrease *(20–25)*. Cyclins D activate

From: *Methods in Molecular Biology, Vol. 91: Flow Cytometry Protocols*
Edited by: M. J. Jaroszeski and R. Heller © Humana Press Inc., Totowa, NJ

Table 1
Cyclins and Their Partner CDKs During the Cell Cycle

Cyclin	Primary CDK partner(s)	Presumed role in cell cycle	Peak of expression	Localization
D type	CDK4 and CDK6	pRB phosphorylation, commitment to S phase	Early in G1	Nucleus
E	CDK2	Initiation of S	G1/S transition	Nucleus
A	CDK2 and CDC2	S and G2 traverse	During G2M	Nucleus
B1	CDC2	G2 traverse entrance to M	Late G2/M	Cytoplasm/ Nucleus[a]

[a]Cyclin B1 is localized in cytoplasm during G2 and undergoes translocation to nucleus during prophase.

CDK4/CDK6 and the complex phosphorylates the retinoblastoma tumor suppressor gene protein RB (pRB) ***(12,26,27)***. Phosphorylation of pRB releases E2F factor that initiates transcription of the components of the DNA replication machinery, thereby committing the cell to S phase ***(26,27)***. This event is the key regulatory point during G1, which appears to be defective in most tumors ***(1,27–29)***.

Immunocytochemical detection of cyclin proteins made it possible to investigate their expression in individual cells by flow cytometry ***(29–32)***. Because of their cell–cycle-phase specificity, analysis of the expression of cyclin proteins can be used, in addition to DNA content, as another marker of a cells' position in the cycle, and provide information about proliferative potential of cell populations. In this chapter we present the optimal conditions for immunocytochemical detection of cyclins B1, A, E, and D applicable for flow or laser scanning cytometry ***(33)***.

2. Materials

1. Cyclin antibodies: Antibodies to cyclin proteins are offered by different vendors. We have tested antibodies from a variety of sources, and have found that only a few were satisfactory for flow cytometric detection of cyclin proteins. Namely, the mouse monoclonal antibodies (MAbs) to cyclin B1 (clone GNS-1), cyclin A (clone BF-683), cyclin D1 (clone G124–326), cyclin D3 (clone G107–565), and to cyclin E (clone HE12; all provided by PharMingen, San Diego, CA; cyclin D1 was also obtained by Immnutech) were tested by us on a variety of cell types of human origin and found to be cell–cycle phase specific in all these cell types (*see* **Note 1**).
2. FITC-conjugated goat antimouse IgG (available from Sigma, St. Louis, MO).
3. Mouse IgG1:isotypic control (available from Sigma).

4. Cell fixative: In 80% ethanol. Keep in the freezer (approx –20°C) prior to use.
5. RNase A (DNase-free RNase, available from Sigma).
6. 1 mg/mL RNase A in phosphate-buffered saline (PBS) (DNase-free RNase, available from Sigma).
7. 1% bovine serum albumin (BSA) in PBS.
8. 0.25% Triton X-100 in PBS.

3. Methods

3.1. Cyclin Detection by Flow Cytometry

1. Fix cells in suspension by pipeting 1–2 × 10^6 cells in 0.5 mL of PBS (phosphate buffered salt solution). Add 5 mL of ice-cold 80% ethanol (cyclins E, A, and B1) or ice-cold 100% methanol (D-type cyclins). The cells may be stored in this fixative in the cold (–20 to –40°C) for 2–24 h (*see* **Note 1**).
2. Centrifuge cells at 300*g* for 5 min. Resuspend cell pellet in 5 mL PBS. Keep for 5 min at room temperature. Spin at 300*g* for 5 min. Repeat once more with PBS containing 1% bovine serum albumin (BSA), centrifuge again, and suspend the cell pellet (<10^6 cells) in 1 mL 0.25% solution of Triton X-100 in PBS. Keep on ice for 5 min. Add 5 mL PBS. Centrifuge at 300*g* for 5 min.
3. Suspend cell pellet in 100 µL 1% BSA in PBS containing the primary antibody at the appropriate dilution to obtain 0.5 µg of antibody per sample. Incubate for 60 min at room temperature with gentle agitation or at 4°C overnight. Control cells should be treated identically; however, the control cells should be incubated with the same batch of isotypic antibody instead of a cyclin antibody (*see* **Note 3**).
4. Rinse the cells with 1% BSA in PBS, centrifuge at 300*g* for 5 min.
5. Suspend the cell pellet in 100 µL 1% BSA in PBS containing FITC-conjugated goat anti-mouse IgG antibody (diluted 1:30). Incubate for 30 min in the dark at room temperature with gentle agitation.
6. Rinse the cells with 1% BSA in PBS; centrifuge at 300*g* for 5 min.
7. Suspend the cell pellet in a solution containing 5 µg PI and 1 mg/mL of DNase-free RNase A in PBS. Incubate 20 min at room temperature in the dark before measurement.
8. Measure cell green (530+20 nm; FITC) and red (>590 nm; PI) fluorescence on the flow cytometer using blue light excitation (argon laser).

3.2. Cyclin Detection by Laser-Scanning Cytometer

The method described above can be adapted to stain cells mounted on microscope slides, to be analyzed by multiparameter LSC. To be analyzed by Laser scanning cytometer (LSC) the cells are initially attached to the slides by cytospinning, fixed, rinsed, and then subjected to the procedures as presented above (**Subheading 3.1.**).

1. To attach cells by cytospinning, 300 µL of suspension in tissue culture medium (with serum) containing approx 20,000 cells are added into a cytospin chamber.

The cells are then cytocentrifuged at 1000 rpm for 6 min without allowing the samples to completely dry and submerged in the respective fixative (in Coplin jars). Small volumes of the respective buffers, rinses or staining solutions, as described in **Subheading 3.** are carefully layered on the cytospin area of the horizontally placed slides. At appropriate times, these solutions are removed with Pasteur pipet (or vacuum suction pipet). Small pieces (1 × 1 cm) of thin polyethylene foil may be layered on slides above the solution drops used for cell incubations to prevent drying. Incubations should be carried out in moist atmosphere.

2. Proceed as described in **Subheading 3.1., steps 3–8**. At the final step of the staining procedure, the cells are mounted in a drop of the respective staining solution, made identical as for flow cytometry. Coverslips may be sealed with warm paraffin or a gelatin-based sealer. Cell fluorescence is measured by LSC; the choice of the fluorescence excitation wavelength and emission filters is the same as described above for flow cytometry.

3.3. Expression of Cyclins B1, A, E, and D Measured by Flow Cytometry and Laser Scanning Cytometer

The scheduled timing of expression for cyclins in relation to the major phases of the cell cycle is reflected by a very characteristic pattern of the bivariate cyclin vs cellular DNA content distributions as shown in **Fig. 1** for normal human proliferating lymphocytes (cyclins B1, A, and E) and fibroblasts (cyclin D1). The expression of cyclin B1 is essentially limited to late S-phase cells and to cells with a G2/M content of DNA, although early- and mid-S-phase cells show a very low level of this protein. Expression of cyclin A is progressively increasing during S-phase and is maximal in cells having a G2/M DNA content; most G1 cells are either cyclin A negative or show minimal level of this protein. Expression of cyclin E can be summarized as follows: the maximal level of this protein is detected in the cells undergoing transition from G1 to S; its level continuously decreases during cell progression through S, with the result that G2+M cells are essentially cyclin E negative; and a distinct threshold in cyclin E expression is apparent at the G1/S transition. The cells have to accumulate cyclin E above the threshold level to enter S phase. Similar patterns of expression of cyclins B1, A, and E to that presented by proliferating lymphocytes is observed in normal fibroblasts and in several tumor cell lines.

The presence of cyclin D1 in exponentially growing normal fibroblasts is limited to cell in G0/1 (**Fig. 1**). Most cells in S and G2/M are cyclin D1 negative, with the exception of a very few cells with a G2/M DNA content. The latter may be G1 cell doublets, since not all doublets can be identified by analysis of the shape (pulse width) of the electronic signal *(34)*. Expression of cyclin D2 in U2–OS-sarcoma and C82 hybridoma cell lines measured by Lukas et al. (28) shows patterns similar to that of cyclin D1 in fibroblasts. Likewise,

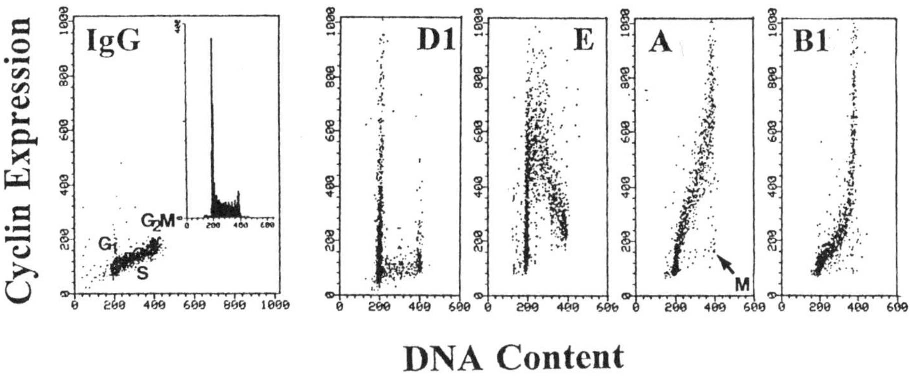

Fig. 1. Typical bivariate cyclin vs. DNA content distributions (scatterplots) showing expression of cyclin D1, E, A, and B1 vs DNA content in normal cells. The first panel shows the level of fluorescence of the respective control cells stained with the isotype IgG prior to fluoresceinated secondary antibody. The remaining panels are bivariate plots from cells stained with cyclin MoAb for D1, E, A, and B1.

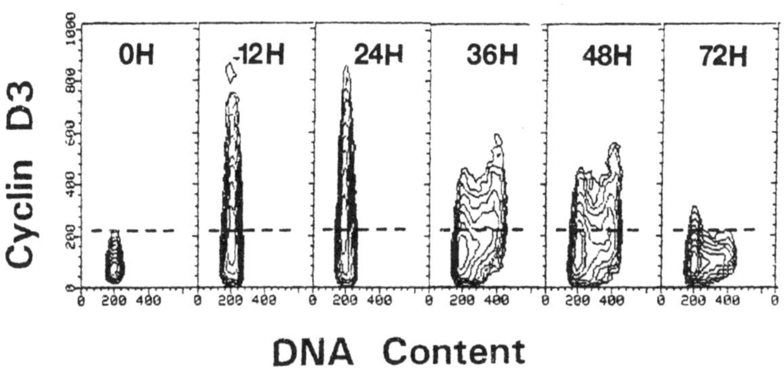

Fig. 2. Expression of cyclin D3 at different times during lymphocyte stimulation by PHA. The isometric contour maps show unstimulated lymphocytes at 0 and PHA-stimulated lymphocytes for 12, 24, 36, 48, and 72 h respectively. The broken horizontal line shows fluorescence level of the respective G0/1 controls (isotypic control) *(35)*.

expression of cyclin D3 by mitogenically stimulated human lymphocytes during exponential growth is also similar, being limited primarily to a subset of G0/1 cells with most of the S and G2/M cells being negative (***32,35***; **Fig. 2**; *see* **Notes 4–6**).

4. Notes

1. Some antibodies available from particular vendors may not be suitable for immunocytochemical detection of the antigen. Often, the antibody which is

very specific and useful for Western blotting, is ineffective for the detection of the antigen *in situ*. This is because the epitope of the denatured protein may react with the antibody, but in the native state, or somewhat altered by fixation, may no longer bind the antibody. The authors should always provide information (the vendor and hybridoma clone number) of the reagent used in their study.

2. The choice of fixative (cold ethanol, methanol, acetone, 1% formaldehyde) appears not to be a critical factor for cyclin detection and, although the absolute level of the immunofluorescence may vary, different fixation protocols yield essentially similar cyclin distributions with respect to the cell cycle position ***(32)***. Each fixative has some undesirable effects (e.g., increased cell clumping in the case of ethanol:acetone mixture, or cell autofluorescence and poor DNA stainability when formaldehyde is used) and one often has to compromise between these effects and the optimal detection of a particular cyclin.
3. The relative cellular content of a particular cyclin plays a role in its detection. For example, the signal-to-noise ratio (ratio of fluorescence intensity of the cyclin-positive cells to the control cells, stained with the isotype immunoglobin) is much higher in the case of cyclin B1 than in the case of cyclins E or A. This is because the cell immunofluorescence after staining with cyclin B1 antibody is much stronger compared to the immunofluorescence of cells stained for cyclins A or E. The level of expression of D-type cyclins varies markedly depending on the cell type and the phase of cell growth. High sensitivity of the instrument and low level of cell autofluorescence, therefore, are of greater importance for the detection of cyclins E or A than of cyclin B1 or D-type cyclins.
4. The change in the growth rate of the cells from exponential phase, such as subconfluence, addition of fresh medium with serum, inhibition of cell growth by antitumor drugs, and so on alters expression of D-type cyclins, both in terms of the absolute level of these proteins, as well as the pattern of their expression *vis a vis* cell cycle position. It is critical that expression of D type cyclins in different cell types is compared under identical conditions of cells growth because of their extreme sensitivity to environmental factors.
5. Perturbation of cell cycle progression such as cell synchronization by agents that interfere with DNA replication, results in significant changes in expression of cyclin proteins. Namely, the presentation of G1 cyclins (i.e., cyclin E) is observed in the cells in G2/M, or G2/M cyclins (i.e., cyclins A and/or B1) in G1 cells ***(34,35)***. This type of change was denoted by us as "unscheduled expression" of cyclins ***(32)***.
6. The majority of the tumor cell lines exhibit patterns of expression of cyclins B1, A, and E similar to the "scheduled" expression, seen for normal fibroblasts or lymphocytes. Some tumor cell lines, however, have distinctly "unscheduled" expression of these cyclins even when observed under conditions of exponential, unperturbed growth. Such phenotype of the tumor may be associated with defective regulation of the cell–cycle progression ***(32)***.

Acknowledgments

This work was supported by NCI Grant CA 28704, and the Chemotherapy Foundation. Gloria Juan, on leave from University of Valencia, Valencia, Spain, is supported by the "This Close" Fellowship from the "This Close" for Cancer Research Foundation. We thank Susan Wormsley of PharMingen for kindly providing the cyclin monoclonal antibodies used in this chapter.

References

1. Cardon-Cardo, C. (1995) Mutations of cell cycle regulators. Biological and clinical implications for human neoplasis. *Am. J. Pathol.* **147,** 545–560.
2. Draetta, F.-G. (1994) Mammalian G1 cyclins. *Curr. Opin. Cell Biol.* **6,** 842–846.
3. Filmus, J., Robles, A. I., Shi, W., Wong, M. J., Colombo, L. L., and Conti, C. J. (1994) Induction of cyclin D1 overexpression by activated *ras*. *Oncogene* **9,** 362,363.
4. Hartwell, L. H. and Kastan, M. B. (1994) Cell cycle control and cancer. *Science* **266,** 1821–1823.
5. Hartwell, L. H. and Weinert, T. A. (1989) Checkpoints: controls that ensure the order in cell cycle events. *Science* **246,** 629–634.
6. Morgan, D. O. (1995) Principles of CDK regulation. *Nature* **374,** 131–134.
7. Murray, A. W. (1992) Creative blocks: cell cycle checkpoints and feedback controls. *Nature* **359,** 599–604.
8. Nigg, E. A. (1995) Cyclin-dependent protein kinases:key regulators of the eukarotic cell cycle. *BioEssays* **17,** 471–480.
9. Norbury, C. and Nurse, P. (1992) Animal cell cycles and their control. *Annu. Rev. Biochem.* **61,** 441–470.
10. Nurse, P. (1994) Ordering S phase and M phase in the cell cycle. Cell **79,** 543–550.
11. Pines, J. (1994) Arresting developments in cell-cycle control. *Trends Biochem. Sci.* **19,** 143–145.
12. Sherr, C. J. (1994) G1 phase progression: cycling on cue. *Cell* **79,** 551–555.
13. Pines, J. and Hunter, T. (1991) Human cyclin-A and cyclin B are differentially located in the cell and undergo cell cycle-dependent nuclear transport. *J. Cell Biol.* **115,** 1–17.
14. Girard, F., Strausfield, U., Fernandez, A., and Lamb, N. J. C. (1991) Cyclin A is required for the onset of DNA replication in mammalian fibroblasts. *Cell* **67,** 1169–1179.
15. Krek, W. and Xu, G., and Livingston, D. M. (1995) Cyclin A-kinase regulation of E2F-1 DNA binding function underlies suppression of an S phase checkpoint. *Cell* **83,** 1149–1158.
16. Pagano, M., Pepperkok, R., Verde, F., Ansorge, W., and Draetta, G. (1992) Cyclin A is required at two points in the human cell cycle. *EMBO J.* **11,** 961–971.
17. Dulic, V., Lees, E., and Reed, S. I. (1992) Association of human cyclin E with a periodic G1–S phase protein kinase. *Science* **257,** 1958–1961.

18. Koff, A., Giordano, A., Desai, D., Yamashita, K., Harper, J. W., Elledge, S., Nishimoto, T., Morgan, D. O., Franza, B. R., and Roberts, J. M. (1992) Formation and activation of a cyclin E-cdk2 complex during the G1 phase of the human cell cycle. *Science* **257,** 1689–1694.
19. Knoblich, J. A., Sauer, K., Jones, L., Richardson, H., Saint, R., and Lehner, C. F. (1994) Cyclin E controls S phase progression and its down-regulation during *Drosophila* embryogenesis is required for the arrest of cell proliferation. *Cell* **77,** 107–120.
20. Baldin, V., Lukas, J., Marcote, M. J., Pagano, M., and Draetta, G. (1993) Cyclin D1 is a nuclear protein required for cell cycle progression in G1. *Genes Dev.* **7,** 812–821.
21. Matsushime, H., Quelle, D. E., Shurtleff, S. A., Shibuya, M., Sherr, C. J., and Kato, J.-Y. (1994) D-type cyclin D kinase activity in mammalian cells. *Mol. Cell Biol.* **14,** 2066–2076.
22. Matsushime, H., Roussel, M. F., Ashmun, R. A., and Sherr, C. J. (1991) Colony stimulating factor 1 regulates novel cyclins during G1 phase of the cell cycle. *Cell* **65,** 701–713.
23. Musgrove, E. A., Hamilton, J. A., Lees, C. S., Sweeney, K. J. E., Watts, C. K., and Sutherland, R. L. (1993) Growth factor, steroid, and steroid antagonist regulation of cyclin gene expression associated with changes in T-47D human breast cancer cell cycle progression. *Mol. Cell Biol.* **13,** 3577–3587.
24. Ohtsubo, M. and Roberts, J. M. (1993) Cyclin-dependent regulation of G_1 in mammalian fibroblasts. *Science* **259,** 1908–1912.
25. Raffeld, M. and Jaffe, E. S. (1991) *bcl*-1 t(11;14), and mantle-cell derived lymphomas. *Blood* **78,** 259–263.
26. Shirodkar, S., Ewen, M., DiCaprio, J. A., Morgan, J., Livingston, D. M., and Chittenden, T. (1992) The transcription factor E2F interacts with the retinoblastoma gene product and a p107-cyclin A complex in a cell cycle-regulated manner. *Cell* **68,** 157–166.
27. Weinberg, R. A. (1995) The retinoblastoma protein and and the cell cycle control. *Cell* **81,** 323–330.
28. Lukas, J., Bartkova, J., Welcker, M., Petersen, O. W., Peters, G., Strauss, M., and Bartek, J. (1995) Cyclin D2 is a moderately oscillating nucleoprotein required for G1 phase progression in specific cell types. *Oncogene* **10,** 2125–2134.
29. Juan, G., Gong, J., Traganos, F., and Darzynkiewicz, Z. (1996) Unscheduled expression of cyclins D1 and D3 in human tumor cell lines. *Cell Prolif.* **29,** 259–266.
30. Gong, J., Traganos, F., and Darzynkiewicz, Z. (1993) Expression of cyclins B and E in individual MOLT-4 cells and in stimulated human lymphocytes during their progression through the cell cycle. *Int. J. Oncol.* **3,** 1037–1042.
31. Gong, J., Traganos, F., and Darzynkiewicz, Z. (1993) Simultaneous analysis of cell cycle kinetics at two different DNA ploidy levels based on DNA content and cyclin B measurements. *Cancer Res.* **53,** 5096–5099.
32. Darzynkiewicz, Z., Gong, J., Juan, G., Ardelt, B., and Traganos, F. (1996) Cytometry of cyclin proteins. *Cytometry* **25,** 1–13.

33. Kamentsky, L. A. and Kamentsky, L. D. (1991) Microscope based multiparameter laser scanning cytometer yielding data comparable to flow cytometry data. *Cytometry* **12,** 381–387.
34. Sharpless, T. K., Traganos, F., Darzynkiewicz, Z., and Melamed, M. R. (1975) Flow cytofluorimetry: discrimination between single cells and cell aggregates by direct size measurements. *Acta Cytol.* **19,** 577–581.
35. Gong, J., Bhatia, U., Traganos, F., and Darzynkiewicz, Z. (1995) Expression of cyclins A, D2 and D3 in individual normal mitogen stimulated lymphocytes and in MOLT-4 leukemic cells analyzed by multiparameter flow cytometry. *Leukemia* **9,** 893–899.

7

Estimation of Cell Viability by Flow Cytometry

James L. Weaver

1. Introduction

The estimation of the viability of a cell population by flow cytometry is based on a simple yet powerful principle: Dead cells leak. Cells that die via the necrosis pathway, in contrast with apoptosis, rapidly lose membrane integrity *(1)*. All of the different methods for evaluating viability are based on either direct leak detection or measurement of a direct consequence of this leakage. Dyes are used that either do not leak into live cells or do not leak out of live cells. Some combination methods use two dyes of different colors and direction of leak detection. The advantage to this is that there are two independent measures of the same phenomenon. This reduces the probability of deceptive results. In addition, these methods more clearly indicate cells that are damaged or dying but are not yet completely dead. The disadvantage is the expenditure of two colors for viability reduces the availability of colors for simultaneous measurement of other cellular parameters. Some methods look at a consequence of cell leakage. For example, cells that leak are not capable of maintaining an electrical potential across the cell membrane. Therefore, potential-sensing dyes can be used to estimate viability, although this is primarily used with bacterial studies *(2,3)*. In addition, cell-associated enzyme activity stops either because of loss of energy or direct leakage of the enzyme; therefore, loss of enzyme activity can be used as a different estimate of cell viability. For example, many dyes used to evaluate cellular ions are nonfluorescent acetomethoxyesters (AM) that are cleaved by nonspecific esterases into the fluorescent ion-sensitive compound. These compounds rapidly leak out of dead cells, giving a secondary measure of the viability of the population *(4)*. Viability measurement can be combined with other measurements of cellular antigen expression or physiology for evaluation or sorting purposes. For example, a viability dye combined with a labeled antibody could

From: *Methods in Molecular Biology, Vol. 91: Flow Cytometry Protocols*
Edited by: M. J. Jaroszeski and R. Heller © Humana Press Inc., Totowa, NJ

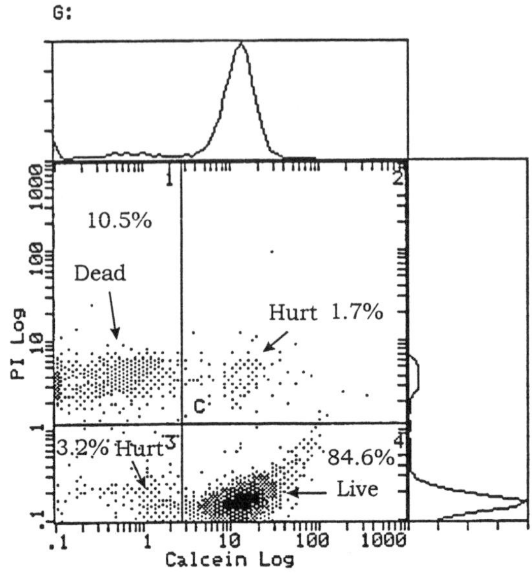

Fig. 1. Fluorescence dot plot of ungated calcein vs PI fluorescence. The data are from a 4-d culture of K562 cells. Cells in quadrant 4 are the healthy, viable cells; quadrant 1 contains dead cells; quadrants 2 and 3 have cells that are thought to be intermediate between healthy and dead. The single parameter histograms are shown along the axes.

3. Add 2 μL of PI stock and mix well.
4. Collect data for 10,000 cells without light-scatter gating. Cells bright for calcein and dark for PI are considered to be viable.

3.3. Labeling with Fura-Red When Green Channel Is Needed for FITC

Another dye that can be used to positively identify viable cells but that has emission in the red (PE/FL2) channel is Fura Red (Molecular Probes). This probe can be used when it is desired to use the FITC channel for a different marker and positive identification of viable cells is also needed.

1. Prepare a single-cell suspension by the method appropriate for the cells being examined. Bring to a final concentration of approx 1×10^6/mL in phenol red-free medium. Use 1 mL per test.
2. Add 2 μL of Fura Red-AM stock solution, mix by inversion, and incubate 30 min at 37°C in a cell–culture incubator.
3. Wash 1X and resuspend in 1 mL of phenol red-free medium and incubate 30 min at 37°C in cell–culture incubator.

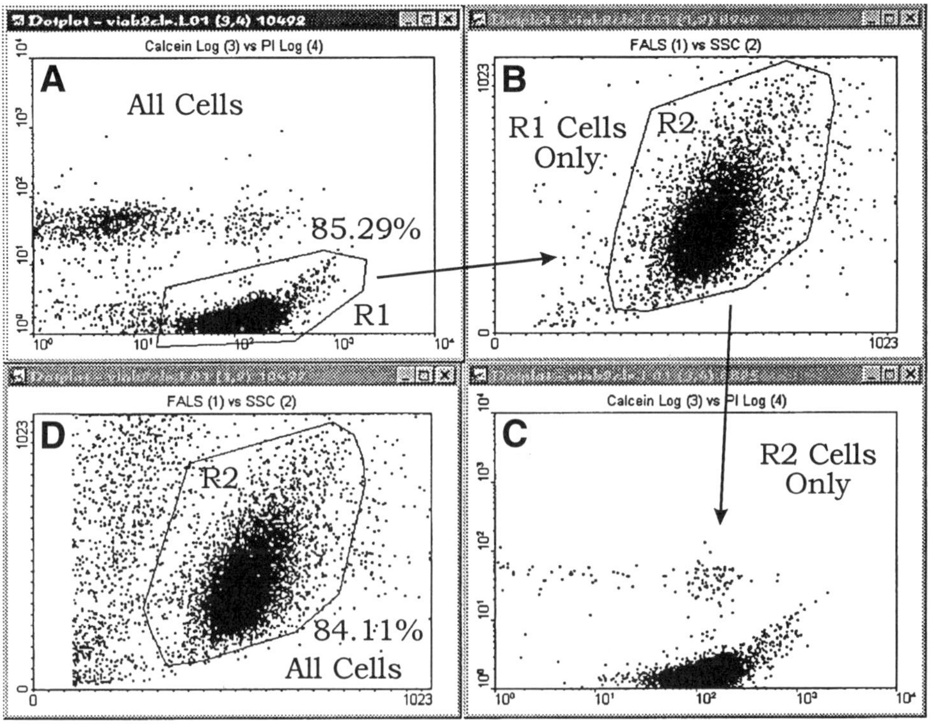

Fig. 2. Evaluation of light scatter gating to estimate viability. **(A)** Calcein by PI plot of all cells. Region 1 (R1) is set around the calcein-bright, PI-dark viable cells. **(B)** A plot of FS vs SS showing only those cells that fall in the R1 gate. A new region (R2) is drawn around the major population displayed. **(C)** A back check plot of calcein vs PI showing all cells that fall in the R2 gate. **(D)** Ungated FS vs SS plot showing R2; 84.11% of all cells fall in R2 as compared to 85.29% of all cells that are found in R1.

4. Spin cells, resuspend in 1 mL of PBS, and transfer to the tube for analysis on the flow cytometer.
5. Collect data for 10,000 cells without light-scatter gating. Cells bright for Fura Red are considered viable.

3.4. Estimating Viability Using Light Scatter

Some cell lines show very distinctive changes in light scatter on loss of viability. In these cases, gating on a specific light-scatter population can be used as a reasonable surrogate for more direct viability measurements (*see* **Note 7**).

1. Collect data as indicated above to determine if this is appropriate for the cell type of interest.
2. Make a two-dimensional plot of ungated calcein vs PI fluorescence and draw a gate around the calcein bright, PI dark population (R1 in **Fig. 2A**).

3. Set up a second two-dimensional plot of FS vs SS, showing only those cells that fall in R1 (**Fig. 2B**). Draw a second region (R2) around the majority of the population.
4. To provide a check of how well R2 distinguishes viable cells, set up another calcein vs two-dimensional plot, this time showing only those cells that fall in R2. **Figure 2C** shows this result. For this data set, <1% of the nonviable cells are included in the R2 light-scatter gate.
5. Compare the percentage of all cells that fall in R1 on the calcein vs PI plot with the percentage that fall in R2 on the FS vs SS plot. For the example shown in **Fig. 2**, these values are close enough for general use.
6. If the percentage of cells in R1 is close to that in R2 and if the back-check fluorescence plot (**Fig. 2C**) shows reasonable accuracy, then the light-scatter gate can be used as a reasonable estimate of the viability of this specific cell type.

4. Notes

1. The indicator calcein is usually an extremely bright label. If fully compensated two (or more) color analysis is needed then it may be necessary to reduce the concentration of calcein-AM used in loading significantly below that given in **Subheading 3.1.** The amount of dilution will vary with the cell type and the intensity of the label in the other channel(s).
2. Data collection should take place within 30 min after the addition of the PI. This label will leak into healthy cells over time so that they should not be exposed to PI longer than needed. Calcein labeling is usually very stable; however, if cells need to be kept for more than 30 min before addition of PI, they should be placed in complete medium for the wait and then washed 1X and resuspended in PBS before addition of the PI.
3. The concentrations of PI reported in the literature vary widely from 0.5 to 32 μg/mL; incubation times range from 5 to 30 min. If the fluorescence is not sufficient with these conditions, a higher PI concentration and/or longer incubation time may be appropriate.
4. Many acetomethoxy esters, including calcein-AM , are substrates for the p-glycoprotein (MDR-1) multi-drug resistance pump. Cells that express this protein will show reduced levels of calcein fluorescence as compared to similar cells that are MDR-1-negative. Cell lines that have been drug treated or drug selected often express MDR-1 and/or other drug resistance mechanisms as well.
5. The times and concentrations given above are typical and should be reasonable for most cell types. However, many cell types are being used for research, some of which have atypical responses. If the results obtained with a specific cell line are not satisfactory, changes in the concentrations of reagents are entirely justified. If needed, cells fixed in ethanol make an excellent positive control as dead cells.
6. Some cell lines express a large anion transporter system (LATS) that can move negatively charged molecules out of the cell very rapidly. If a time-course experiment shows rapid drops in calcein fluorescence, try adding 2.5 m*M* probenecid before the calcein-AM is added.

7. Light scatter is accepted as a way of gating on generally viable cells for further evaluation of other properties. However, if viability is an experimental outcome, then use of a specific fluorescent indicator, such as PI and/or calcein, will be expected by most editors and reviewers.

References

1. Darzynkiewicz, Z., Bruno, S., Del Bino, G., Gorcyzyca, W., Hotz, M. A., Lassota, P., and Traganos, F. (1992) Features of apoptotic cells measured by flow cytometry. *Cytometry* **13,** 795–808.
2. Mason, D. J., Power, E. G., Talsania H., Phillips, I., and Grant, V. A. (1995) Antibacterial action of ciprofloxacin. *Antimicrob. Agents Chemotherap.* **39,** 2752–2758.
3. Porter, J., Pickup R., and Edwards, C. (1995) Membrane hyperpolarization by valinomycin and its limitations for bacterial viability assessment using rhodamine 123 and flow cytometry. *FEMS Microbiol. Lett.* **132,** 259–262.
4. Rotman, B. and Papermaster, B. W. (1966) Membrane properties of living cells as studied by enzymatic hydrolysis of fluorogenic esters. *Proc. Natl. Acad. Sci. USA* **55,** 134–141.
5. Burgess, S. M., Delannoy, M., and Jensen, R. E. (1994) MMM1 encodes a mitochondrial outer membrane protein essential for establishing and maintaining the structure of yeast mitochondria. *J. Cell Biol.* **126,** 1375–1391.
6. Horan, P. K. and Wheeless, L. L. (1977) Quantitative single cell analysis and sorting. *Science* **198,** 149–157.
7. Shapiro, H. M. (1995) *Practical Flow Cytometry,* 3rd ed., Wiley-Liss, New York, pp. 307,308.
8. Garner, D. L. and Johnson, L. A. (1995) Viability assessment of mammalian sperm using SYBR-14 and propidium iodide. *Biol. Reprod.* **53,** 276–284.
9. Becker, B., Clapper, J., Harkins, K. R., and Olson, J. A. (1994) In situ screening assay for cell viability using a dimeric cyanine nucleic acid stain. *Anal. Biochem.* **221,** 78–84.
10. Riedy, M. C., Muirhead, K. A., Jensen, C. P., and Stewart, C. C. (1991) Use of a photolabeling technique to identify nonviable cells in fixed homologous or heterologous cell populations. *Cytometry* **13,** 795–808.
11. Bhuyan, B. K., Loughman, B. E., Fraser, T. J., and Day, K. J. (1976) Comparison of different methods of determining cell viability after exposure to cytotoxic compounds. *Exp. Cell Res.* **97,** 275–280.

8

Flow Cytometric Detection of Fluorescent Redistributional Dyes for Measurement of Cell Transmembrane Potential

Michael K. Tanner and Samuel R. Wellhausen

1. Introduction

Eukaryotic cells are electronegative when compared to the surrounding environment. This negative charge is called the transmembrane potential (TMP) and is partly caused by concentration gradients of K^+, Na^{2+}, and Cl^- ions across the cell membrane. For purposes of cell osmolarity and pH balance, Na^{2+} and Cl^- ion concentrations are kept lower inside the cell relative to the external environment, whereas K^+ ions are maintained at a high intracellular concentration (relative to outside the cell) balancing negative charges on cytoplasmic organic molecules (**Fig. 1**). These ion gradients are maintained by the relative impermeability of the plasma membrane to charged particles and by membrane-bound, energy-dependent ion pumps. The net negative TMP is generated by a combination of K^+ ion loss through leak channels (down the K^+ concentration gradient), Na^{2+} ion loss through the Na^{2+}/K^+ ATPase, and negatively charged organic molecules trapped in the cytoplasmic compartments of the cell *(1,2)*.

Historically, measurement of TMP could be performed using microelectrodes implanted into cells *(5)* or the patch clamp technique *(6)*. The creation of charged lipophilic fluorescent probes *(9)* provided a noninvasive means of assessing TMP that was first applied with spectrophotometry and bulk measurements *(10)*. These probes also facilitated multiparametric TMP analysis of single-cell events by flow cytometry *(11)*. This was a significant advance because it was the first analytical method that allowed for detection of TMP heterogeneity within a population of cells. The first dyes to be tested with flow cytometry were the cationic carbocyanine dyes. These positively charged

From: *Methods in Molecular Biology, Vol. 91: Flow Cytometry Protocols*
Edited by: M. J. Jaroszeski and R. Heller © Humana Press Inc., Totowa, NJ

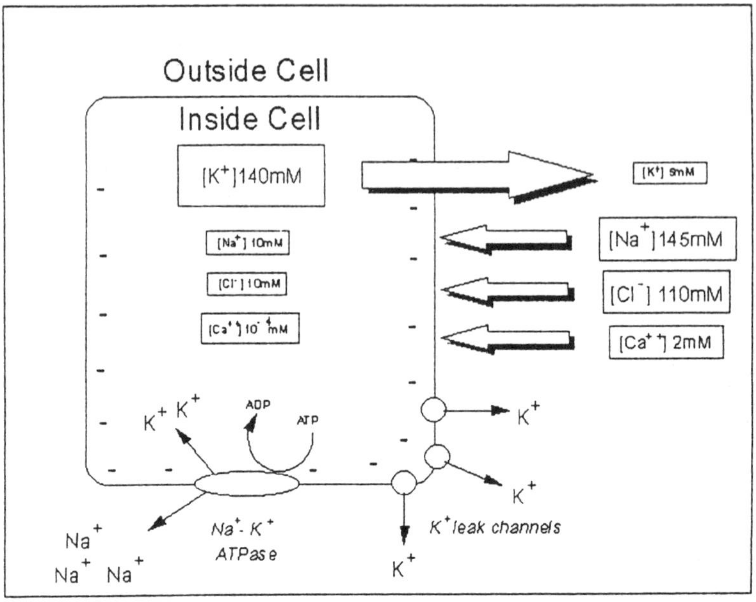

Fig. 1. Forces that regulate and determine membrane potential. Approximate intracellular and extracellular ion concentrations for mammalian cells are represented in boxes. White arrows represent direction and strength of electrochemical gradients for each ion. Negative signs represent charge of the cell membrane.

distributional probes bind readily to the negatively charged cells. When cells become hyperpolarized (increased negative charge) they take up more of the cationic distributional probes, whereas depolarization is indicated by a decrease in probe fluorescence ***(2)***.

Although the carbocyanine dyes have been widely used, a major drawback is the contribution of the cell's highly electronegative mitochondria and lipophilic sites in the cytoplasm to the TMP as measured with these probes ***(12,13)***. In response to this concern, a family of anionic fluorescent lipophilic distributional probes, the oxonols, have been applied to flow cytometric TMP evaluation. These negatively charged probes do not stain the negatively charged cells as intensely as do the carbocyanines and will distribute exactly opposite to the cyanine dyes in response to TMP changes ***(13,14)***. That is, hyperpolarization results in decreased oxonol fluorescence and depolarization results in increased fluorescence.

TMP has been shown to be important in immune cell function. Mitogenic stimulation of lymphocytes ***(3,4)***, monocytes ***(5,6)***, and neutrophils ***(7)*** are associated with changes in TMP. Experimental manipulation of lymphocyte

TMP can prevent lymphocyte responses to mitogens ***(3,8)***. Distributional fluorescent probes of TMP and flow cytometry have contributed significantly to investigations of a wide variety of these systems. Immune-cell activation, natural killer cell cytotoxicity ***(15)***, response to cytokines ***(16)***, response to bacterial toxins ***(17)***, and the effects of immunosuppressive agents ***(18)*** have been associated with changes in TMP as measured by cyanines and oxonols. A relatively recent application of this method has correlated bacterial TMP with viability in studies of antibiotic resistance ***(19)***. Although there will be differences between organisms, experimental systems, and distributional probes, the basic steps for TMP measurement by flow cytometry are similar:

1. Equilibrate TMP probe with cells.
2. Assess health of cells with hyperpolarizing and depolarizing controls ***(2)***.
3. If controls work, analyze experimental variables in the same manner as the controls.

We describe methods here for using cyanine and oxonol dyes to measure transmembrane potential in human peripheral mononuclear cells.

2. Materials

2.1. Stains and Ionophores

1. DIOC6(3) (dihexyloxacarbocyanine) may be purchased from Molecular Probes (Eugene, OR. cat. no. D-273). A 0.5 m*M* stock solution of this dye is made in dimethyl sulfoxide (DMSO) and stored wrapped in foil (light sensitive) at room temperature. This stock solution is stable up to 1 mo.
2. $DiBAC_4$ (bis oxonol) may be purchased from Molecular Probes (cat. no. B-438). A 300 μ*M* stock solution is made in ethanol, wrapped in foil, and stored at room temperature.
3. FCCP (carbonyl cyanide p-trifluoromethoxyphenyl-hydrazine) can be purchased from Sigma (St. Louis, MO, cat. no. C-2759). Prepare a 4 m*M* stock in DMSO fresh before each experiment.
4. A23187 (calcium ionophore) may be purchased from Molecular Probes (cat. no. A-1493). Prepare a 10 μ*M* working solution in DMSO fresh before each experiment.
5. Valinomycin (potassium ionophore) may be purchased from Sigma (cat. no. V-0627). Prepare a 6 μ*M* working solution fresh in DMSO before each experiment.
6. Gramicidin (depolarizing control) may be purchased from Sigma (cat. no. G-5127). Prepare a 20 μ*M* working solution fresh in DMSO before each experiment.

2.2. Buffers

The media in which the cells are resuspended before TMP experiments should be protein free because any protein in the media will effect the equilibration of the distributional probes ***(2)***. RPMI 1640 with 25 m*M* HEPES works well because the pH remains constant outside a CO_2 incubator for the short time course of the experiment and there is no protein.

In order to provide hyperpolarizing and depolarizing controls for experiments with bis oxonol, the external potassium concentration of the cell environment needs to be varied from 5.4–122.4 m*M* using buffers ***(13)***. These are prepared as follows:

1. High potassium buffer: 4.56 g KCl, 1 g $NaHCO_3$, 0.1 g $MgSO_4$, 0.07 g NaH_2PO_4, add 100 mL 18 m*M* $CaCl_2$ (2 g/L) and 100 mL 200 m*M* HEPES (52 g/L) (*see* **Note 8**) and distilled water to 1 l. Adjust to pH 7.4.
2. Low potassium buffer: 0.2 g KCl, 3.42 g NaCl, 1 g $NaHCO_3$, 0.1 g $MgSO_4$, 0.07 g NaH_2PO_4, add 100 mL 18 m*M* $CaCl_2$ (2 g/L), and 100 mL 200 m*M* HEPES (52 g/L) (*see* **Note 8**) and distilled water to 1 l. Adjust to pH 7.4.

3. Methods

3.1. Cell Preparation

1. Obtain peripheral blood using acid citrate dextrose as an anticoagulant. Heparin and EDTA anticoagulants may inhibit functional studies of mononuclear cells.
2. Dilute blood at least twice with RPMI/HEPES to insure optimal separation.
3. Isolate the mononuclear cell fraction by centrifugation over Histopaque 1077 (Sigma, St. Louis, MO) at 500*g* for 20 min.
4. Remove the mononuclear fraction and wash twice with RPMI using centrifuge speeds of 100*g* to pellet then gently resuspend the cell pellets with a pipet.
5. Resuspend the cells in RPMI/HEPES at a concentration of 5×10^5 cells/mL.
6. Assess the viability of the cell preparation microscopically using dye (e.g., trypan blue) exclusion and only use the cells if the viability >95%.

3.2. Use of Carbocyanine Dye DIOC6(3)

The carbocyanine dye DIOC6(3) gives a measure of TMP by its distribution across the cell membrane. This is a dynamic process so it is imperative to keep dye incubation times and conditions constant. Experiments can be accomplished faster (important for cells without serum nutrients) by lagging dye and reagent addition to new samples every 5 min and using a multiple event timer to monitor sequential incubations.

DIOC6(3) fluorescence can be detected using a standard filter setup for fluorescein isothiocyanate (FITC) with 488 nm excitation and gating events on the basis of light scatter. TMP is then directly related to DIOC6(3) mean fluorescent channel (MFCH) on a linear histogram.

Condition of the cell preparation is one of the most important factors in determining the success or failure of using DIOC6(3) to measure TMP. Viability should be >95% microscopically using dye exclusion. Even viability does not guarantee cell function in this assay, so each cell preparation must be tested using hyperpolarizing (valinomycin) and depolarizing (gramicidin) controls. Valinomycin is a K^+ ionophore that allows K^+ to pass out of the cell because of the

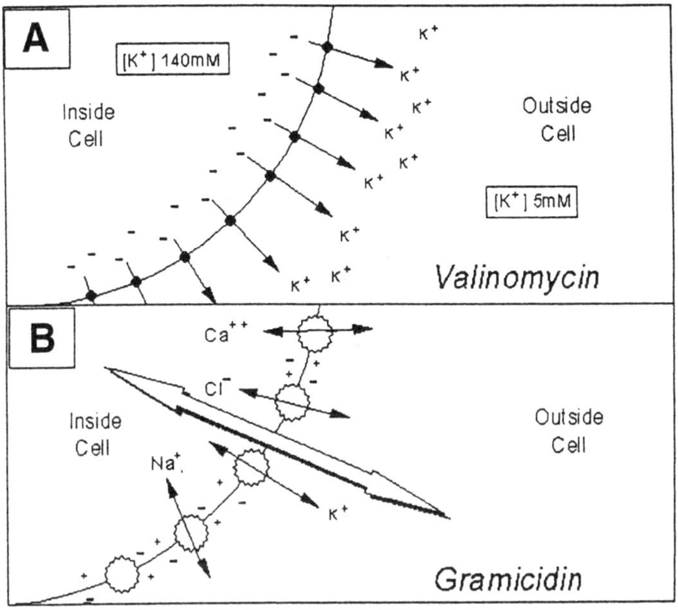

Fig. 2. Ionophore deregulation of membrane potential. **(A)** Potassium ionophore valinomycin hyperpolarizes the cell membrane by allowing K^+ ions to pass freely through cell membrane down their electrochemical gradient. **(B)** Gramicidin forms a pore that allows all ions to pass freely through cell membrane eventually causing total depolarization (no charge on membrane).

concentration gradient (**Fig. 2A**). The afflux of these positively charged ions increases the net electronegative charge on the cell membrane (hyperpolarization). Gramicidin forms pores in the cell membrane that allow all monovalent and divalent cations and anions to pass through the membrane effectively negating all ion concentration gradients and depolarizing the cell *(2)* (**Fig. 2B**). These reagents can be used in the following variation of a method described by Shapiro *(2)*:

1. Prepare 5 μ*M* working dilution from 0.5 m*M* stock of DIOC6(3) in DMSO (1:100 dilution). Incubations and sample acquisition on flow cytometer can be conducted at room temperature unless reagents need to be tested at 37°C.
2. Minimize contamination from sample to sample by flushing lines sequentially with bleach, ethanol, and media containing the final concentration of DIOC6(3) before the experiment and between samples.
3. Add 5 μL of DIOC6(3) working dilution to 1 mL cell suspension in RPMI with 10 m*M* HEPES (final concentration of dye is 25 n*M*) and 5 μL of DMSO to establish a nontreated control histogram. Wait 15 min for the dye to equilibrate at room temperature and run on flow cytometer adjusting the green PMT so that events are centered on the linear histogram.

4. Analyze a sample that contains 25 n*M* DIOC6(3) and 5 µL of 6 µ*M* valinomycin (final concentration 30 n*M*) as a hyperpolarizing control. After the 15-min incubation this tube should increase MFCH compared to the untreated histogram 100–200 channels.
5. Next analyze the depolarizing control that contains 5 µL of 20 µ*M* gramicidin (final concentration 100 n*M*) and 25 n*M* DIOC6(3), which after 15 min dye equilibration should decrease MFCH compared to the untreated histogram 100–300 channels.
6. Proceed in the same manner (incubation time and dye concentration) with test reagents and cell populations if hyperpolarizing and depolarizing controls work. If the controls do not work, it may be necessary to determine optimal dye concentration (*see* **Note 1**) and/or incubation time (*see* **Note 2**). Keep in mind that test reagents should be added in concentrations so as to keep the total solvent equal to or less than 1% of the total volume of the sample. Taking into consideration that 5 µL is used by the dye, the test reagents must be concentrated so that they can be added in 5 µL into a 1-mL sample.

There are different ways to analyze fluorescent data from these experiments. The MFCH of each tube may be directly compared within a single experiment. Alternatively, relative membrane potential (*see* **Note 7**) may be calculated by dividing the MFCH of the test reagent or cells by the MFCH of the gramicidin depolarized control (*see* **Note 3**):

$$\text{Relative membrane potential or } \Psi = (\text{MFCH test})/(\text{MFCH gramicidin}) \quad (1)$$

This calculation may be used to evaluate the effects of reagents on the TMP of single cell type and to compare the resting TMPs among various cell types ***(21)***. Relative TMP (not quantitative TMP, *see* **Note 6**) can be compared between experiments.

One criticism of cyanine dyes is that they become concentrated in the mitochondria. The mitochondria are negatively charged with respect to the cell plasma membrane ***(13)*** and work from our lab ***(18)*** has shown this could effect the ability of DIOC6(3) to give a true measure of the electrochemical potential at the plasma membrane. One way to eliminate the mitochondrial influence on TMP is to uncouple oxidative phosphorylation in the mitochondria using the cyanine analog FCCP ***(20)***. The experiments can be done exactly as described above except for the addition of 20 µ*M* FCCP (5 µL of 4 m*M* FCCP in DMSO) to the cells at the beginning of the 15 min incubation with the cyanine dye (*see* **Note 4**).

3.3. Use of Oxonol Dye di-BA-C_4

The problem mentioned above with the cyanine dyes is caused by their positive charge. The cyanines enter the cell and bind to internal hydrophobic sites and negatively charged organelles, such as mitochondria ***(12,13)***. One way to compensate for this problem is to use a negatively charged redistributional probe for TMP such as di-BA-C_4 (bis oxonol). Bis oxonol should not enter negatively charged cells to the extent that the positively charged cyanines do ***(5,6,18)*** and therefore may only redistribute because of the electrochemical gradient at the cell plasma membrane ***(13)***.

Bis oxonol can be analyzed on a flow cytometer with 488 nm excitation and the same collection optics used for FITC (550 nm dichroic long pass and a 525 nm band pass). We have used bis oxonol to give a relative measure of TMP in the following variation of a method described by Wilson and Chused ***(13)***. The bis oxonol procedure requires the use of different controls from those used with cyanines. Valinomycin and gramicidin cannot be used as hyperpolarizing and depolarizing controls because they may bind di-BA-C_4. Calcium gated potassium channels are activated with Ca^{2+} ionophore A23187 in media of differing potassium concentrations to provide these controls as follows:

1. Bring cells up to 5×10^6/mL (10X normal concentration) in RPMI/HEPES. The cells are then diluted 1:10 in media or high or low K^+ buffers just before adding bis oxonol or ionophores. This method prevents clumping that we observed with long-term storage of cells in high and low K^+ buffers.
2. Dilute 300 μ*M* bis oxonol stock solution 1:10 in RPMI/HEPES to reach a 30 μ*M* working solution.
3. Add 100 μL of 10X cell preparation to 900 μL of RPMI/HEPES for the first sample. Add 5 μL of the bis oxonol working solution for a final concentration of 150 n*M* di-BA-C_4 and allow dye to equilibrate with cells for 10 min. This tube is then run on flow cytometer without analysis in order to equilibrate sample tubing with bis oxonol. In our experiments, we noticed the first sample we ran was always significantly dimmer than the following tubes. Rinse sample tubing with bleach and water after this and all subsequent samples. This tube may therefore be considered a "dummy" sample.
4. Repeat sample preparation in **step 3**. After 10-min incubation this control sample is run on flow cytometer and the green PMT should be adjusted so that light scatter-gated bis oxonol fluorescence falls in the center of the green histogram.
5. Resuspend a 100-μL aliquot of 10X prep in 900 μL of low K^+ buffer. Add bis oxonol and 5 μL of 10 μ*M* stock of the calcium ionophore A23187 (final concentration 50 n*M*). After the 10-min incubation, cells should show a 50–100 channel decrease in fluorescence compared to the RPMI/HEPES control indicating a relative hyperpolarization (*see* **Note 5**). Remember, this is opposite the increase in fluorescence that cyanine dyes exhibit with a hyperpolarization.
6. Resuspend an aliquot of the cell prep in high K^+ buffer and add bis oxonol and A23187. After the 10-min incubation cells should show a 100–200 channel increase indicating a depolarization of the cell membrane.
7. Proceed with cells resuspended in RPMI/HEPES and incubated with bis oxonol as above (if cells respond appropriately with the controls) remembering to allow

Table 1
Determination of Optimal Working Concentration of DIOC6(3)[a]

Dye concentration, n*M*	Treatment	DIOC6(3) fluorescence
12.5	DMSO	596
	Valinomycin	911
	Gramicidin	447
25	DMSO	432
	Valinomycin	665
	Gramicidin	187
50	DMSO	449
	Valinomycin	589
	Gramicidin	203
250	DMSO	437
	Valinomycin	457
	Gramicidin	140

[a]Mononuclear cells were incubated with different concentrations of DIOC6(3) for 15 min at room temperature. The TMP of 1000 monocytes was expressed as mean DIOC6(3) fluorescence. Values represent one experiment.

a 10 min equilibration of dye after addition of test reagents. As with DIOC6(3), if the controls do not work it may be necessary to determine optimal dye concentration (*see* **Note 1**) and/or incubation time (*see* **Note 2**).

4. Notes

1. Numerous variables such as cell type and media may effect oxonol and cyanine redistribution. Therefore, it is essential to titrate the concentration of each of these dyes to the experimental conditions in which they will be used. This can be accomplished with the cyanine dyes using gramicidin and valinomycin controls to determine which dye concentration provides the largest range of MFCH deflection for measuring depolarizations and hyperpolarizations. This type of analysis **(Table 1)** indicated that in our system higher concentrations of DIOC6(3) decreased the ability to determine hyperpolarizations, whereas lower concentrations decreased the ability to determine depolarizations ***(18)***. The same type of analysis should be done with bis oxonol using the hyperpolarizing and depolarizing controls mentioned in **Subheading 3.2**.
2. Because of the redistributing nature of these TMP probes, investigation of the optimal equilibration time for a specific system is time well-invested. In our work with both DIOC6(3) and bis oxonol we saw that these dyes are continually entering cells. In the case of cyanine, uptake plateaus somewhat after 10 min, but is still increasing at 30 min. Therefore, although maintaining consistent incubation periods is required between samples, determining the plateau

equilibration time will assure that measurements are made at a time when dye uptake will be less variable.

3. Although positioning the untreated control at the midpoint of the green histogram started each experiment at approx the same MFCH, we found that trying to compare MFCH's between experiments was too much variation for statistical analysis. So, all test values within a given experiment were normalized to the depolarized control with the reasoning that the totally depolarized measure should be consistent between experiments. This is calculated as the relative potential as described in **Subheading 3.2., Eq. 1**.
4. Although FCCP treatment of cells provided good information on the mitochondrial contribution to the total TMP, there were two major drawbacks in using it. First it caused a relative depolarization of approx 100 channels in resting cells, and therefore lessened the ability to detect depolarizations. Second, the valinomycin control depolarized the cells in the presence of FCCP so mitochondria may play some role in valinomycin's hyperpolarizing effect. Valinomycin is still required to assess cell fitness in these experiments, but is of no use as a hyperpolarizing control for FCCP treated cells.
5. Bis oxonol is useful as a means of eliminating mitochondrial influences on TMP, however in our hands, the dye seemed much more sensitive to depolarizations than hyperpolarizations based on the fact that we observed variable control hyperpolarizations. The hyperpolarizing control was obtained by activating calcium-gated potassium pump with the calcium ionophore A23187 in a buffer in which the potassium concentration was physiologic (5.4 m*M*).
6. Fluorescence intensity cannot be quantitatively related to the mV value of TMP *(22)*. A calibration curve may be established by setting the membrane potential in a series of buffers of varying K^+ concentration in the presence of valinomycin *(23)*. This is accomplished by replacing NaCl with KCL in the medium in order to maintain isotonicity. The set TMP is given by the Nernst equation in which Vm = –60 mV log [Ki]/[Ko]. [Ki] is the internal K^+ level of the cell and [Ko] is that of the medium. In low [Ko] the cells will hyperpolarize. The level of hyperpolarization will decrease as [Ko] increases and cells should depolarize when [Ko] exceeds [Ki]. This procedure assumes that the K^+ diffusion potential in the presence of valinomycin greatly exceeds the contributions of other ion conductances and ignores the lipophilicity of the cyanines or any other non-TMP-related binding properties of TMP dyes.
7. Measurement of TMP by flow cytometric analysis of fluorescent redistributional dyes is indirect and relative in nature and therefore is not the method of choice for physical-chemical determinations of TMP *(2)*. The advantage of flow cytometric methods is the ability to rapidly make many measurements in mixed cell populations for relative comparisons of TMP among different cell types and the modulating effects of reagents.
8. $CaCl_2$ and HEPES are made in separate solutions because they are difficult to dissolve.

References

1. Alberts, B., et al. (1994) Membrane transport of small molecules and the ionic basis of membrane excitability, in *Molecular Biology of the Cell,* Garland, New York, pp. 507–549.
2. Shapiro, H. M. (1988) Optical probes of cell membrane potential, in *Practical Flow Cytometry* Liss, Inc., New York, pp. 181–189.
3. GeFland, E. W., Cheung, R. K., Mills, G. B., and Grinstein, S. (1987) Role of membrane potential in the response of human T lymphocytes to phytohemagylutanin. *Immunology* **138,** 527–531.
4. Tsien, R. Y., Pozzan, T., and Rink, T. J., (1982) T cell mitogens cause early change in cytoplasmic free Ca^{2+} and membrane potential in lymphocytes. *Nature* **295,** 68–71.
5. Gallin, E. K. (1986) Ionic channels in leukocytes. *J. Leuk. Biol.* **39,** 241–254.
6. Gallin, E. K. and Livengood, D. R., (1981) Inward rectification in mouse macrophages: evidence for a negative resistance region. *Am. J. Phys.* **214,** C9–C17.
7. Segilmann, B., Chused, T. M., and Gallin, J. I. (1981) Human neutrophil heterogenetic identified using flow microfluorimetry to monitor membrane potential. *J. Clin. Invest.* **68,** 1125–1131.
8. Oettgen, H. D., Terhorst, C., Cantley, L. C., and Rosoff, P. M. (1985) Stimulation of T3-T cell receptor complex induces a membrane potential sensitive calcium influx. *Cell* **40,** 583–590.
9. Cohen, L. D. and Salzberg, B. M. (1978) Optical measurement of membrane potential. *Rev. Physiol. Biochem. Pharmacol.* **83,** 85.
10. Sims, P. J., Waggoner, A. S., Wang, C. H., et al. (1974) Studies on the mechanism by which cyanine dyes measure membrane potential in red blood cells and phosphatidylcholine vesicles. *Biochemistry* **13,** 3315–3330.
11. Shapiro, H. M., Natale, D. J., and Kamentsky, L. A. (1980) Estimation of membrane potential of individual lymphocytes by flow cytometry. *Proc. Natl. Acad. Sci. USA* **76,** 5728–5730.
12. Johnson, L. V., Walsh, M. L., Bockus, B. J., and Chen, L. B. (1981) Monitoring of relative mitochondrial membrane potential in living cells by fluorescence microscopy. *J. Cell. Biol.* **88,** 526–535.
13. Wilson, H. A., Seligmann, B. E., and Chused, T. M. (1985) Voltage sensitive cyanine dye fluorescence signals in lymphocytes: plasma membrane and mitochondrial components. *J. Cell. Physiol.* **125,** 57–71.
14. Wilson, H. A. and Chused, T. M. (1985) Lymphocyte membrane potential and Ca^{2+} sensitive potassium channels described by oxonol dye fluorescence measurements. *J. Cell. Physiol.* **125,** 72–81.
15. Radosevic, K., Bakker Schut, T. C., Van Graft, M., deGrooth, B., and Greve, M. (1993) A flow cytometric study of the membrane potential of natural killer and K562 cells during the cytotoxic process. *J. Immunol. Methods* **161,** 119–128.
16. Balint, E., Cheng, M., Rupp, B., Grimley, P. M., and Aszalos, A. (1992) Cytoskeletal modulation of plasma membrane events induced by Interferon-α. *J. Interferon Res.* **12,** 249–255.

17. Taichman, N. S., Masayasu, I., Lally, E. T., Shattil, S. J., Cunningham, M. E., and Korchak, H. M. (1991) Early changes in cytosolic calcium and membrane potential induced by *Actinobacillus actinomycetem-comitans* leukotoxin in susceptible and resistant target cells. *J. Immunol.* **147,** 3587–3594.
18. Tanner, M. K., Wellhausen, S. R., and Klein, J. B. (1993) Flow cytometric analysis of altered mononuclear cell transmembrane potential induced by cyclosporin. *Cytometry* **14,** 59–69.
19. Ordenez, J. V. and Wehman, N. M. (1993) Rapid flow cytometric antibiotic susceptibility assay for *Staphylococcus aureus. Cytometry* **14,** 811–818.
20. Hasmann, M., Valet, G. K., Tapiero, H., Trevorrow, K., and Lampidis, T. (1989) Membrane potential differences between adriamycin-sensitive cells and resistant cells by flow cytometry. *Biochem. Pharm.* **38,** 305–312.
21. Kessel, D., Beck, W. T., Kukuruga, D., and Schulz, V. (1991) Characterization of multidrug resistance by fluorescent dyes. *Cancer Res.* **51,** 4665–4670.
22. Waggoner, A. S. (1979) Dye indicators of membrane potential. *Ann. Rev. Biophys. Bioeng.* **8,** 47–68.
23. Ishida, Y. and Chused, T. M. (1993) Lack of voltage sensitive potassium channels and generation of membrane potential by sodium potassium ATPase in murine T lymphocytes. *J. Immunol.* **151,** 610–620.

9

Flow Cytometric Analysis of Macrophage Oxidative Metabolism Using DCFH

Amy Imrich and Lester Kobzik

1. Introduction

Macrophages throughout the body function to phagocytose pathogens, tissue debris, or foreign particulates. For example, lung macrophages have frequent contact with inhaled pathogens or inert particles and have developed interesting and diverse responses to these challenges. One response of particular interest is the respiratory burst (RB), which involves converting oxygen into reactive oxygen intermediate species (ROI). Depending on the amount and site of production, these oxidants can contribute to intracellular signaling, participate in killing of ingested microorganisms, or cause injury to the cell itself or adjacent tissues *(1,2)*. Indiscriminate production of oxidants by alveolar macrophages (AMs) interacting with otherwise innocuous environmental particles could have deleterious consequences for the lung. Flow cytometric analysis of particle uptake and intracellular RB has enhanced our understanding of AM–particle interactions and determinants that lead to a RB or down-regulation of potentially harmful AM responses *(3–6)*.

Intracellular production of oxidants within macrophages from the lung or other sites can be measured using the fluorescent reporter molecules, dichlorofluorescin (DCFH), hydroethidine (HE), or dihydrorhodamine 123 (DHR). These probes share a common principle of action, but differ in their final cellular fate. All three are nonfluorescent, reduced precursor molecules that easily enter the cell and then become oxidized during basal metabolism or during a RB event. On oxidation, the reporter is converted to a fluorescent molecule whose concentration can be quantitated by flow cytometric (or fluorometric) techniques. For DCFH, the product is a dichlorofluorescein (DCF), present diffusely in the cytoplasm and having the spectral properties of fluorescein.

From: *Methods in Molecular Biology, Vol. 91: Flow Cytometry Protocols*
Edited by: M. J. Jaroszeski and R. Heller © Humana Press Inc., Totowa, NJ

Oxidation of HE (the sodium-borohydride-reduced derivative of ethidium bromide) produces ethidium bromide, which intercalates with nuclear DNA and thereby enhances its red fluorescence *(7)*. DHR is oxidized to rhodamine 123, a red fluorescent molecule accumulated in mitochondria *(8)*. The relative specificity of these agents for different reactive oxidant species is unsettled. Initial characterization suggested that DCFH was specific: for H_2O_2, whereas HE and DHR have been advanced as agents specific for O_2^- *(7,9)*. In fact, DCFH can also be oxidized by nitric oxide or products of NO interaction with O_2^- *(10)*. HE shows sensitivity to both O_2^- and H_2O_2, as identified by the need for both super oxide dismutase and catalase to abrogate increased intracellular fluorescence in HE-loaded cells exposed to xanthine/xanthine oxidase as a source of ROI *(3)*. Hence, we consider DCFH (and the other fluorescent agents) to be most useful as reporters of changes in the intracellular redox state of macrophages. Moreover, combination of these reporters with other functional or phenotypic analysis can provide important information.

By using flow cytometric analysis of AMs loaded with DCFH we have studied the RB of cells in response to a variety of particulates. Our goal in this chapter is to provide a comprehensive protocol for using DCFH in AMs and to describe methods of cell handling, experimental design, and execution that will produce consistent results.

2. Materials

2.1. Buffers

Add the following reagents to 1 L distilled water, adjust to pH 7.3, then sterile filter (0.2-μm bottle-top filter, Costar, Cambridge, MA) into clean, autoclaved bottles. Store at 4°C.

1. PBS-EDTA (ravage buffer): 8 g NaCl, 2.16 g Na_2PO_4 (7 H_2O), 0.2 g KCl, 0.2 g KH_2PO_4, 0.023 g $EDTA(Na)_2$.
2. BSS (assay buffer): 7.25 g NaCl, 0.43 g KCl, 1.8 g dextrose, 4.76 g HEPES, 0.044 g $CaCl_22 \cdot 2H_2O$, 0.203 g $MgCl_2 \cdot 6H_2O$.

2.2. Reagents

1. Dichlorofluorescin-diacetate DCFH-DA (Molecular Probes, Eugene, OR): To prepare a 100X stock solution, dissolve 31 mg of DCFH-DA powder in 1 mL 100% ethyl alcohol and gradually add to 30 mL BSS at room temperature while gently stirring with a magnet. The DCFH-ethanol solution will turn cloudy as it is added to the buffer and should go into solution as it gets mixed in. The result is a clear-opaque solution of DCFH at 2 m*M* in 3% ethanol that will be used at a final concentration of 20 μ*M*. Aliquot 0.5 mL into microfuge tubes and freeze at –80°C, then transfer to –20°C for storage. DCFH is light sensitive (photo-oxidation), so mini-

mize exposure during all handling and storage, keep on ice after thawing, and use a fresh aliquot each day.

2. Propidium iodide PI, Sigma (St. Louis, MO) cat. no. P4170: To prepare a 20X stock solution, dissolve 10 mg in 50 mL BSS to get 200 µg/mL stock that is stored at 4°C and used at a final concentration of 10 µg/mL. PI is poisonous, so follow the precautions detailed in the accompanying safety materials.
3. Phorbol myristate acetate PMA (Sigma cat. no. P8139): Prepare a 10^{-3} *M* stock solution in DMSO, aliquot 20 µL in microfuge tubes and store –20°C or colder. Because PMA can easily stick to surfaces when diluted in aqueous buffer, our working solution is made from 100% DMSO and is used within 30 min. PMA is our standard positive control for induction of a respiratory burst in macrophages and is used at a final concentration of 10^{-7} *M*.
4. Antimycin-A (Sigma cat. no. A8674): Make a 10^{-3} *M* stock in DMSO, aliquot 20 µL in microfuge tubes and store at –20°C or colder. Follow PMA guidelines for handling and use. Antimycin-A inhibits mitochondrial respiration and is one of several pretreatments that can be used to investigate macrophage oxidative metabolism.

2.3. Particles for Phagocytosis

1. Latex spheres: We use 1-µm diameter red fluorescent latex beads (Duke Scientific, Palo Alto, CA). To create immunoglobulin (IgG) coated spheres, latex beads are first coated with BSA (100 µL of beads at 3×10^{10}/mL are incubated for 1 h, 37°C, in 10 mg/mL BSA in BSS buffer, followed by three centrifugation washes at 10,000*g*). Subsequent incubation with rabbit anti-BSA antibody (Capper, Malvern, PA; resuspend bead pellet in 1 mL of 200 µg/mL antibody in BSS) produces antibody-opsonized latex beads, capable of stimulating a macrophage respiratory burst. They are stored in aliquots at –20°C and after thawing, they are held at 4°C for 2–3 d then discarded. Other sources of latex beads include PolySciences (Warrington, PA) and IDC (Portland, OR).
2. Zymosan-A (Sigma cat. no. Z4250): Zymosan-A is a biological particulate consisting of hollow yeast cells. Incubation of zymosan in serum results in surface deposition of complement and IgG, facilitating AM uptake via corresponding surface receptors. To make opsonized zymosan (Z-OPS), incubate 10 mg, of zymosan at 37°C in 5 mL BSS containing 20% autologus serum. Gentle motion is recommended during opsonization. After 1 h, add 40 mL of BSS, spin 10 min at 400*g*, and resuspend in 3 mL BSS. Dilute 10 µL into 1 mL, sonicate, and count in a hemocytometer. We get approx 4×10^8 Z-OPS per 5 mg and use one batch for a group of related experiments. Adjust stock to 2×10^8 Z-OPS/mL, store in aliquots at –20°C, and use fresh each day (*see* **Note 1** about particle sonication).

2.4. Flow Cytometer

We use an ORTHO 2150 Cytoflourograf equipped with a 15 mW 488-nm emitting air-cooled argon laser (Cyonics/Uniphase, Sunnyvale, CA), and a commercially available data acquisition and analysis system (Cyclops Software, Cytomation, Fort Collins, CO). DCF fluorescence is collected through a 530 band-pass filter, whereas PI fluorescence and red fluorescent latex spheres are measured through a 630 long-pass filter. All parameters are collected on a linear scale.

2.5. Miscellaneous

1. 1.7-mL microfuge tubes with caps attached (VWR cat. no. 20170-546).
2. End-to-end rotation device (Labquake Shaker cat. no. 400-110, Labindustries, Berkeley, CA).
3. 5 × 10 microfuge tube storage box.
4. 5 × 16 microfuge tube rack.
5. Incubator at 37°C.
6. Probe sonicator (Model W-200P, Ultrasonics, Plainview, NY).

3. Methods

3.1. Cell Isolation

1. Female Syrian golden hamsters are euthanized with a lethal IP injection of sodium pentobarbital (75 mg/kg).
2. After dissection, tubing is inserted into the upper trachea through a small incision just below the cricothyroid cartilage and secured with suture. We use an 18-gage canula extended with plastic tubing (PE-190, 1.19-mm id, Clay-Adams, VWR).
3. Using a 5-mL syringe and the canula tubing, bronchoalveolar lavage (BAL) is performed by gently injecting 3–4 mL of cold PBS-E, then withdrawing it. The chest is massaged simultaneously as fluid goes in and out. This is repeated 8–10 times while the recovered cell-containing fluid is pooled and stored on ice (*see* **Note 2**). The volume recovered from the first two ravages is <100% but after this, all fluid that goes in should come out.
4. Pellet cells by centrifugation for 10 min at 250*g*.
5. Pour off PBS-E and gently resuspend in 5 mL cold BSS.
6. Count cells with a hemocytometer and adjust to 500,000/mL. Keep on ice and let sit 20 min prior to use (*see* **Note 3**).

3.2. Experimental Design and Procedure

The DCFH assay includes three major steps. First, the experiment is designed and the sample number is determined. Second, reagents are prepared, cells are preincubated with treatments and then exposed to buffer containing DCFH-DA and the stimuli of interest. Finally, cell samples are analyzed by flow cytometry. A typical experiment will include untreated control cells to establish a baseline, cells treated with PMA as a positive control to cause increased oxidative metabolism (RB), and cells treated by incubation with phagocytic particles or soluble compounds of interest (*see* **Note 4**).

We will illustrate a protocol designed to test the effect of the mitochondrial respiration inhibitor antimycin A on lung macrophage RB response to PMA or opsonized zymosan particles. Our sample experiment includes the following treatment groups: control, PMA (10^{-7} *M*), and Z-OPS (40:1) +/– pretreatment with antimycin A (10^{-7} *M*), all in duplicate, total = 12 (*see* **Note 5**). For 12 samples, 0.5 mL each, we need 6 mL of cells (ravaged AMs) at 500,000/mL BSS.

1. Reagent preparation I: Prior to starting the assay, prepare reagents to be added (as 50-μL aliquots) in **step 4**. Make 0.5 mL of each of the following: buffer control (BSS); 10^{-6} *M* PMA in BSS; zymosan (Z-OPS) at 2×10^8 per mL (50 μL = 40:1 for 250,000 cells). Prepare 10^{-5} *M* antimycin A in BSS, thaw an aliquot of 2 m*M* DCFH, and keep all reagents on ice.
2. Begin preincubation. First add 5 μL of antimycin A (10^{-5} *M*) to appropriate microtubes, then aliquot 0.5 mL of cells per tube. Rotate end-to-end at 37°C for 10 min. (*see* **Note 6** for details of securing tubes, and so on).
3. Reagent preparation II: Sonicate Z-OPS (0.5 mL prepared in **step 1**) for 15 sec. Add 55 μL of 2 m*M* DCFH to the 0.5 mL solutions of BSS, PMA, and sonicated Z-OPS. This achieves a final DCFH concentration of 200 μ*M* in each reagent stock.
4. Add reagents to cells. After 10 min of preincubation +/– pretreatment (antimycin A in our example), add 50 μL of DCFH-containing reagents prepared in **step 3** (this includes stimuli or control buffer) to appropriate tubes as quickly as possible. Close caps tightly, invert to mix, and rotate end-to-end for 30 min at 37°C in dark (*see* **Note 7**).
5. Stop the assay. After 30 min, remove each row of tubes (still taped together, as per **Note 6**) and place directly on ice. Keep in the dark, cool for 10 min, take off the tape, and add PI at 10 μg/mL final (25 μL of 20X 200 μg/mL stock). Invert to mix, wait 5 min, then begin flow cytometric analysis.

3.3. Flow Cytometric Analysis

We begin sequential gating using a FWD vs RED (forward light seater vs red fluorescence) bivariate histogram. All viable, PI-excluding cells having low red fluorescence are gated and sent to a FWD vs RAS (right angle scatter) bivariate histogram where AM are identified and selected based on their light scattering properties (*see* **Figure 1**). The green fluorescence (GFL) of these cells is then displayed as a GFL univariate histogram.

All samples are run at a flow rate of 300–500 cells/s. Microfuge tubes containing cells are kept on ice throughout the analysis. Initial cell scatter parameters are set with unused leftover ravaged cells (kept on ice). Next, the GFL gain setting is established by running the positive control, PMA-treated cells, and adjusting the gain so as to place the resulting histogram midway on the GFL intensity scale, having no more than 5% off scale. For experiments in which a substantial increase over the PMA-treated sample is expected, the

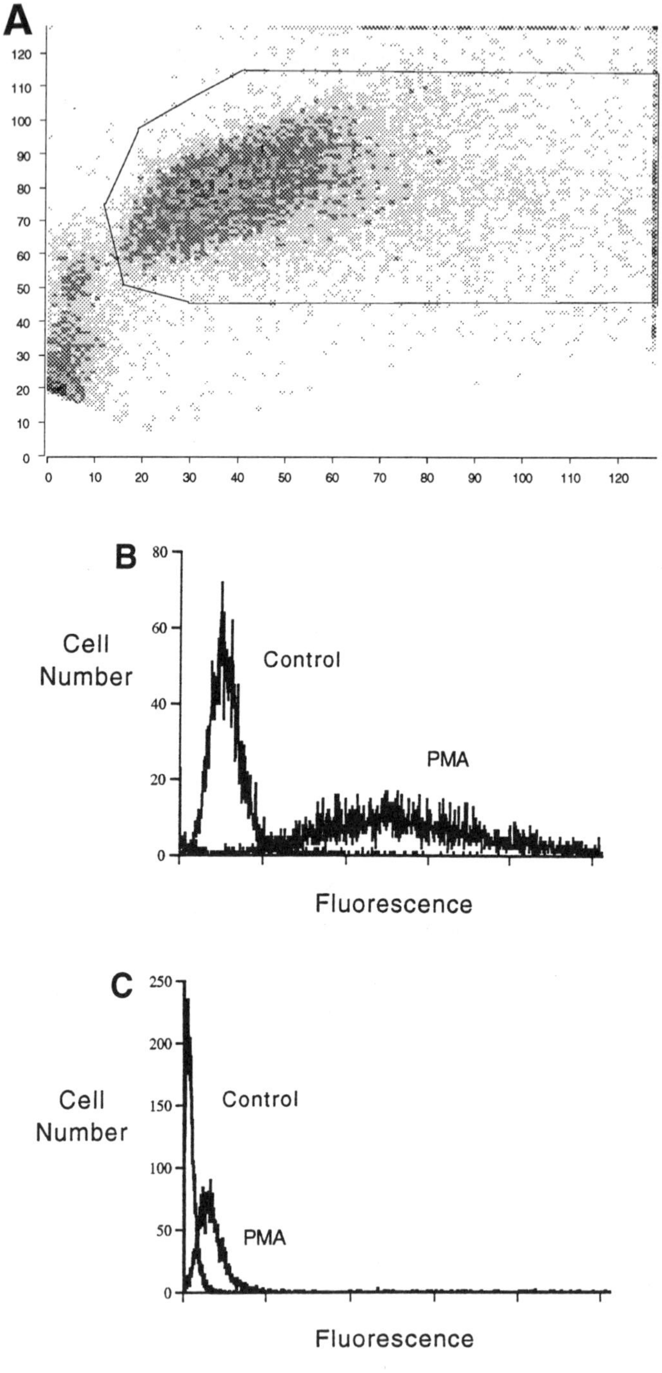
A
120
110
100
90
80
70
60
50
40
30
20
10
0
0
10
20
30
40
50
60
70
80
90
100
110
120
B
80
60
40
20
0
Cell Number
Control
PMA
Fluorescence
C
250
200
150
100
50
0
Cell Number
Control
PMA
Fluorescence

Fig. 1.

gain should be lowered accordingly to move the histogram closer to the ordinate and prevent off scale events and/or a log scale could be used. The remaining samples are run at the set GFL gain using uniform sample pressurization and equal run time before data acquisition (*see* **Note 8** about pressure effects).

When analyzing cells loaded with DCFH for uptake of red fluorescent latex beads, PI red fluorescence cannot be simultaneously measured and is excluded from the assay. AMs are gated directly from FWD vs RAS into Red and GFL histograms. The RED gain was set by running a dilute solution of bead stock and setting the fluorecsence of one bead at the low end of the scale (e.g., we target channel 50 of 1024).

3.4. Expected Results and Interpretation

Typical results following the above procedures are illustrated in **Fig. 1**. **Figure 1A** shows a bivariate histogram (forward vs right angle light scatter) of PI-negative (viable) AMs. Gating of the AM cloud is used to select cells for display of fluorescence histograms. **Figure 1B** shows typical green fluorescence histograms of unstimulated (control) and PMA-stimulated AMs loaded with DCFH. **Figure 1C** shows the effects of inhibiting intracellular mitochondrial respiration by pretreatment with antimycin A. Both basal and stimulated AM oxidative metabolism are inhibited by antimycin. A small, antimycin-insensitive RB is detected and reflects the antimycin insensitive NADPH oxidase system *(3)*.

Untreated AMs should have about the same baseline respiration and RB response to PMA (±15%) on any given day (*see* **Notes 3** and **6** about cell activation). When all related experiments are analyzed using the same GFL gain setting, data can be expressed as the mean fluorescence channel averaged from several animals. Another option is to show results as a percent increase over control values. This is a valid way to compare data from experiments that were analyzed using different GFL gain settings for reasons such as

---

Fig. 1. *(previous page)* Typical histograms obtained during flow cytometric analysis of macrophage oxidative metabolism using DCFH. (**A**) A bivariate histogram displays alveolar macrophage (AM) right angle light scatter (X-axis) vs forward light scatter (Y-axis). The encircled cells are gated for further analysis of their fluorescence (**B**,**C**), whereas smaller cells and debris are excluded. (**B**) Histogram of relative green fluorescence channel number (1-1023) in DCFH-loaded AMs treated with control buffer or PMA; the last channel that displays all fluorescence values ≥1024 and typically contains <5% of events is not illustrated. (C) Same as (B), except AMs were pretreated with antimycin A to inhibit mitochondrial respiration. Note substantial inhibition of both resting and stimulated DCFH oxidation; a small component of antimycin-resistant, nonmitochondrial AM respiratory burst is demonstrated.

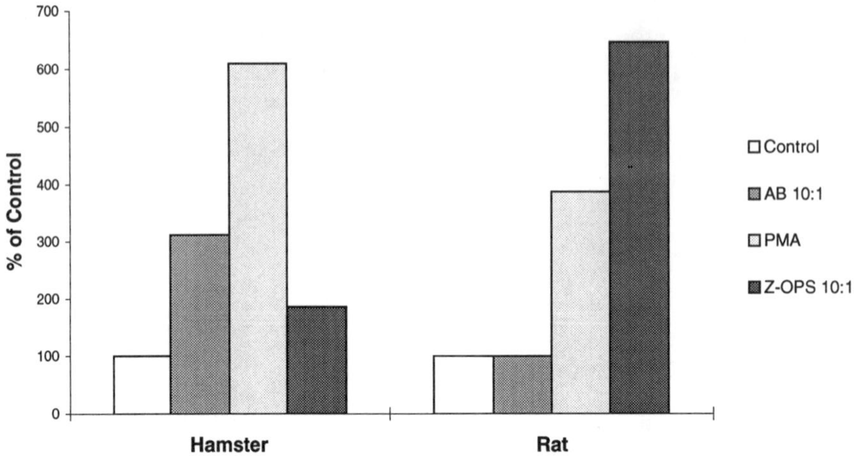

Fig. 2. Increases in mean fluorescence of DCFH-loaded AM populations in response to various stimuli. The results are expressed as percent of control (AB = antibody opsonized latex beads at 10:1 bead:cell ratio, PMA = phorbol myristate acetate 10^{-7} *M*, Z-OPS = serum opsonized zymosan at 10:1 particle:cell ratio). The results demonstrate that species differences in macrophage respiratory burst responses occur (e.g., hamster vs rat, response to PMA, Z-OPS) that should be considered in experimental design. These are caused, in part, by differences in phagocytic uptake (i.e., average particle per cell, hamster vs rat: 3.3 vs 0 for latex beads, 1.1 vs 3.2 for opsonized zymosan).

instrument/laser variability, unexpected cell responses, or the presence of many RBCs (*see* **Note 8**). For most experiments, this will suffice to accurately test hypotheses. If subpopulations are observed within the histograms, then further analysis by gating and/or analysis of the mean of the relevant subpopulations may be warranted.

Figure 2 shows results comparing responses of hamster vs rat AMs to PMA, opsonized zymosan or antibody-coated latex beads. Note the differences in species response to PMA (hamster >> rat) and Z-OPS (rat >> hamster). For evaluation of the effect of phagocytic particles, one must also compare relative uptake of the particle. This can be done most easily when using red fluorescent particles (e.g., latex beads) or by parallel analysis of particle uptake using microscopy (e.g., counting cell-associated zymosan particles in cytospin preparations).

4. Notes

1. All working dilutions of phagocytic particles are probe sonicated for 30 s/mL immediately prior to use. This provides a uniform preparation of single particles for each assay. Efficacy can be confirmed by microscopy. We have found that

repeated sonication of the same stock adds variation in the surface integrity of opsonized particulate. Therefore, aliquots are sonicated only once, then discarded.

2. This procedure is easily adapted for use in other laboratory animals. Lavage of a healthy hamster will yield 7–9 million cells, 95% being AMs. Findings characteristic of lung inflammation include a 30% or more increased macrophage yield, the presence of 5% or more inflammatory cells (neutrophils, eosinophils, or lymphocytes), or high numbers of red blood cells (RBCs) in the first few ravages. RBCs that appear after the first few ravages are usually introduced by ravage technique. Before starting an assay, make a cytospin or smear of each ravage sample, stain the cells with DiffQuik (modified Wright-Giemsa, VWR, Boston, MA), and perform a cell differential count to judge the health of the animal before starting. Sitting on ice for more than 3–4 h can effect AM-oxidative responses, so work within this time frame.
3. Harvesting of tissue macrophages or adherent macrophages cultured in vitro often requires isolation procedures like enzyme digestion, density gradient centrifugation, RBC lysis, and/or mechanical dissociation. Many of these procedures will cause some level of cell activation or impairment in function which, in turn, can adversely effect ROI production and DCFH assay results. Before measuring baseline and RB responses of such cells, consider short-term (hours or overnight) suspension culture of macrophages to give them time to settle down or replace surface receptors. Preliminary experiments should compare baseline and RB capacity at time intervals after macrophage purification. We recommend CoStar low-binding plates for overnight suspension culture of AM (in RPMI-2% fetal bovine serum). To prepare for the DCFH assay, cool cultured cells on ice 10 min, spin at 250*g*, then resuspend in cold BSS.
4. Preliminary experiments are usually needed to establish assay systems. For example, **Fig. 3** illustrates the results of testing different concentrations of DCFH to detect hamster AM respiratory burst to PMA. A second optimization step usually involves dose-response testing of the stimuli or agents of interest at two or three time points. Particle:cell ratios as well as cell concentration are variables one should consider initially adjusting to improve phagocytic uptake. We often include inhibitors (e.g., antimycin A, mitochondrial respiration poison) as pretreatments in order to learn more about the nature of AM RB responses. Experimental design can include comparisons of AM function after overnight suspension culture +/– endotoxin, cytokines, or any drugs of interest. To study AM activation in vivo, cells from control vs treated or diseased animals can be compared, but these should be carried out on the same day in parallel as much as possible.
5. Duplicate samples are recommended as good experimental practice and because they provide confidence when observing small RB changes (i.e., 15–30%). We often add more controls when large sample numbers are being tested and run a fresh one every 10–15 samples. This assures that samples sitting on ice are compared to the baseline value seen in a comparably treated control sample. This basal fluorescence drift is rather small (approx 5–10% over 30–60 min) and does not usually affect interpretation of most experiments where increases of 200–300% are observed.

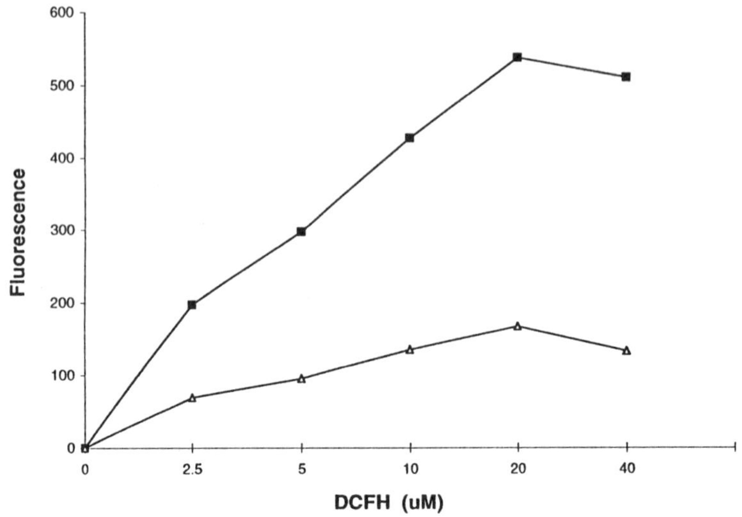

Fig. 3. Dose-response analysis of PMA-induced AM respiratory burst. Optimization of DCFH assay conditions will often include a dose-response analysis of loading concentrations vs maximal response, as illustrated here. We commonly use the 20 μ*M* concentration based on these results. PMA; Δ control.

6. We use the following protocol to minimize variability and cell activation resulting from assay conditions. Set up microtubes in a 80-well microtube (5 × 16) rack, label, and group together all tubes that will receive the same subsequent treatments (no more than 10 tubes in one row). Add pretreatment(s) and cells. Close lids tightly and apply a strip of 1/2-in. wide tape across the front of each tube to create a tape-secured grouping. Do not tape over the lids. Pick up the excess tape ends, transfer and secure the grouping of tubes into a 50-slot 5 × 10 microfuge tube-storage box. Keep the cover off to provide even air circulation. Gently invert the box to help start the fluid moving from end to end then strap it onto the rotator with a rubberband. Rotate as detailed in the protocol. Make sure the box does not interfere with 360° rotation.

   The end-to-end format is preferred over shaking for the following reasons. The baseline tends to be lower, particle-cell interaction is more uniform, and AMs remain more viable, especially during incubation periods longer than 30 min. We have noticed that some styles of microfuge tubes work well, whereas in others, movement is irregular and the fluid gets stuck at one end. If this is happening, adding 0.2% FBS to BSS will help. Also, by taping the tubes into a box, rather then clipping them on the rotator, more tubes can be set up at one time and adding reagents is quicker.
7. In our experiments, we have found that adding DCFH along with stimulants, as opposed to preloading DCFH (e.g., for 5–10 min) then adding stimulants, generally results in a higher percent increase over baseline DCF production, especially for short incubation times. We have observed no significant differences in over-

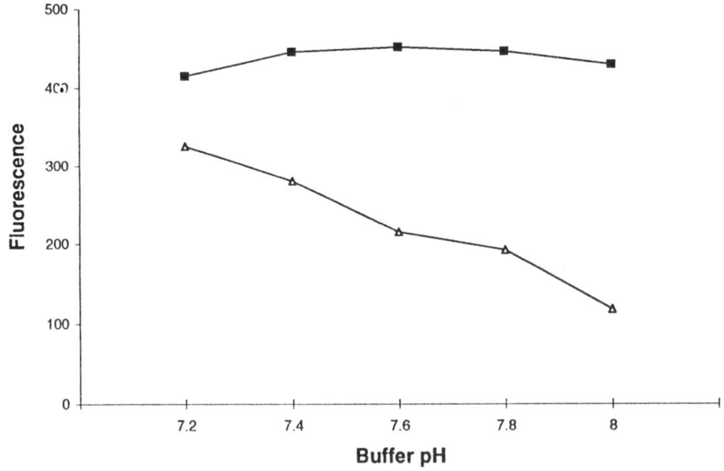

Fig. 4. Effects of pH and BSA on AM DCFH oxidation. Alkaline pH, in the presence of BSA, will inhibit AM basal (not shown) and PMA-stimulated DCFH oxidation. Hence pH should be monitored when adding significant amounts of basic reagents in assays that include BSA in the buffer. PMA; Δ PMA + BSA.

all results using this approach. Investigators developing new assays may wish to compare results using preloading to exclude the possibility of alterations in DCFH uptake as a potential confounding effect of experimental reagents. It may also be easier to preload DCFH (add it to the stock of cells in **step 2**) when there are many treatments to be added in **step 4**.

8. We have observed that some common reagents and assay conditions can markedly affect DCF fluorescence. These include effects of RBCs in the samples, effects of pH in the presence of BSA, and handling of samples during flow analysis.

   The presence of RBC 10–20X more than other cells will decrease the intensity of GFL among all samples containing them. We have not investigated the basis for this problem in detail. We speculate that some partitioning or uptake of DCFH by the RBCs may be a factor. In practice, we try to avoid introduction of RBCs when harvesting macrophages. If RBC lysis solutions are used, these have intrinsic effects on macrophage RB and appropriate controls (e.g., treat all samples with lysis buffer) should be considered. Brief suspension culture after RBC lysis is also an option (*see* **Note 3**).

   We have found that BSA can compete with DCFH for a portion of the ROI produced during cell respiration, e.g., act as a free radical scavenger ***(11)***. If BSA is added (e.g., to improve cell viability, to minimize aggregation of phagocytic particles), this effect must be considered. All samples must have the same amount of BSA and comparison experiments in the absence of BSA may be needed. In addition we have observed that BSA promotes a marked decrease in DCF fluorescence when small pH increases are introduced (**Fig. 4**). This effect can be an important confounder when using reagents as free bases and/or in sufficient concentration to shift pH.

Finally, during flow cytometric analysis all samples should be subjected to the same initial pressurization to the degree that is reasonably possible. We have observed that excessive pressurization of samples will skew results (increased DCF fluorescence), so we try to introduce all samples into the instrument under the same conditions and run them for the same amount of time before collecting data. We usually begin data acquisition 5–10 s after the flow rate becomes stable.

References

1. Fantone, J. C. and Ward, P. A. (1982) Role of oxygen-derived free radicals and metabolites in leukocyte-dependent inflammatory reactions. *Am. J. Pathol.* **107,** 397–418.
2. Halliwell, B. and Gutteridge, J. (1993) *Free Radicals in Biology and Medicine,* Oxford University Press, Oxford, UK.
3. Kobzik, L., Godleski, J. and Brain, J. (1990) Oxidative metabolism in the alveolar macrophage: analysis by flow cytometry. *J. Leuk. Bio.* **47,** 295–303.
4. Kobzik, L., Godleski, J., and Brain, J. (1990) Selective down-regulation of alveolar macrophage oxidative response to opsonin-independent phagocytosis. *J. Immunol.* **144,** 4312–4319.
5. Kobzik, L., Huang, S., Paulauskis, J. D., and Godleski, J. J. (1993) Particle opsonization and lung macrophage cytokine response. In vitro and in vivo analysis. *J. Immunol.* **151,** 2753–2759.
6. Kobzik, L. (1995) Lung macrophage uptake of unopsonized environmental particulates: role of scavenger-type receptors. *J. Immunol.* **155,** 367–376.
7. Rothe, G. and Valet, G. (1987) Use of hydroethidine (HE) and 2,7-dichlorofluorescin (DCFH) for the flow-cyotmetric measurement of NADPH-oxidase and mitochondrial oxygen radical formation in phagocytes. *Cytometry* **(Suppl. 1),** 77.
8. Chen, L. B. (1989) Fluorescent labelling of mitochondria. *Methods Cell Biol.* **29,** 103–123.
9. Rothe, G., Emmendorffer, A., Oser, A., Roesler, J., and Valet, G. (1991) Flow cytometric measurement of the respiratory burst activity of phagocytes using dihydrorhodamine 123. *J. Immunol. Methods.* **138,** 133–135.
10. Rao, K., Padmanabhan, J., Kilby, D., Cohen, H., Currie, M., and Weinberg, J. (1992) Flow cytometric analysis of nitric oxide production in human neutrophils using dichlorofluorescein diacetate in the presence of a calmodulin inhibitor. *J. Leuk. Biol.* **51,** 496–500.
11. Holt, M., Ryall, M., and Campbell, A. (1984) Albumin inhibits human polymorphonuclear leucocyte luminol-dependent chemiluminescence: evidence for oxygen radical scavenging. *Br. J. Exp. Pathol.* **65,** 231–241.

10

Measurement of Environmental Particulate Uptake by Lung Cells Using Flow Cytometry

Bradley Stringer and Lester Kobzik

1. Introduction

Inhaled particles can cause a spectrum of responses ranging from simple clearance to pathologic reactions, including acute injury and chronic fibrosis. In vivo and in vitro studies of particulate interaction with lung cells has greatly advanced our understanding of the mechanisms of particle health effects *(1,2)*. Methods for studying particulate interaction with lung cells include simple light microscopy, electron microscopy, and flow cytometry *(3)*. Many investigators have used flow cytometry and fluorescent latex beads to study particle–cell interactions. These particles are useful in that particle uptake by lung cells can be exactly quantified by taking advantage of their fluorescent properties *(4–7)*. However, these latex beads can only serve as a surrogate for actual "real world" environmental particulates.

Recent epidemiological data have shown an association between respirable air pollution particulates, and morbidity and mortality in humans *(8,9)*. New methods to study lung cell interactions with nonfluorescent, real world particles are needed in order to facilitate investigations of the epidemiologic observations. In this chapter, we describe a flow cytometric method that allows one to study particulate-lung cell interactions using the light-scattering characteristics of cells after association with particulates *(10)*.

The assay uses cellular light scatter as measured by the flow cytometer (i.e., forward scatter indicating cell size, and side or right angle scatter, indicating cell granularity) *(11)*. As particulates bind to the cells, the granularity of the cell increases, which in turn increases right angle light scatter (RAS). This increase in RAS is the basis for quantitation of particle uptake by lung cells. Please note that precise quantitation of particle number is not feasible using this technique, in contrast to assays using uniform fluorescent latex beads. In contrast

From: *Methods in Molecular Biology, Vol. 91: Flow Cytometry Protocols*
Edited by: M. J. Jaroszeski and R. Heller © Humana Press Inc., Totowa, NJ

to fluorescent latex beads, this method is designed to provide information about the *relative* and not *absolute* numbers of particulates that are cell associated.

This assay is designed to measure particulates that are cell associated; the method measures particulates that are both ingested and bound to the cells surface. The scatter assay alone, just as using fluorescent latex beads, can not distinguish between the two components without using inhibitors of particulate ingestion (i.e., cytochalasin d) ***(12)***. We therefore will use the broad term "uptake" when referring to particulate–cell association, which includes both bound and ingested particulates.

2. Materials

The basic assay requires only cells, buffer, and particles. The investigator can add other components to the basic buffer depending on the experiment in question. The following materials are needed:

1. Lung macrophages or other lung cell types can be used. Either cell lines or freshly isolated primary cells work well. Particulate uptake experiments can also be done using adherent cells. However, it should be noted that analysis is performed on single cells in suspension. If the cells are adherent, a nonenzymatic method of cell removal from the culture dish or flask should be used so as to preserve particulate receptors and/or particulate–cell attachment. For the A549 lung epithelial cell line, we replace the cell culture media with cold balanced salt solution (BSS) (*see* **item 2**) and incubate the cells at 37°C for 45 min; the cells are then detached by vigorous up and down pipeting.
2. BSS: 124 m*M* NaCl, 5.8 m*M* KCl, 10 m*M* dextrose, 20 m*M* HEPES; titrate with NaOH to pH 7.4 before use (*see* **Note 1**)
3. 1.5-mL microcentrifuge tubes.
4. Platform shaker.
5. 37°C incubator (need not be CO_2). The incubator should be large enough to accommodate a small platform shaker.
6. Particles: The type of particles used can vary depending upon the investigators studies (*see* **Notes 2** and **3**). The following particles can be used as control particulates:
   a. Titanium dioxide (TiO_2) (Baker Chemical [Philipsburg, NJ], cat. no. 1-4162).
   b. α-Quartz (SiO_2) (US Silica [Berkeley Spring, WV], cat. no. 14808-60-7).

These particles are representative of an inert particulate (TiO_2) and a pathogenic particulate (SiO_2). The particle types are given here as a guide. Other types can be used depending on the experimenters protocol.

3. Methods

3.1. Instrumentation and Gating

The instrumentation setup, including the scatter plots and gating are the basis for the assay. It is important that these parameters be set properly in order to produce meaningful data. Since all flow cytometers work essentially the same, instrument setup will be described in broad terms and not in terms of commands for specific software packages. The investigator will need to adapt the assay to their particular software package and flow cytometer. In the following methods description, the nomenclature specific for the Ortho 2150 Cytofluorograph will be used. In addition, gating procedures outlined will use the nomenclature of the Cicero computer interface system (Cytomation, Ft. Collins, CO).

3.1.1. Setting a Bivariate Scatter Histogram

1. Set a bivariate right angle scatter (RAS) vs forward scatter histogram. The RAS should be displayed on the abscissa and forward scatter displayed on the ordinate (*see* Chapter 9, Fig. 1A). This histogram selects for the cells of interest. This becomes important if one is using a mixed population of cells (i.e., whole blood, macrophage/neutrophil mixtures, and so on). Normally, a homogeneous population of cells is studied using this technique, and only one cluster of cells will be observed.
2. Set the channel number from 0–120 for both the ordinate and abscissa. This will be the initial scatter-plot used to assess the RAS of the cells being studied (*see* **Note 4**)

3.1.2. Setting a Univariate Scatter Histogram

1. Set a univariate histogram that plots cell number on the ordinate and RAS on the abscissa. The abscissa channel number should be set from 0–1024. RAS should be gated from the cell population in the bivariate plot (*see* **Subheading 3.1.3.** for details).

3.1.3. Gating Cells from the Bivariate Scatter Plot

1. Set an electronic gate inside the bivariate plot in order to isolate the cells being studied (*see* Chapter 9, Fig. 1A).
2. Adjust the width (RAS parameter) of the gate so that the RAS of the cell population can be measured over the entire range of channels (usually 120). This is needed in order to facilitate measurement of RAS in the presence of particulates, since particulate interaction with the cells will result in the shift of the cell RAS to the right in the bivariate histogram. The gate needs to encompass the last channel of RAS data in order not to underestimate the RAS of cells after particulate uptake (*see* **Note 5**).
3. Adjust the height (forward scatter) of the gate to just encompass the cell population, since particle-cell interaction will not normally alter the forward scatter characteristics. A narrow forward scatter gate facilitates the electronic removal of free, unbound particles. This step minimizes increases in RAS caused by free particles and generating erroneous data (*see* **Note 7**).

3.2. Uptake Assay

3.2.1. Cell–Particle Preparation

A stepwise protocol for the measurement of titanium dioxide (TiO_2) uptake by hamster alveolar macrophages (AMs) will be illustrated. This or other cell types can be studied; the cells can be treated with numerous inhibitors, receptor ligands, or fluorochromes needed for a given experimental question.

1. Obtain hamster AMs by repeated bronchoalveolar lavage (BAL).
2. Centrifuge BAL fluid at 150*g* for 10 min at 4°C and resuspend cells in BSS.
3. Count cells and determine viability by trypan blue dye exclusion. Modified Wright-Giemsa staining of cytospins or smears can be done to insure >95% AMs.
4. Adjust cell suspension to 1×10^6 cells/mL in BSS.
5. Add 400 µL of the cell suspension to 1.5-mL microcentrifuge tubes.
6. Add particulates (TiO_2) at 50 µg/mL in a 100 µL volume. Vortex the suspension to mix.
7. Incubate 30 min at 37°C on a platform shaker set on a low to moderate speed (approx 300–500 rpm).
8. Remove cell suspensions to ice, and analyze immediately by flow cytometry.

3.2.2. Flow Cytometry

1. Measure RAS of control cell (no particulates) and set the RAS channel number to approx 60 (of 1024 channels, *see* **Note 6**).
2. Measure RAS of cell–particulate sample with highest concentration of particles and verify that the gate set in the bivariate RAS vs forward scatter histogram (*see* **Subheading 3.1.3.**) excludes free particles. (*see* **Note 7**).
3. Analyze 3000–5000 cells per sample at a flow rate of approx 250 cells/s. Record the mean channel RAS for the cell populations from the univariate histogram set in **Subheading 3.1.2.**

3.3. Data Analysis

Typical results from the assay for two different particles types are shown in **Figs. 1** and **2**. Data analysis is done by comparing the RAS of the control cells (no particles) to the RAS of cells incubated with particulates. The RAS mean channel number calculated from the univariate RAS histogram should be used for comparisons. Data are generally expressed as fold changes in RAS for a given particulate/cell sample as compared to RAS of control cells

4. Notes

1. The basic BSS buffer can be altered to contain numerous ions, ligands, and so on, in order to study their effects on particulate uptake. We add $CaCl_2$ (0.3 m*M*) and $MgCl_2$ (1.0 m*M*) to facilitate particle binding, since unopsonized particle binding is commonly (but not exclusively) calcium dependent ***(12)***.
2. This assay is designed to measure *relative* numbers of particles per cell. Each particle type has a unique scatter pattern with regard to the magnitude of change between control cells and treatments; particle size and aggregation will contribute to these differences. With these caveats in mind, it is recommended that caution be used when attempting to compare uptake of two *different* types of particulates. The assay works best when comparing changes in particle uptake between different treatment groups (i.e., ligands, ions, and so on) using the same particle type. If comparisons between different particle types are desired, it is

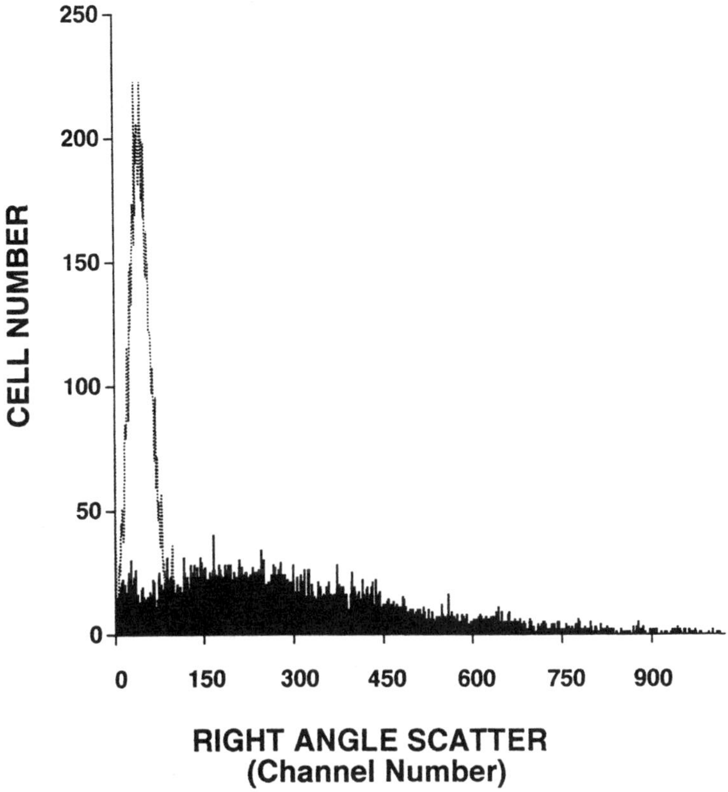

Fig. 1. TiO_2 uptake by hamster alveolar macrophages (AMs). Univariate histogram showing right angle light scatter (RAS) for AMs in the absence of particulates (open histogram) and after addition of 40 μg/mL TiO_2 (closed histogram, 30 min incubation). RAS increased from 59.6 for control cells to 308.3 after incubation with TiO_2, a fivefold increase. Light microscopy confirmed an increase in cell-associated particulates (data not shown).

recommended that particles be added to the cell suspension based on number rather than concentration. Also, the RAS scatter pattern should be similar (i.e., size homogeneity between particle types one wishes to compare).

3. Any type and size of particulate can be studied as long as it will flow through the orifice of the instrument. Generally, particulates at or below 2–3 μm (respirable size) are studied in the context of lung cell research and present no problem. The volume of particulates added should be no more than approx 100 μL so as not to dilute the cell suspension; the particulate concentration should be adjusted accordingly.
4. A second bivariate scatter plot may be included at this point. It is often desirable to only analyze live cells. This is accomplished by using propidium iodide ***(13)*** to exclude nonviable cells, and setting a forward scatter (ordinate) vs red fluorescence

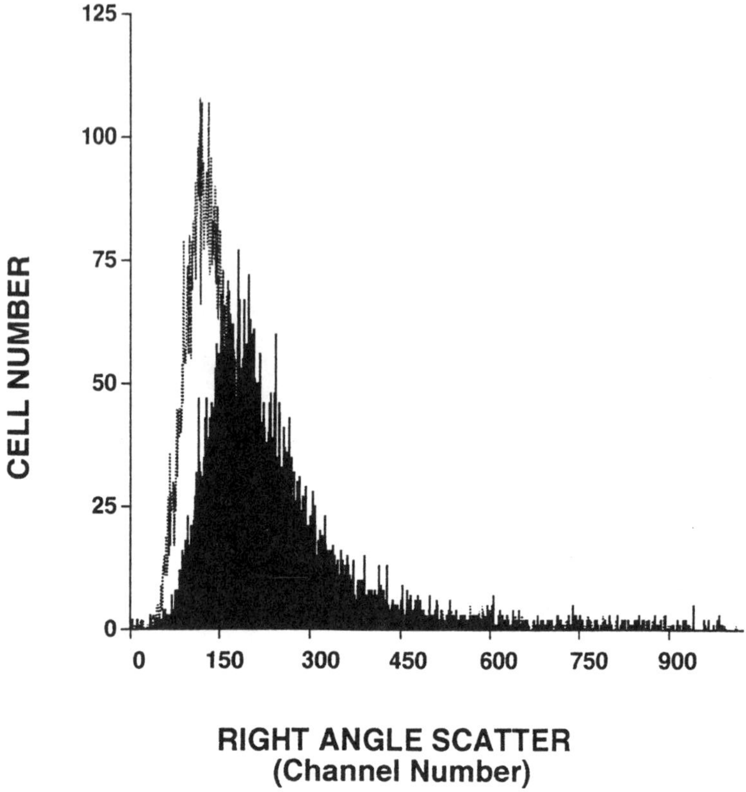

Fig. 2. Uptake of α-quartz by hamster alveolar macrophages. Univariate histogram showing right angle light scatter (RAS) for AMs in the absence of particulates (open histogram) and after addition of 100 μg/mL α-quartz (closed histogram, 30 min incubation). The initial RAS value for control AMs (no particles) was set to 164.8, since the RAS changes resulting from uptake of α-quartz is not as great as for TiO_2 (*see* **Fig. 1**). After incubation with 100 μg/mL α-quartz for 30 min, RAS increased to 244.8, a 1.5-fold increase.

(abscissa) histogram. The nonflourescent cell population is then gated into the forward vs right angle histogram and the assay proceeds as described in **Subheading 3.**

5. The univariate plots of cell RAS should resemble those shown in **Figs. 1** and **2**. A narrow RAS histogram of the cell population, as well as a small mean RAS channel number (50–100), is advantageous in order to maximize RAS differences in the presence of particulates. However, this depends on the particulate being studied. As shown in **Fig. 1**, uptake of TiO_2 by AMs results in a large RAS change as compared to the control. Therefore, the RAS gain can be lower. In the case of α-quartz, a higher RAS gain must be used in order to maximize differences in scatter between control and treatment (**Fig. 2**).

6. The initial RAS gain settings should be determined before the actual experiment is to be done by analyzing a cell–particle sample at the highest concentration one wishes to use. This will allow maximal RAS differences to be measured between control and experimental groups. The RAS for control cell (no particles) can be set at any value one wishes (usually between 60–150 out of 1024). The value of 60 is given as a guide and can be adjusted upward depending on the differences between the control and cell–particle sample. It should be noted that once the RAS gain settings are made, the settings should not be adjusted during the course of the experiment unless a new sample of control cells is analyzed at the new setting.
7. When cell–particle suspensions are analyzed, it is normal to observe a population of free particles with a small forward scatter, and a moderate RAS, depending on the size homogeneity of the particles being studied. RAS of a particle–cell sample should be measured when initially making the instrument settings in order to determine the optimum height (forward scatter) of the gate so as to minimize free particulate interference.

Acknowledgment

This work was supported by grant no. NIENS-00002 from the National Institute of Environmental Health.

References

1. Adamson, I. Y. R. and Bowden, D. (1981) Dose response of the pulmonary macrophagic system to various particulates and its relationship to transepithelial passage of free particles. *Exp. Lung Res.* **2,** 165–175.
2. Brain, J. D. (1985) Macrophages in the respiratory tract. *Handbook Physiol.* **I,** 447–471.
3. Absolom, D. R. (1986) Basic methods for the study of phagocytosis. *Methods Enzymol.* **132,** 95–180.
4. Hed, J., Hallden, G., Johansson, S., and Larsson, P. (1987) The use of fluorescence quenching in flow cytofluorometry to measure the attachment and ingestion phases in phagocytosis in peripheral blood without prior cell separation. *J. Immunol. Methods* **101,** 119–125.
5. Kobzik, L., Huang, S., Paulauskis, J., and Godleski, J. (1993) Particle opsonization and lung macrophage cytokine response: in vitro and in vivo analysis. *J. Immunol.* **151,** 2753–2759.
6. Oda, T. and Maeda, H. (1986) A new simple fluorometric assay for phagocytosis. *J. Immunol. Methods* **88,** 175–183.
7. Stewart, C. C., Lehnert, B. E., and Steinkamp, J. A. (1986) In vitro and in vivo measurement of phagocytosis by flow cytometry. *Methods Enzymol.* **132,** 183–192.
8. Dockery, D., Pope III, C., Xu, X., Spengler, J., Ware, J., Fay, M., Ferris, B., and Speizer, F. (1993) An association between air pollution and mortality in six U.S. cities. *N. Engl. J. Med.* **329,** 1753–1759.

9. Pope, C. A., Bates, D. V., and Raizenne, M. E. (1995) Health effects of particulate air pollution: time for reassessment? *Envir. Health Persp.* **103,** 472–480.
10. Stringer, B., Imrich, A., and Kobzik, L. (1994) Lung macrophage uptake of environmental particulates: a flow cytometric assay. *Cytometry* **20,** 23–27.
11. Salzman, G. and Mullaney, P. P., B (1979) Light scattering approaches to cell characterizations. *Flow Cytometry Sorting* **I,** 105–124.
12. Parod, R. and Brain, J. (1983) Uptake of latex particles by pulmonary macrophages: role of calcium. *Am. J. Physiol.* **245,** C227–C234.
13. Kobzik, L. (1995) Lung macrophage uptake of unopsonized environmental particulates: role of Scavenger-type receptors. *J. Immunol.* **155,** 367–376.

11

Estimation of Drug Resistance by Flow Cytometry

Adorjan Aszalos and James L. Weaver

1. Introduction

Drug resistance is a subject that covers two areas. In one case it refers to the ability of microbes to defeat antibiotics. In a different setting it refers to the ability of cancer cells to resist chemotherapy. In the first case, the drug resistance is evaluated on the basis of the percentage of viable cells after treatment with a given drug concentration. This is then simply an application of viability testing, the reader interested in this specific area is referred to Chapter 7.

The second area, multidrug resistance in cancer includes several different mechanisms by which tumor cells are able to evade or defeat the effects of chemotherapeutic drugs. Several types of resistance have been characterized. This includes MDR1 (P-glycoprotein/P170) and MRP *(1,2)* as well as mechanisms that increase the production of a drug target and elevate drug metabolism *(3)*. This list is not intended to be comprehensive and it is nearly certain that additional mechanisms of drug resistance remain to be described.

There are several efflux pumps, two of which have been reasonably well characterized, MDR1 and MRP. These are both members of a larger family of proteins that include a number of bacterial transport proteins and the CFTR chloride channel *(2)*. MDR1 and MRP both have the ability to move neutral molecules from the cytoplasm to the exterior of the cell and require ATP for this activity *(4)* but differ in that MDR1 can also move positively charged molecules *(5)*, whereas MRP appears to be able to move negatively charged substrates *(6)*. They also differ in the tissues in which they are normally expressed and in their comparative substrate and blocker specificities. For example, probenecid is a blocker for MRP *(7)*, but has no effect on MDR1 activities *(5)*. However there is significant overlap in the molecules that can be transported

From: *Methods in Molecular Biology, Vol. 91: Flow Cytometry Protocols*
Edited by: M. J. Jaroszeski and R. Heller © Humana Press Inc., Totowa, NJ

by these two proteins. A more recently described efflux pump, LRP, is able to affect intracellular doxorubicin levels and can be blocked by brefeldin-A ***(8)***, but is otherwise still in the process of being characterized.

Flow cytometry has been widely used in the characterization of the activity of efflux pumps, but is not currently the method of choice for characterizing other mechanisms of drug resistance such as increased levels of glutathione. In the case of efflux pumps, the standard strategy is to incubate cells with an inherently fluorescent efflux pump substrate such as daunorubicin or rhodamine 123 (R123). Expression of efflux proteins can also be measured using standard antibody-labeling methods ***(9)***.

2. Materials

1. Probenecid: 30 mg/mL in DMSO (Sigma, St. Louis, MO).
2. Verapamil: 10 m*M* in DMSO (Sigma).
3. Genistein: 20 m*M* in DMSO (Sigma).
4. Daunorubicin: 100 μg/mL in DMSO (Sigma).
5. Calcein-AM: 50 μ*M* in DMSO (Molecular Probes, Eugene, OR).
6. Rhodamine 123: 2.6 m*M* (1 μg/mL) in ethanol (Sigma).
7. Cyclosporin A: 1 mg/mL in DMSO (Sigma).
8. Flunarazine: 10 mg/mL in DMSO (Sigma).
9. MRK-16, a mouse monoclonal antibody to MDR1 (Signet Labs, Dedham, MA).
10. Anti-Mouse-IgG-FITC (Sigma).
11. MRPPr1, a rat monoclonal antibody to MRP (Signet).
12. Anti-rat-IgG-FITC (Sigma).
13. FACS Permeabilizing Solution: 10% (v/v) in PBS (Becton Dickinson, San Jose, CA).

3. Methods

3.1. Evaluation of Potential Blockers of MRP Pumping Activity

1. For this purpose, parental and MRP-expressing cells are collected (with or without trypsinization) and are resuspended in phenol red-free medium at 0.5×10^6 cells/mL (*see* **Note 1**). After equilibration for 2–5 min, both parental and MRP cells are treated with potential blockers, solvents, positive controls, or no treatment. As positive controls for blocking MRP activity, any of the following compounds could be used: probenecid (100–300 μg/mL), verapamil (10 μg/mL), or genistein (200 μ*M*).
2. Cells are incubated with test compounds or controls for 10–20 min. If genistein is used, the incubation period before addition of substrate should be at least 1 h since this agent acts by blocking the phosphorylation needed for MRP activity.
3. Next the substrate is added, recommended substrates are daunorubicin (0.3 μg/mL) or calcein-AM (250 n*M*). Incubation with the substrate is for 1 h at either 22 or 37°C in the dark. This time has been shown to be sufficient to allow maximal accumulation of substrates in parental and MRP cell lines.
4. After incubation, cells are collected by centrifugation and kept as a pellet without supernatant until flow cytometry (*see* **Note 2**). Pellets keep their intensity for

>1 h if kept in the dark (*see* **Note 3**). One-parameter histograms are collected for the substrate (calcein: 525–535 nm [FL1]; daunorubicin: 565–605 nm [FL2]). Data are collected for 3000–10,000 cells.

5. Evaluation of the blocking ability of a specific compound is done by comparing the relative percent fluorescence of untreated parental vs MRP cells to the drug treated pair. Correct evaluation of a given drug should include two experiments on different days as well as dose response studies.

3.2. Evaluation of Potential Blockers of MDR1 Pumping Activity

Testing for MDR1 activity is similar to that for MRP, since MDR1 has different substrate preferences, the substrates and blockers are different from those used for the MRP method.

1. Parental and MDR1-expressing cells are collected (with or without trypsinization) and are resuspended in phenol red-free medium at 0.5 × 10^6 cells/mL (*see* **Note 1**). After equilibration for 2–5 min, both parental and MDR1 cells are treated with potential blockers, solvents, positive controls, or no treatment. As positive controls for blocking MDR1 activity, the following compounds could be used: cyclosporin A (1 µg/mL), or flunarazine (5 µ*M*).
2. Cells are incubated with test compounds or controls for 10 min.
3. Next the substrate is added, recommended substrate is R123 (5.2 µ*M*). Incubation with the substrate is for 20 min at 37°C in the dark.
4. After incubation, cells are collected by centrifugation and kept as a pellet without supernatant until flow cytometry (*see* **Note 2**). Pellets keep their intensity for >1 h if kept in the dark (*see* **Note 3**). One parameter histograms are collected for 10,000 cells in the FITC channel.
5. Evaluation of the blocking ability of a specific compound is done by comparing the relative percent fluorescence of untreated parental vs MDR1 cells to the drug-treated pair. Correct evaluation of a given drug should include two experiments on different days as well as dose response studies.

An example of the results of this type of experiment is shown in **Fig. 1**. Here L5178Y cells from parental and MDR1–transfected cell lines were exposed to no treatment or 1 µ*M* cyclosporin A (CsA). The inhibition of MDR1 by CsA prevents the efflux of R123 and results in a large increase in R123 fluorescence. In contrast, CsA has almost no effect on R123 fluorescence in the MDR1–negative parental cells. The mean fluorescence values are: parental: 30.0, parental/CsA: 26.4, MDR: 0.329, MDR/CsA: 22.2.

3.3. Testing for Expression of MDR1 (P-Glycoprotein)

Testing for MDR1 gene expression on MDR1-positive and MDR1-negative cells can be needed to differentiate between cells expressing MDR1, MRP, or both. This is very important in evaluation of potential blockers since interpretation can be very complex if cells express both proteins. Expression of MDR1

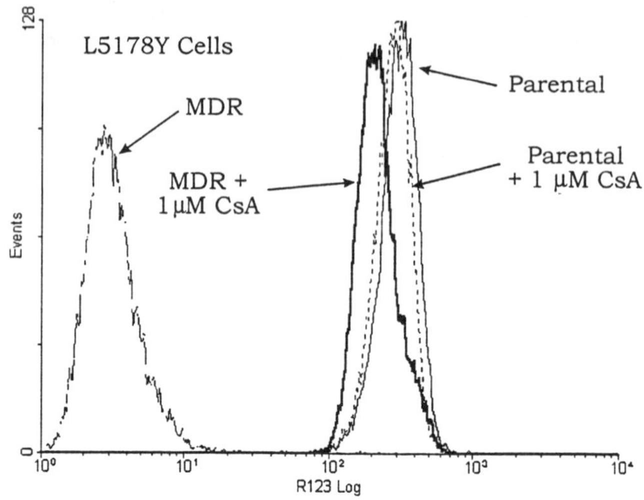

Fig. 1. The effect of cyclosporin A on rhodamine 123 fluorescence in parental and MDR1–transfected mouse L5178Y lymphoma cells. Thin solid line: parental, dashed line: parental with 1 µ*M* CsA, interrupted line: MDR, strong solid line: MDR with 1 µ*M* CsA.

can be evaluated using the monoclonal antibodies such as MRK-16 or 4E3.16, which bind to a extracellular epitopes of MDR1. Other antibodies are available to internal epitopes, but these require fixation and permeabilization to allow the MAb access to the epitope. Any of these antibodies can be used either as a direct-fluorochrome conjugate or with a secondary labeled antibody.

1. Collect at least 10^5 cells/tube by centrifugation and resuspend in 100 µL of PBS. Inclusion of controls including isotype control for the primary, is recommended. For MDR1 expression, add 2 µL of: MRK-16 from a stock of 0.5 mg/mL, or an equivalent amount of MRK-16-FITC. Mix and incubate on ice for 20–30 min.
2. If unlabeled primary antibody is used, add 1 mL of ice-cold PBS, spin, and resuspend in 100 µL of cold PBS. Add 5 µL of antimouse IgG-FITC, mix and incubate on ice for 20–30 min. If labeled primary antibody is used, proceed directly to **step 3**.
3. After all antibody labeling steps are completed, wash the cells 2X in PBS and resuspend in 0.5 mL PBS. Single parameter histograms are collected on 10,000 cells. Labeling is evaluated as increase in fluorescence as compared to isotype controls (*see* **Note 4**).

An example of the results of this type of experiment is shown in **Fig. 2**. Here, transfected NIH3T3MDR cells were exposed to either directly labeled MRK-16-FITC or unlabeled MRK-16 followed by antimouse-IgG-FITC. Note the significant nonspecific binding of the antimouse-IgG-FITC.

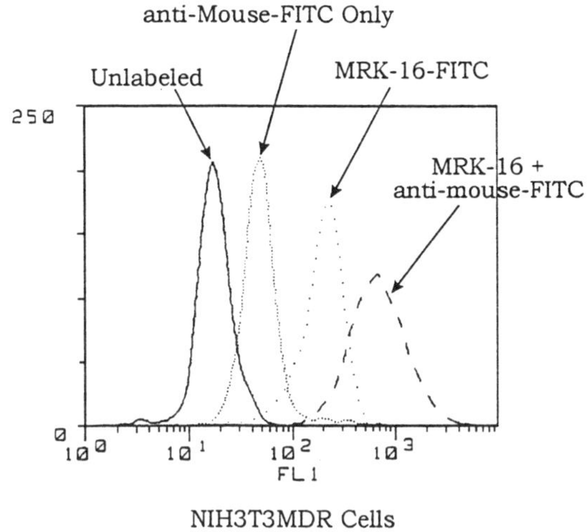

Fig. 2. Binding of FITC-labeled or unlabeled MRK-16 to NIH3T3MDR cells. Cells were treated with MRK-16-FITC, MRK-16 followed by antimouse-IgG-FITC, or antimouse-IgG-FITC only.

3.4. Testing for Expression of MRP

The antibody to MRP detects an intracellular epitope, therefore the cells must be permeabilized to allow the antibody access to its target.

1. Collect 0.5×10^6 cells by centrifugation and resuspend in 1 mL of FACS-permeabilizing stock, incubate for 10 min at room temperature.
2. Add 1.5 µg of MRPPr1 and incubate 60 min at room temperature.
3. Wash cells 1X, resuspend in 1 mL PBS, add 10 µL of antirat-IgG-FITC, and incubate for 30 min.
4. Wash 2X and resuspend in 0.5 mL PBS. Single parameter histograms are collected on 10,000 cells. Labeling is evaluated as increase in fluorescence as compared to isotype controls (*see* **Note 4**).

4. Notes

1. We have suggested the use of cell–culture medium for incubations for efflux experiments since both MDR1 and MRP require good levels of ATP for proper function. Extended incubations in nonglucose-containing solutions such as PBS may result in loss of efflux activity.
2. In some cell lines, it is necessary to allow the cells an efflux period between exposure to the fluorescent substrate and evaluation by flow cytometry. Typically, after the incubation with the substrate, cells would be washed 1X and resuspended in phenol red-free medium, and incubated for an additional time period that can range between 20 and 60 min. If no change in fluorescence is seen after 60 min, efflux activity is probably not biologically significant.

3. Note that many fluorescent substrates have a significant "spontaneous" leak rate. Be sure to use controls with either non-MDR/MRP expressing cells or cells treated with strong blockers to differentiate between efflux and leakage.
4. Levels of expression of these efflux pumps varies significantly among various cell lines. For example, four different MDR transfected cell lines showed expression varying from 8000–55,000 antibody binding sites/cell *(9)*.

References

1. Goldstein, L. J., Pastan, I., and Gottesman, M. M. (1992) Multidrug resistance in human cancer. *Crit. Rev. Oncol./Hematol.* **12,** 245–253.
2. Cole, S. P. C., Bhardwaj, G., Gerlach, J. H., Mackie, J. E., Grant, C. E., Almquist, K. C., Stewart, A. J., Kurz, E. U., Duncan, A. M. V., and Deely R. G. (1992) Overexpression of a transporter gene in a multidrug-resistant human lung cancer cell line. *Science* **258,** 1650–1654.
3. van der Zee, A. G., Hollema, H. H., de Bruijn, H, W., Willemse, P. H., Boonstra, H., Mulder, N. H., Aalders, J. G., and deVries, E. G. (1995) Cell biological markers of drug resistance in ovarian carcinoma. *Gynecol. Oncol.* **58,** 165–178.
4. Ambudkar, S., Lelong, I. H., Zhang, J., Cardarelli, C. O., Gottesmann, M. M., and Pastan, I. (1992) Partial purification and reconstitution of the human multidrug resistance pump: characterization of the drug-stimulatable ATP hydrolysis. *Proc. Natl. Acad. Sci. USA* **267,** 21,020–21,026.
5. Weaver, J. L., Szabo, G., Jr., Pine, P. S., Gottesman, M. M., Goldenberg, S., and Aszalos, A. (1993) The effect of ion channel blockers, immunosuppressive agents, and other drugs on the activity of the multidrug transporter. *Int. J. Cancer* **54,** 456–461.
6. Hollo, Z., Homolya, L., Hegedus, T., and Sarkadi, B. (1996) Transport properties of the multidrug resistance-associated protein (MRP) in human tumour cells. *FEBS Lett.* **383,** 99–104.
7. Feller, N., Broxterman, H. J., Wahrer, D. C., and Pinedo, H. M. (1995) ATP-dependent efflux of calcein by the multidrug resistance protein (MRP): no inhibition by intracellular glutathione depletion. *FEBS Lett.* **368,** 385–388.
8. Verovski, V. N., Van den Berge, D. L., Delvaeye, M. M., Scheper, R. J., De Neve, W. J., and Storme, G. A. (1996) Low-level doxorubicin resistance in P-glycoprotein-negative human pancreatic tumour PSN1/ADR cells implicates a brefeldin A-sensitive mechanism of drug extrusion. *Br. J. Cancer* **73,** 596–602.
9. Weaver, J. L., McKinney, L., Schoenlein, P. V., Goldenberg, S., Gottesman, M. M., and Aszalos, A. (1996) MDR1/P-glycoprotein function. I. Effect of hypotonicity and inhibitors on rhodamine 123 exclusion. *Am. J. Physiol.* **270** (*Cell. Physiol.* **39**), C1447–C1452.

12

Flow Cytometric Monitoring of Cellular Pharmacokinetics

Margaret S. Dordal

1. Introduction

The discovery that one drug transporter, P-glycoprotein, decreases intracellular concentrations of many anticancer drugs explains how drug resistance could simultaneously develop to structurally unrelated compounds *(1)*. These discoveries have focused attention on the kinetics of drug distribution within cells. Flow cytometry can be easily adapted to collect kinetic data on fluorescent compounds by recording time as one parameter.

In clinical pharmacokinetics, inferences are drawn about the movement of drugs throughout the body based on mathematical modeling of serum drug levels *(2)*. In the same fashion, modeling drug uptake and efflux from intact cells can reveal the presence of internal kinetic compartments. This data can then be correlated with microscopic observations to provide insight into mechanisms of drug distribution.

We have equipped a flow cytometer with a rapid injector that adds drug to the sample compartment during a run without disturbing flow *(3)*. Using this system, we measure cell-associated dye at selected times, then model the hypothesized mechanisms of uptake and efflux. Doxorubicin and rhodamine, both P-glycoprotein substrates, diffuse passively into the cell *(4,5)*. Once within the cell, both compounds accumulate irreversibly, doxorubicin within the nucleus *(6)* and rhodamine in mitochondria *(5)*. The simplest model describing both diffusion and sequestration is shown in **Fig. 1A**. Data gathered via the flow cytometric assay was used to obtain membrane permeabilities and sequestration rate constants.

P-glycoprotein is hypothesized to transport drugs outward across the cell membrane, theoretically increasing efflux as shown in **Fig. 1B**. In comparison

From: *Methods in Molecular Biology, Vol. 91: Flow Cytometry Protocols*
Edited by: M. J. Jaroszeski and R. Heller © Humana Press Inc., Totowa, NJ

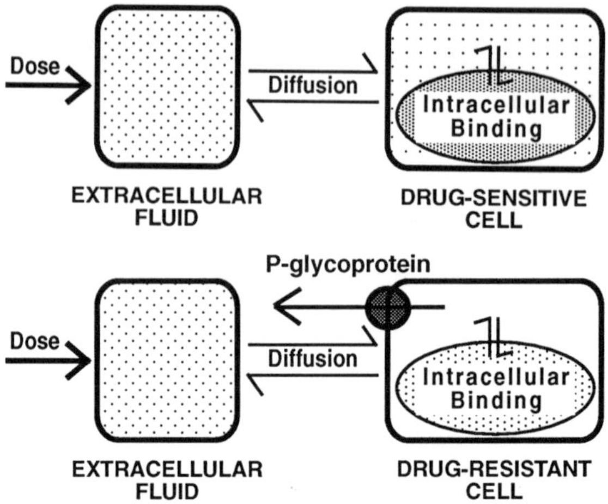

Fig. 1. Regulation of drug distribution in sensitive and resistant cells. Lipophilic drugs diffuse across the cell membrane, eventually reaching an equilibrium between the extracellular fluid and the cytosol. When drugs bind within an internal compartment, such as the nucleus or mitochondria, the equilibrium concentration within that compartment is much higher than that of either cytosol or extracellular fluid. Total cellular uptake is the sum of all such processes. In drug-sensitive cells **(A)** drug will continue to accumulate where irreversibly bound until the sites are saturated. In cells expressing P-glycoprotein **(B)**, the equilibrium between cytosol and extracellular fluid is altered by outward transport. However, drug should continue to accumulate where bound unless the transporter also regulates intracellular distribution of drug.

to nonresistant cell lines, those cells expressing P-glycoprotein showed small differences in rhodamine efflux, but much larger differences in sequestration *(3,7)*. These results suggest that drug transporters may regulate intracellular distribution as well as transmembrane efflux.

The protocol below describes the kinetic analysis of rhodamine transport used to measure permeability and sequestration in the presence of a P-glycoprotein inhibitor. This method can be used to directly study the cellular pharmacokinetics of fluorescent compounds or to screen nonfluorescent inhibitors of drug transporters for which a fluorescent substrate exists.

2. Materials

1. Flow cytometer and fluorescence microscope: Our studies were conducted on a Coulter (Hialeah, FL) Epics model 752 dual-beam laser cytometer equipped with argon and xenon lasers. For the majority of studies we used a mirrored biohazard sample tip with a 200 μ aperture. We confirmed the intracellular localization of fluorescence with a Zeiss epifluorescence microscope.

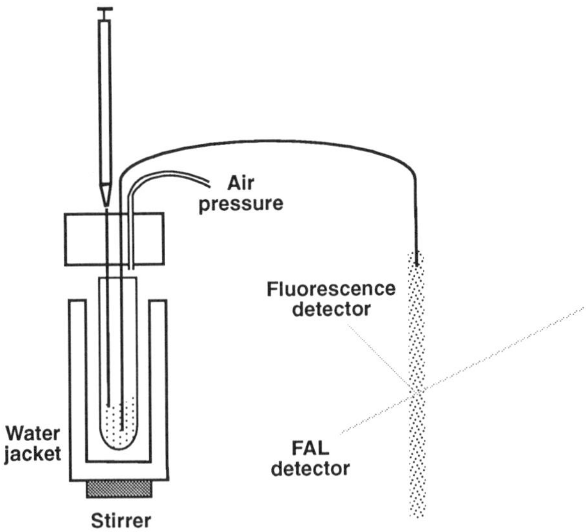

Fig. 2. Rapid injection system for flow cytometry. The rapid injection sample holder (Cytek) permits injection of up to 250 μL of drug into the sample at any point during accumulation of data without removing the cuvet from the instrument. The drug may be injected either through a septum (shown) or through a right-angle stopcock. A sudden increase in the air pressure clears the sample line of untreated cells.

2. Special equipment: To permit rapid injection and mixing of drug and sample, the cytometer must be equipped with a stirred, temperature-controlled rapid injection system, such as the Cytek Time Zero Module (Fremont, CA) (**Fig. 2**). The following data should be collected for each particle observed: time, forward angle scatter, 90° scatter, volume, if available, and the emitted fluorescence on a linear scale (**Note 1**).

   If the cytometer is not equipped with a sizing channel, mean cell volumes (MCV) may be obtained using a Coulter counter model ZBI equipped with a Channelyzer. A calibration constant K is calculated from a standard curve generated with beads of three different sizes (Coulter); volumes are calculated from the following equation:

   $$V(\mu^3) = K * \text{current} * 1/\text{amplitude} * (\text{mean channel of sample} - \text{threshold}) \quad (1)$$

   The fluorescence response factor is determined by quantitating intracellular dye using a high-performance liquid chromatograph. For rhodamine, a 25-cm Zorbax ODS column provides good quantitation.
3. Cell lines: We have successfully studied drug uptake in a number of human cell lines growing in suspension, including those derived from leukemias (HL-60, K562), lymphoma (SU-4), and myeloma (8226). Drug-resistant subclones expressed P-glycoprotein (HL-60R_{Winter}, K562/$_{\text{INF}}$, SU-4R) ***(3,7–10)***, LRP (8226_{mitox40}) ***(10)***, or MRP (HL-60R_{Center}) ***(10)***. Fresh human red blood cells,

mononuclear cells, and dispersed lymph-node biopsies were studied following Ficoll separation *(8)*. Cells should be in exponential growth stage; culture medium should not be allowed to become acidic (*see* **Note 2**). The standard procedure will be illustrated with the SU-4 and SU-4R cell lines.

4. Drugs and dyes: Compounds with sufficient fluorescence may be monitored directly. We have successfully monitored the uptake of doxorubicin, daunomycin, rhodamine, and irinotecan and its metabolite SN-38 in a variety of cells (*see* **Note 3**). Mi et al. have used a similar technique to study topotecan binding ***(11)***. We study potential inhibitors of P-glycoprotein using 2 μ*M* rhodamine. Rhodamine will be used as an example throughout the procedure.

   The morning of an experiment, prepare a 20-μ*M* working solution of rhodamine in PHG buffer from the 1 mg/mL methanolic stock solution, which is stable stored at –20°C for several months.
5. PHG buffer: commercial phosphate-buffered saline, 0.5 m*M* $MgCl_2$, 0.9 m*M* $CaCl_2$, supplemented with 25 m*M* HEPES and 1 g/L glucose, pH 7.45, osmolality 290 mosm (will require approx 50 mL water), filter-sterilized (*see* **Note 2**).
6. Mobile phase: 0.05 *M* sodium phosphate, pH 2.7, 40% acetonitrile.
7. Methanol: for rinsing tubing and syringes between runs.

3. Methods

3.1. Flow Cytometric Monitoring of Drug Uptake

1. Two days prior to drug-uptake analysis, seed a sufficient volume of culture medium with each cell line (SU-4 and SU-4R) to grow 4×10^6 cells for each condition to be tested (usually approx 100,000 cells/10 mL medium).
2. One hour prior to analysis, collect the cells by centrifugation at 8°C (**Note 4**). Immediately resuspend cells in 37°C PHG buffer, measure cell number and volume, and adjust the suspension to 1×10^6 cells/mL. During the experiment, store the cell suspension at 37°C with frequent agitation.
3. With the flow cytometer properly aligned and filters adjusted to collect the desired wavelengths (>515 nm for single-color analysis of rhodamine), adjust the photomultiplier tube voltage to fix the fluorescence of DNA check beads (Coulter, Hialeah, FL) in channel 200 at gain 1. Adjust the boost pressure of the injector to clear the uptake tubing of residual particles within 4–8 s (**Note 5**).
4. The final volume of the cell suspension after addition of all drugs and injection of dye should be 1.2 mL. Typically, 100 μl of test drug is added to 1.0 mL of cells 12 min before analysis. The injection syringe is overfilled with rhodamine working solution (130 μL), the excess expressed through the injection needle, and the 100 μL column of fluid is withdrawn slightly to prevent premature mixing of dye and cells. Wipe off any dye adhering to the outside of the needle (**Note 6**).
5. At the start of analysis, the cell suspension is placed in the sample holder. When stable flow of 200–300 cells/s is achieved, data collection begins with recording the baseline fluorescence. At 30 s, the dye is injected, the boost pressure is activated to clear the tubing of unstained cells, and data collection continues for an additional 10 min (**Fig. 3A**).

6. When data collection is complete, representative samples should be examined under the fluorescence microscope to ensure that cells are intact and to confirm the intracellular location of fluorescence.
7. In between each sample, needle, stir bars, and syringes are rinsed with methanol to minimize carryover. Fresh tubes are used for each sample (**Note 6**).

3.2. Data Analysis

1. Draw a gate tightly around the cell population as displayed in the histogram of forward scatter and 90° scatter (**Fig. 3A inset**). Construct a histogram of cell number vs time from the cells within this gate. Place gates (A–E) corresponding to the desired 3-s-wide time slices the histogram (**Fig. 3A**).

   Determine the mean fluorescence intensity of cells at each time point by building rhodamine-fluorescence histograms gated on one of A–E (**Fig. 3B** and **C**). Record the mean channels for time and fluorescence intensity (**Fig. 3D**). If additional analysis times are desired, adjust the ranges for gates A–E and rebuild the histograms. Repeat for each uptake curve. Frequently it is most convenient to perform data analysis after all data collection is complete.
2. Sampling times should include the baseline and 10–14 additional points to fully characterize an uptake curve. Data can reliably be captured after flow has returned to a stable pattern, approx 10 s after dye injection.
3. Autofluorescence (given by the baseline mean fluorescence) can be subtracted using a spreadsheet, during the kinetic analysis, or if the flow cytometry program permits, automatically.

3.3. Calibration for Quantitative Analysis

1. Incubate 2×10^6 cells with each of four rhodamine concentrations (0.25, 0.75, 2, or 4 μ*M*) for 10 min, wash with 2 mL PHG, and resuspend in 2.0 mL PHG. Layer 1 mL of each concentration above silicone oil SF-1250 in a microfuge tube.
2. Using the identical flow cytometer settings as for kinetic analysis, collect a histogram of 20,000 cells from the remaining sample. At the midpoint of each data collection, centrifuge the parallel sample at 16,000*g* for 10 s. Repeat in triplicate.
3. Discard the extracellular fluid, cut the tip off the microcentrifuge tube and extract the rhodamine in 10 mL ethyl acetate.
4. Rhodamine concentrations may be measured by high-pressure liquid chromatography after drying the organic phase and resuspending in mobile phase. Chromatograph on a 25 cm Zorbax ODS column at a flow rate of 1.0 mL/min. Detect rhodamine fluorescence at 530 nm following excitation at 488 nm and compare to a standard curve prepared with cell lysate spiked with known rhodamine concentrations to determine the intracellular rhodamine (**Note 7**).
5. Construct a plot of mean fluorescence intensity vs picomoles of intracellular rhodamine. The response factor is the slope of this line. Multiply the mass of rhodamine injected by the response factor to determine the "dose" in relative fluorescence units for the kinetic analysis.

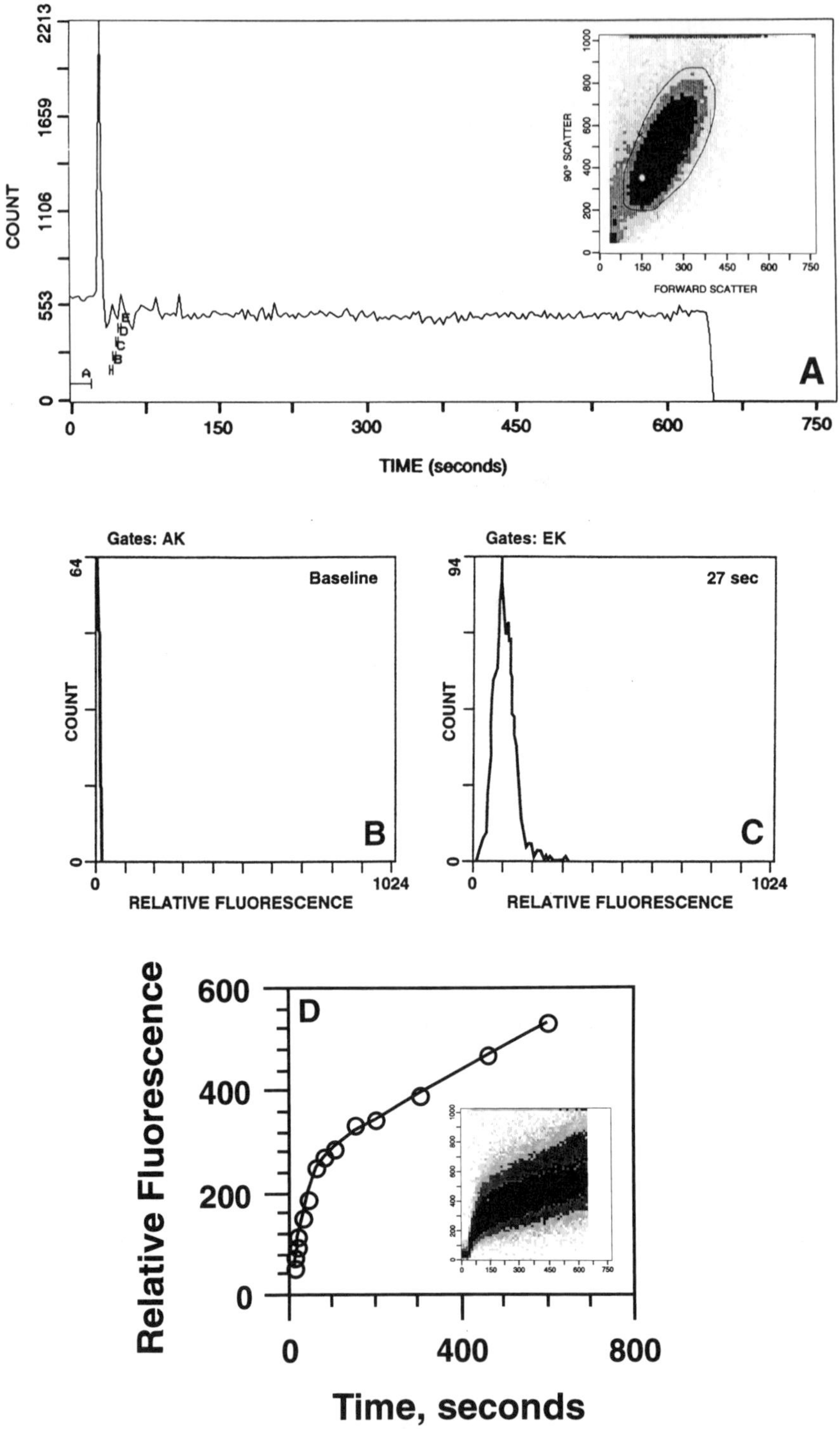

COUNT
TIME (seconds)
90° SCATTER
FORWARD SCATTER
A
Gates: AK
Baseline
COUNT
RELATIVE FLUORESCENCE
B
Gates: EK
27 sec
COUNT
RELATIVE FLUORESCENCE
C
Relative Fluorescence
Time, seconds
D

3.4. Kinetic Analysis

1. We use the SAAM program, available for both PC and Macintosh (SAAM Institute, Redmond, WA), to analyze the data and determine whether the hypothesized model fits.
2. Choose the desired model. The three-compartment model diagrammed in **Fig. 3** is the SAAM schematic representing the system shown in **Fig. 1**. Enter the known volume of the extracellular fluid (1.2 mL), the measured cell volume, and the dose of dye injected (in fluorescence units). Make estimates of the rate constants for uptake and release of the dye (**Note 8**).
3. Enter the times and mean fluorescence values observed in the data window as instructed in the user's guide. Integrate your estimates to determine their accuracy, then allow the computer to fit the model to the data. An adequate fit is determined by visual inspection, by random fluctuation of errors around the predicted curve, and by a minimal residual variation.
4. The permeability coefficient P for a given condition is given by

$$P = (\lambda_{12} \times V_1)/A \qquad (2)$$

where V_1 is the known volume of extracellular buffer and A is the average surface area of the cell membrane. A is calculated from the cell volume, assuming that the cell is a sphere.
5. Sequestration, the rate constant for nonexchangeable pool inflow S_{pool}, is calculated by

$$S_{pool} = \lambda_{2,3} * V_{2app}, \qquad (3)$$

where V_{2app} is the apparent volume of distribution of compartment 2. V_{2app} is calculated as shown below. When transporter-mediated efflux is present in addition to diffusion, the calculated value for V_{2app} will be decreased.

Fig. 3. *(previous page)* Analysis of kinetic data by flow cytometry **(A)** Histogram of particles counted per second vs time. When the boost pressure is activated at 30 s, sample flow peaks. Unstained cells are cleared from the tubing in approx 8 s. Gate A collects cells that have not yet been exposed to drug. Gates B through E collect cells observed during 3-s time slices at 9, 12, 18, and 27 s after injection. These gates should be moved to the desired time points. The inset shows a histogram of forward scatter vs 90° scatter; living cells are selected in the gate shown. **(B)** Distribution of fluorescence intensities of cells selected in both the forward scatter vs 90° scatter (living cells) and A (unstained cells) gates. **(C)** Distribution of fluorescence intensities of cells selected in the living cell gate and E (cells exposed to rhodamine for 27 s). **(D)** The mean fluorescence intensity at selected times (Y-axis, circles) is plotted against time (X-axis). The solid line is the least-squares-fitted curve generated by the SAAM program. The inset shows the individual data points. The fluorescence intensity of living drug-sensitive cells increases rapidly upon injection of rhodamine for the first 2 min, then increases linearly. The range of drug uptake at a given time is predominantly a function of cell volume.

$$V_{2app} = (\lambda_{12} / \lambda_{21}) \cdot V_1 \quad (4)$$

We normalize for changes in cell size to give a dimensionless constant, F, which is also decreased in the presence of transporter-mediated efflux (**Note 9**).

$$F = V_{2app} / \text{MCV} \quad (5)$$

6. The total outward clearance of dye across a membrane is the sum of the clearances produced by diffusion and transporters. In sensitive cells,

$$\text{CL}_{out} = \text{Cl}_{diffusion} = \lambda_{in} * V_1 \quad (6)$$

$$\lambda_{diff,out} * F * \text{MCV} = \lambda_{in} * V_1 \quad (7)$$

$$\lambda_{diff,out} = (\lambda_{in} * V_1) / (F * \text{MCV}) \quad (8)$$

In resistant cells

$$\text{CL}_{out} = \text{CL}_{MDR} + \text{Cl}_{diffusion}. \quad (9)$$

Although the permeability of sensitive and resistant SU-4 cell membranes differs, intracellular binding of dye, Φ, is identical in sensitive and resistant cells. By substitution,

$$\text{CL}_{MDR} = \lambda_{2,1} * \text{MCV} - ((\lambda_{1,2} * V_1) / \text{F}_{WT}) \quad (10)$$

3.4. Michaelis-Menten Analysis

Whereas a full explanation of transport kinetics is beyond the scope of this chapter, the parameters calculated above may be used to determine the apparent Km and Vmax of transport and the Ki for inhibitors (**Note 10**). Plot $1/\text{CL}_{MDR}$ or $1/\text{S}_{pool}$ vs rhodamine to obtain plots that are analogous to Eadie-Hofstee plots.

4. Notes

1. The quantity of dye in a cell is linearly related to its fluorescence intensity. The concentration of dye is the fluorescence intensity divided by the volume of the cell, times a "response factor" which must be determined for each instrument, dye, and wavelength (*see* **Subheading 3.3.**). Frequently, a rapidly changing signal is kept on scale by using a logarithmic channel. However, this conversion is usually not mathematically precise, so that data from logarithmic channels may not be used for quantitative studies. To be sure that the full range of data is collected accurately, we collect the fluorescence on two linear channels with a two- or fivefold difference in gains.
2. Most drug transporters use ATP; therefore it is essential that the cells be healthy. Fluid shifts shortly before or during the experiment will change the volume of the cell. Changes in pH during the experiment will alter the distribution of ionic dyes. Proteins in serum may adsorb the dye, altering uptake. Therefore osmolal-

ity is carefully controlled, a stable buffer is selected, and protein conditions must be specified. If a Ficoll separation is necessary, wash the cells in PBS, then culture them in medium (RPMI plus 10% fetal calf serum) overnight at 37°C in a CO_2 incubator to restore normal metabolism.

3. Doxorubicin and daunomycin (United States Pharmacopeia, Rockville, MD) stock solutions are prepared at a concentration of 1 mg/mL in methanol and stored at –20°C. Irintotecan and SN-38 are unstable and must be prepared the day of use. Irinotecan and SN-38 are excited at 350 nm and emit at 425–500 nm (CPT-11) and >590 nm (SN-38). Chemotherapy agents may cause local irritation and are carcinogenic; they should be handled with gloves and disposed of as hazardous chemicals.
4. Cells must remain at or above room temperature for the hour prior to use to prevent alteration of membrane fluidity. We find that cells take 30–60 min to return to normal uptake patterns once they have been chilled. We centrifuge cells at 8°C to prevent excessive cooling.
5. The pressure in the sample chamber is transiently increased ("boosted") to remove unstained cells from the uptake tubing. Higher pressures remove cells more quickly but cause unstable flow, clogging, and erroneous fluorescence measurements. Inadequate increases in sample pressure leave residual unstained cells in the tubing for up to 70 s. When present, these unstained cells decrease the average fluorescence intensity.
6. The highly lipophilic rhodamine adsorbs readily to tubing. PEEK is preferred over silastic tubing if possible. Rhodamine adsorption to polypropylene varies considerably from manufacturer to manufacturer and sometimes from lot to lot. Rainin cryostat tubes both fit the injector and give reproducible results.
7. Alternatively, uptake of radiolabeled dye, if available, may be used to quantitate the intracellular concentration of free drug.
8. Within the SAAM program, rate constants are named "receiving compartment, comma, donating compartment" rather than the format common in biology. We change formats at the time of publication.
9. In our earliest studies, we used the change in F to detect P-glycoprotein-mediated efflux. However, this approach assumes that the permeability of sensitive and resistant cell membranes is identical. This assumption is not correct for the diffusion of rhodamine into SU-4 and SU-4R cells. A more general approach is to calculate the outward clearance due to diffusion (CL_{diff}) for the resistant cells as derived in **Subheading 3.4., step 6**.
10. If the Ki of an inhibitor is desired, it is most convenient to study 4–5 inhibitor concentrations at a single dye concentration on a given day, varying the substrate concentration on subsequent days because large differences in dye concentration may require use of neutral density filters and recalibration.

Acknowledgments

I thank Charles Goolsby and Kenneth Bauer, Yi Fan Fu, and the staff of the Northwestern Flow Cytometry Facility for their advice and assistance in the development of these procedures.

References

1. Gottesman, M. M. and Pastan, I. (1988) The multidrug transporter, a double-edged sword. *J. of Biol. Chem.* **263,** 12,163–12,166.
2. Gibaldi, M. and Perreir, D. (1982) *Pharmacokinetics.* Marcel Dekker, New York.
3. Dordal, M. S., Ho, A. C., Jackson-Stone, M., Fu, Y. F., Goolsby, C. L., and Winter, J. N. (1995) Flow cytometric assessment of the cellular pharmacokinetics of fluorescent drugs. *Cytometry* **20,** 307–314.
4. Dalmark, M. and Storm, H. H. (1981) A Fickian diffusion transport process with features of transport catalysis. Doxorubicin transport in human red blood cells. *J. Gen. Physiol.* **78,** 349–364.
5. Summerhayes, I. C., Lampidis, T. J., Bernal, S. D., Nadakavukaren, J. H., Nadakavukaren, K. K., Shepherd, E. L., and Chen, L. B. (1982) Unusual retention of rhodamine 123 by mitochondria in muscle and carcinoma cells. *Proc. Natl. Acad Sci. USA* **79,** 5292–5296.
6. Dessypris, E. N., Brenner, D. E., Baer, M. R., and Hande, K. R. (1988) Uptake and intracellular distribution of doxorubicin metabolites in B-lymphocytes of chronic lymphocytic leukemia. *Cancer Res.* **48,** 503–506.
7. Dordal, M. S., Jackson-Stone, M., Ho, A. C., Winter, J. N., and Atkinson, A. J. J. (1994) Decreased intracellular compartmentalization of doxorubicin in cell lines expressing P-glycoprotein. *J Pharmacol Exp. Ther.* **271,** 1286–1290.
8. Jackson-Stone, M., Fu, Y. F., Ho, A. C., Winter, J. N., and Dordal, M. S. (1994) Flow cytometric measurement of altered intracellular drug distribution in multidrug resistant cells. *Clin. Res.* **42,** 393A.
9. Dordal, M. S., Fu, Y. F., and Ho, A. C. (1996) P-glycoprotein decreases mitochondrial sequestration of rhodamine as well as increasing efflux. *Proc. Am. Assoc. Cancer Res.* **37,** 312.
10. Dordal, M. S. and Fu, Y. F. (1996) Cellular pharmacokinetics of irinotecan in drug-sensitive and multidrug resistant cell lines. *Proc. Am. Assoc. Cancer Res.* **37,** 429.
11. Roy, D., Munshi, C. B., Mi, Z., and Burke, T. G. (1995) Alteration of uptake and cytotoxicity of camptothecin drugs in MCF-7 breast cancer cells by human serum albumin. *Proc. Am. Assoc. Cancer Res.* **36,** 444.

13

Quantification of Xenoprotein Electroinsertion in Mammalian Cells

Khalid El Ouagari and Justin Teissié

1. Introduction

Recent studies reported the so-called "electroinsertion" of proteins having a membrane-spanning sequence into mouse red blood cell or Chinese hamster ovary cell membrane by exposing the cell suspension to electrical field pulses *(1–6)*. The viability of the pulsed cells is not affected. New receptors are then present on the cell surface. It is proposed to use such a technology in clinical applications such as AIDS therapy *(7)*. Genetically engineered CD4s are inserted in red blood cells (RBC) and act as lures to trap the virus. A critical parameter for the practical use of such a methodology is to know the number of inserted xenoproteins.

In the present chapter, glycophorin A, an integral erythrocyte-glycosilated membrane protein, is described to be stably back inserted into the plasma membrane of Chinese hamster ovary cells by electropulsing a mixture of mammalian cells and proteins. Quantification of the number of resulting modified cells and of the number of inserted proteins is obtained by flow cytometry after immunodetection of the xenoproteins. This procedure is observed to be more sensitive than using fluorescein derivatives of the protein.

1.1. Methodology

The number of binding antigens on a cell surface is obtained by quantitating the fluorescent antibodies' associated intensity on a given cell, which is directly related to the number of bound antibodies.

The effective fluorescence ratio of the antibody (F/P) indicates the average fluorescence intensity per antibody molecule. The number of bound antibodies could be quantitated by determining the average fluorescence intensity of a

From: *Methods in Molecular Biology, Vol. 91: Flow Cytometry Protocols*
Edited by: M. J. Jaroszeski and R. Heller © Humana Press Inc., Totowa, NJ

cell population and dividing it by the average fluorescence intensity of the conjugated antibody molecule.

The problem of quantitating antigens is resolved by using a specially designed microbead standard called Simply Cellular™ in conjunction with the Quantitative Fluorescent Microbead Standard (Becton Dickinson) (flow cytometer calibration).

2. Materials

1. Pulsing buffer (PB): 250 m*M* sucrose, 1 m*M* $MgCl_2$, 10 m*M* potassium phosphate buffer, pH 7.4.
2. Glycophorin A type MN can be obtained from Sigma (St. Louis, MO). It should be emphasized that although glycophorin is an integral protein, it is soluble in water and may be under a micellar form.
3. CHO cells (WTT clone) were grown in suspension by using a spinner flask (500 mL) driven by a magnetic stirrer (400 rpm). The culture temperature was set at 37°C. The culture medium was Eagle's minimum essential medium (MEM 0111, Eurobio, France) supplemented with 6% new-born-calf serum (Boehringer, Mannheim, Germany), penicillin (100 Ul/mL), streptomycin (100 mg/mL) and L-glutamin (0.58 mg/mL). The cell density was maintained in exponential growth phase by daily dilution in culture medium. Cells were harvested by centrifugation (100*g*, 4 min at room temperature), and washed twice by centrifugation in pulsing buffer.
4. The procedure for cell cytofluorimetry described in the instruction manual of the FACScan (Becton Dickinson) was followed. The lysis II software was run on a HP 9153C microcomputer connected to a HP Paintjet printer. The filter set suitable for fluorescein was choosen.

3. Methods

The experimental protocol is as follows. Electroinsertion is first operated. After immunolabeling, two steps are needed to get a quantitative determination of the modification of the cell surface:

1. Calibration of the flow cytometer using Quantitative Fluorescent Microbead Standard.
2. Quantification using Simply Cellular (Becton Dickinson) microbeads.
   a. Determination of the effective fluorescence F/P ratio.
   b. Determination of the number of available antibody binding sites per cell.
   c. Determination of the number of fluorescent cells

3.1. Electropulsation

This should be carried out at room temperature using a CNRS electropulser (Jouan PS1, St. Herblain, France) able to deliver square-wave pulses with parameters (voltage, pulse duration, number and frequency of pulses) that are all independently adjustable. Pulses were monitored using a 15 MHz oscilloscope (Enertec, St. Etienne, France). Stainless steel electrodes were parallel and flat with an anode-cathode distance of 1.5 mm.

3.2. Electric Field-Mediated Membrane–Protein Association

1. Incubate 10^7 cells in PB at 37°C during 15 min, with 50 μg of glycophorin A in a total volume of 18 μL.
2. Apply five pulses of 7 ms with an intensity of 0.9 kV/cm at a frequency of 1 Hz.
3. Wash the cells, 10 min after pulse application, three times in PBS (100*g*, 4 min at room temperature) and resuspended in a final volume of 0.05 mL.
4. Reveal insertion by immunofluorescence.
5. Run a control, in which the sample was subjected to all these steps except that no pulse was applied (*See* **Notes 1–4**).

3.3. Immunofluorescence Assay

1. Incubate the cell suspension at 4°C during 30 min, with 2 μg of antihuman glycophorin mouse monoclonal antibody (MAb) (Immunotech, Marseille, France) in a total volume of 0.1 mL.
2. Wash cells three times in PBS.
3. Incubate cells at 4°C for 30 min, with 12.5 μg of secondary antibody (Fab2) goat antimouse IgG FITC (Immunotech) in a total volume of 0.1 mL.
4. Wash cells three times in PBS.
5. Analyze cells by flow cytometry (*see* **Notes 5–12**).

3.4. Immunofluorescence Calibration

Simply Cellular microbeads (Becton Dickinson), with antigenic activity that would bind mouse MAbs, were used to quantify inserted glycophorins. This antigen activity was derived from goat anti-IgG antibodies covalently bound to the surface of microbeads. These microbeads were calibrated in terms of the number of monoclonal mouse IgG molecules they would bind.

1. Incubate Simply Cellular microbeads at 4°C for 30 min, with 2 μg of antihuman glycophorin mouse MAb (Immunotech) in 0.1 mL PBS.
2. Wash the microbeads three times in PBS.
3. Incubate the beads at 4°C during 30 min, with 12.5 μg secondary antibody (Fab2) goat antimouse IgG FITC (Immunotech) in 0.1 mL PBS.
4. Wash the microbeads three times in PBS.
5. Analyze them by flow cytometry (*see* **Note 13**).

3.5. Calibration of the Flow Cytometer

Calibrating a flow cytometer is determining the instrument response with different calibrated standard samples. Samples of unknown fluorescence intensity can be measured on the resulting calibration plot.

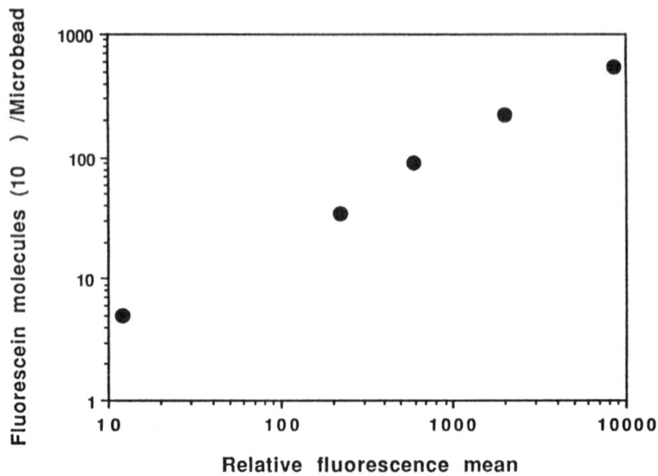

Fig. 1. Data plot for flow cytometer calibration.

1. Determine the relative fluorescence means for the calibrated microbeads (Quantitative Fluorescent Microbead Standard) using the flow cytometer.
2. Plot the fluorescence mean values against the number of fluorescein molecules/microbeads indicated on the bottle of each microbead population.
3. Plot the data on a log-log scale (**Fig. 1**, *see* **Notes 14** and **15**).

3.6. Determination of the Number of Glycophorin A Molecules per Cell

3.6.1. Effective Fluorescence F/P Ratio

1. Immunolabel the Simply Cellular microbeads to saturation by indirect method with MAbs as indicated above.
2. Determine the relative fluorescence mean of the saturated Simply Cellular microbeads using the flow cytometer *(Our microbeads gave a value of 435)*.
3. Determine the number of fluorescein molecules/microbead against the calibration curve constructed above using this relative fluorescence mean (435). (**Fig. 1**) *(We find a value of 6.5 × 10^4 fluorescent molecules/microbead)*.
4. Divide this value by the number of calibrated antibody-binding sites on the microbeads indicated on the bottle of the Simply Cellular microbeads (7 × 10^4 fluorescein molecules/microbead).

The result (0.93) represents the effective fluorescence F/P ratio.

3.6.2. Number of the Glycophorin A Molecules/Cell

1. Immunolabel CHO cells, carrying glycophorin molecules, to saturation by indirect method with the same MAbs as used for Simply Cellular microbeads, as described in **Subheading 3.3.**

2. Determine the relative fluorescence mean of the saturated cells using the flow cytometer. (Our cells gave a value of 260).
3. Determine the number of available antibody sites per cell against the calibration curve constructed above using this relative fluorescence mean value (260), (**Fig. 1**) (We find a value of 4.75×10^4).
4. Divide this value (4.75×10^4) by the effective fluorescence F/P ratio determined above (0.93).

The result (5.1×10^4) represents the average number of glycophorin A molecules per cell, i.e., approx 100 inserted proteins per square micron of the cell surface. The distribution of bound molecules is between 3.2 and 5.8×10^4.

3.7. Quantification of the Number of Labeled Cells

This is classically obtained by getting the number of cells belonging to the subpopulation of fluorescent cells. A threshold must be arbitrarily (but easily) defined from the fluorescence histogram given by the Facscan-associated computer.

4. Notes

1. The viability of the pulsed cells was checked to be only slightly affected and remained larger than 50%.
2. Insertion is obtained under electrical conditions that are specific to the host cells. Preserving the cell viability is a key requirement. Pulse durations must last several milliseconds (square wave pulse) to detect a high number of inserted proteins.
3. The low ionic content of the pulsing buffer reduces the joule heating associated with the electric field pulse, but cells cannot be kept in such a buffer for a long period.
4. The field sensitivity of CHO cells depends on the age of the culture ***(10)***. The present conditions are for a fresh culture.
5. Cell autofluorescence is a problem when the level of the immunofluorescence is low ***(8)***. Furthermore, cells fixed with glutaraldehyde provide a high level of autofluorescence intensity. Paraformaldehyde is normally used.
6. We observed that protein adsorption was present on the cell surface when no pulse was applied. The reversibility of the process was slow. By culturing the cells during one night after pulsing, this artifact was minimized. But, as cell growth was occuring, the pool of inserted proteins was shared between daughter cells.
7. As the fluorescent labels are photosensitive, bleaching may take place. Most irradiated fluorochromes undergo some degree of photodecomposition. This phenomenon may be easily observed under a fluorescence microscope (so-called fadding). Oxygen molecules are converted by the excitation source to high energy singlet oxygens, as well as other radicals, and react with the fluorochrome molecules, converting them into nonfluorescent products ***(9)***. Photobleaching can furthermore occur in room light. Therefore, fluorescent samples must be kept in the dark until they are analyzed. Antioxidants such as glycerol can be used to stop photobleaching.

8. Be careful to disperse the cells, clumping may be a problem in this quantification but is shown by the increase in scattering.
9. Because the number of inserted proteins per cell is low, there is no internal quenching resulting from a high density of fluorescent moieties on the membrane.
10. Direct observation of the cells under a fluorescent microscope shows a fluorescent ring at the cell periphery in agreement with an insertion of glycophorin in the membrane. It was almost impossible to get a quantitative assay by such an approach where a strong fading is present.
11. If the immunoassay was run 24 h after electropulsing, the same fluorescence pattern was observed under the microscope proving the stability of the insertion.
12. To be detected, proteins must show the determinant on the external cell surface. This will give an underestimation if the insertion occurs under two different topologies.
13. If further adjustments are needed to the alignment of the instruments after establishing a calibration plot, then a new calibration plot has to be established.
14. Quantitative Fluorescent Microbead Standards (Becton Dickinson) were used to calibrate the flow cytometer in terms of fluorescence intensity. These microbeads eliminate errors resulting from the daily fluctuation of instrument laser power or alignment. We used five populations with 5×10^3, 3.4×10^4, 9×10^4, 2.2×10^5, and 5.5×10^5 fluorescein molecules/microbead, respectively.
15. The calibration plot is not linear and a log-scale is needed.

References

1. Moumeine, Y., Tosi, P.-F., Gazitt, Y., and Nicolau, C. (1989) Electroinsertion of Xenoglycophorin into the red blood cell membrane. *Biochem. Biophys. Res. Comm.* **159,** 34–40.
2. Moumeine, Y., Tosi, P. F., Barhoumi, R., and Nicolau, C. (1990) Electroinsertion of full length recombinant CD4 into red blood cell membrane. *Biochim. Biophys. Acta.* **1027,** 53–58.
3. Moumeine, Y., Tosi, P. F., Barhoumi, R., and Nicolau, C. (1991) Electroinsertion of xenoproteins in red blood cell membranes yields a long lived protein carrier in circulation. *Biochim. Biophys. Acta.* **1066,** 83–89.
4. Zeira, M., Tozi, P. F., Moumeine, Y., Lazarte, J., Sneed, L., Volsky, D. J., and Nicolau, C. (1991) Full length CD4 electroinserted in the red blood cell membrane as a long-lived inhibitor of HIV infection. *Proc. Natl. Acad. Sci. USA* **88,** 4409–4413.
5. El Ouagari, K., Gabriel, B., Benoist, H., and Teissié, J. (1993) Electric field-mediated glycophorin insertion in cell membrane is a localized event. *Biochim. Biophys. Acta.* **1151,** 105–109.
6. El Ouagari, K., Benoist, H., Sixou, S., and Teissié, J. (1994) Electropermeabilization mediates a stable insertion of glycophorin A with Chinese hamster ovary cell membranes. *Eur. J. Biochem.* **219,** 1031–1039.
7. Hannig, J., Dawkins, C., Tosi, P.-F., and Nicolau, C. (1995) Stability and immunological reactivity of recombinant membrane CD4 electroinserted into the plasma membrane of erythrocytes. *FEBS Lett.* **359,** 9–14.

8. Aubin, J. (1979) Autofluorescence of viable cultured mammalian cells. *J. Histochem. Cytochem.* **27,** 36–43.
9. Johnson, G. D., Davidson, R. S., Russell, G., Goodwin, R. S., and Holborow, E. J. (1982) Fading of immunofluorescence during microscopy: a study of the phenomenon and its remedy. *J. Immunol. Methods* **55,** 7–12.
10. Rols, M. P., Coulet, D., and Teissié, J. (1992) Highly efficient transfection of mammalian cells by electric field pulses. *Eur. J. Biochem.* **206,** 115–121.

14

Flow Cytometry Quantification of Electropermeabilization

Marie-Pierre Rols and Justin Teissié

1. Introduction

Cell plasma membranes act as highly impermeable barriers for the exchange of molecules between cells and external medium. Transfer of compounds can only take place via the existence of specific transport systems that allow a limited number of molecules to cross membranes. Alteration in membrane selective permeability can nevertheless be achieved by several drastic methods such as microinjection, particle bombardment, polyethylene glycol, and viruses. In the early 1970s, a physical method was described. This method was termed "electropermeabilization" ("electroporation") *(1)*. It is used to gain access to the cytosol *(2)*. It is now routinely used to transfect cells with DNA, and is applicable to different systems including mammalian cells, plant protoplasts, intact bacteria, and yeast cells *(3)*. More recently, clinical applications have been reported *(4)*.

When an external electric field is applied to the cells, it induces a transmembrane potential which is added to the natural potential of cells. This potential is given by the following equation *(5)*:

$$\Delta V = f.g.R.E.\cos\theta \quad (1)$$

It depends both on physiological parameters as cell shape f, membrane conductivity g, cell radius R, and on physical parameters as the applied electric field intensity E, and the angle θ between the electric field vector and the normal at the surface where the potential is measured. When the resulting potential is higher than a threshold of 200–300 mV, the membrane becomes permeable *(6,7)*. This permeation, induced for electric field values greater than a threshold Ep, can be transient only under controlled conditions that utilize electrical parameters, and do not affect cell viability.

From: *Methods in Molecular Biology, Vol. 91: Flow Cytometry Protocols*
Edited by: M. J. Jaroszeski and R. Heller © Humana Press Inc., Totowa, NJ

Applications of electropulsation need control over the number of exogeneous molecules that are transfered in the cytoplasm of the host cell. Very few studies have indeed provided quantitative determinations of permeabilization vs the different electric field parameters of the technique. Quantitative analysis of the phenomenon based on flow of released ATP by chemiluminescence ***(8)***, and on molecular transport across erythrocyte ghost membranes by flow cytometry ***(9)*** have been reported. Quantitative flow ϕ of molecules S diffusing across the permeabilized membrane is given by:

$$\phi\,(S,t) = K \cdot P_S \cdot \Delta S \cdot (1 - Ep/E) \cdot X(N,T) \cdot e^{-k(N,T)t}, \qquad (2)$$

where K is a constant, P is the permeability coefficient of the electropermeabilized membrane for S, X is the probability of permeabilization of the membrane that depends on both pulse duration T and pulse number N, and t is the time after the pulse ***(8)***.

Quantitative determination at the individual cell level is needed to test theoretical models and to guide applications of the technique in basic and applied research. Although digitized image analysis of individual cells provide high spatial and temporal resolutions, they are restricted to measure only a few cells at a given time, without taking into account possible population distributions of cell behavior. Flow cytometry provides quantitative data based on optical measurements from large numbers; of individual cells at rates up to 10^4 cells/s. Molecular uptake and cell damage can be independently assessed. Light scattering is linked to morphology and fluorescence intensity gives quantitative measurements of the number of fluorescent molecules that are trapped in the cell as a result of its transient permeabilization ***(9)***. In the present chapter, we described the protocol for quantitative electropermeabilization of Chinese hamster ovary (CHO) cells grown either in suspension or on Petri dishes.

2. Materials

1. Growth medium: minimum essential medium (MEM 01 11) containing 8% fetal calf serum, glutamine (0.584 mg/mL), and antibiotics (penicillin, streptomycin 100 µg/mL).
2. Electropulsation buffer (PB): 10 m*M* phosphate buffer, adjusted to pH 7.2, 250 m*M* sucrose, 1 m*M* $MgCl_2$ dissolved in double-distilled deionized water (*see* **Notes 1** and **2**).
3. Phosphate-buffered saline (PBS): 10 m*M* phosphate buffer, 150 m*M* NaCl, adjusted to pH 7.2.
4. Propidium iodide (PI): 100 µ*M* in electropulsation buffer. This solution can be stored in the dark at 4°C for several weeks. As it may cause heritable genetic damage, it has to be handled with care.
5. Trypsin (0.5 g/L)-EDTA (0.2 g/L) solution, (TE) in phosphate-buffered saline.

6. Electropulsation apparatus: the CNRS cell electropulsator (Jouan PS1, St. Herblain, France) is used, connected to two flat, parallel stainless-steel electrodes and to an oscilloscope (Enertec, Saint-Etienne, France) to monitor the shape of the pulses (**Fig. 1**; *see* **Note 3**)
7. Flow cytometry apparatus: a Becton Dickinson FACScan is used. Laser excitation is at 488 nm, detection of fluorescence is performed above 650 nm.

3. Method

Electropermeabilization of cells is quantified by the penetration of the unpermeant dye PI. The main interest of this fluorescent dye is that its fluorescence quantum yield is increased 1000-fold when Pl is inserted between nucleotide base pairs. It is then not necessary to wash cells to eliminate nonincorporated PI as compared to other fluorescent dyes. Cells, pulsed in the presence of PI, are analyzed by flow cytometry. This avoids the negative consequences of pipeting and centrifuging pulsed cells.

3.1. Electropermeabilization of Cells

1. Grow cells in suspension or on Petri dishes. Cells grown in suspension can be replated readily on Petri dishes. Keep them at 37°C in a 5% CO_2 incubator. Cells in suspension should be centrifuged for 5 min at 100*g* and resuspended at a concentration of 10^7 cells per mL in electropulsation buffer PB containing 100 µ*M* PI. (*see* **Notes 4** and **5**).
2. Electropermeabilize cells in suspension or cells attached to Petri dishes. A pulsation chamber is formed by creating contact between the electrodes and the bottom of a Petri dish (35 mm in diameter). In the case of cells growing on suspension, 100 µL of the suspension are placed between the electrodes (**Fig. 1**). In the case of cells growing on Petri dish, the culture medium is discarded, contact of the bottom of the dish with electrodes created, and 1 mL of the PB buffer is added to the cells in the presence of PI (100 µ*M*). Voltage pulses are then applied for different electric voltages, pulse duration, and number. Controls have to be performed (no electric field, no PI).
3. Check sterility: All these experiments are performed at room temperature and can be carried out under sterile conditions by working under a laminar flow hood. Electrodes should be sterilized by immersion for 2–3 s in 70% ethanol, and kept 10 min in a laminar flow to dry.
4. Remove electrodes 5 min after pulsation. In the case of cells in suspension, transfer 50 µL of the cell solution into a sterile 2058 Becton Dickinson centrifuge tube containing 1 mL of PBS. In the case of cells growing on Petri dish, it is necessary to put them in suspension by trypsinization for flow cytometry. Discard the pulsation buffer, wash once with 1 mL PBS buffer, and add 10 µL TE solution at the area of electropulsation. Then observe the cells under a microscope. When they begin to round up and move, add 10 µL of culture media and transfer the suspension to a tube containing 1 mL of PBS (*see* **Note 6**).

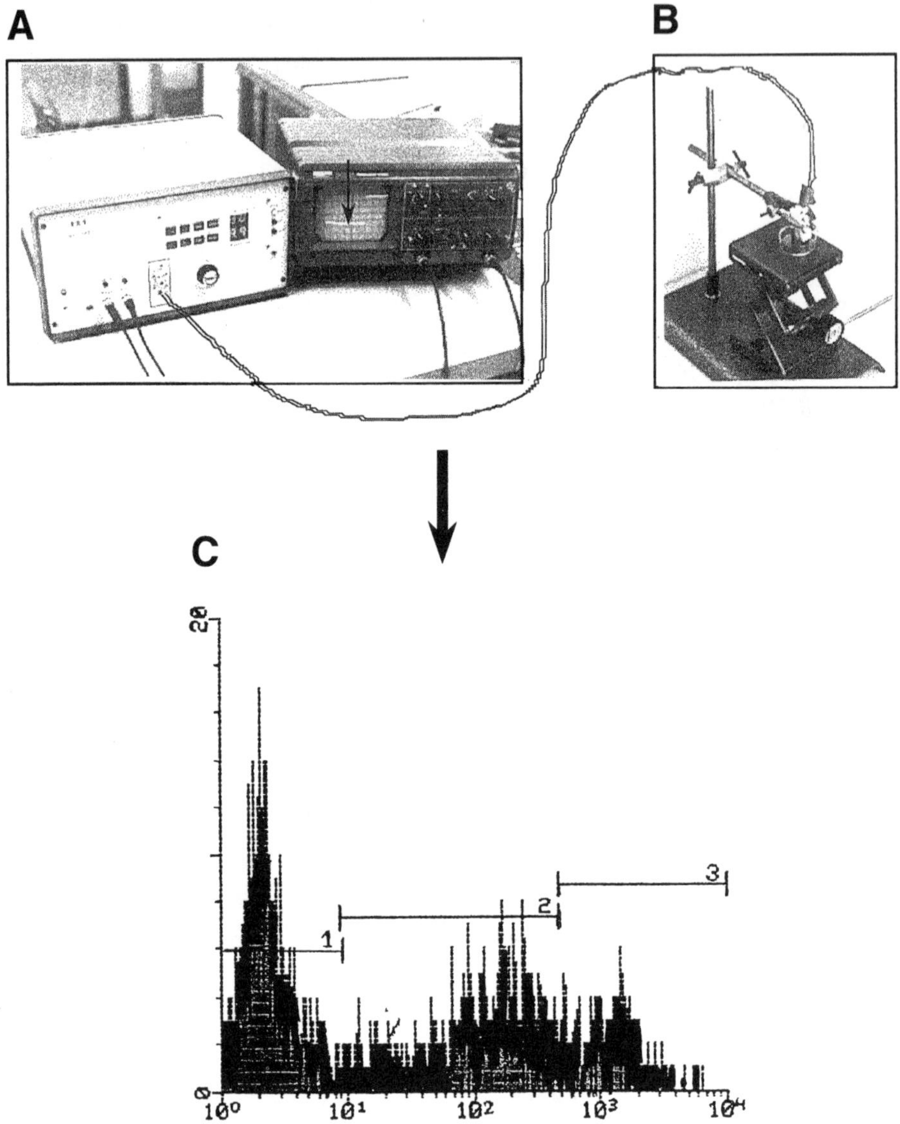

Fig. 1. **(A)** Cell electropulsator: The arrow shows the square wave shape of the electric pulse. **(B)** Experimental chamber: formed by the electrodes and a Petri dish. **(C)** Fluorescence histograms showing uptake of propidium iodide. The vertical axis gives the number of events; the horizontal axis gives the fluorescence intensity. Three classes of cells can be defined: unpermeable cells (class 1), reversibly permeabilized cells (class 2), and irreversibly permeabilized cells (dead cells) (class 3).

5. Perform a postpulse permeabilization assay for reversibility. Pulse cells as before in the pulsing buffer in absence of PI. Add 100 µL of the dye containing PB after the indicated time span. Incubate cells 5 min thereafter; dye penetration is used to monitor the induced permeabilization of the membrane and not to assess the cell viability, as routinely described ***(8)***.

3.2. Quantification of Electropermeabilization by Flow Cytometry

Flow cytometry gives access both to the percentage of fluorescent cells and to the level of fluorescence associated with each of these cells. In the case of electropermeabilization, it leads to assessment of permeabilization efficiency (i.e., percentage of cells that have been permeabilized and to their mean fluorescence), and allows detection of dead cells.

1. Resuspend the cells by gently pipeting them (in order to avoid possible aggregation of cells that occurs at long delays between pulsation and flow cytometric analysis).
2. Record data (i.e., fluorescence of cells) from each tube for approx 1000–2000 cells.
3. Use statistical tools to determine three classes of cells: control, nonfluorescent cells (class 1); reversibly permeabilized, fluorescent cells (class 2); and irreversibly permeabilized, highly fluorescent cells (dead cells; class 3) (*see* **Fig. 1**). Normally, control cell population is localized at low values of fluorescence (class 1), but sometimes a few percent of the cell population can belong to class 3. These cells are dead cells (*see* **Note 7**).
4. By definition, the percentage of permeabilized cells is 100 times the ratio of the number of fluorescent cells to the total number of cells, i.e., the number of cells belonging to classes 2 and 3 to the number of cells belonging to classes 1, 2, and 3. Usually, increasing pulse duration, number or intensity leads to an increase in the percentage of fluorescent cells (classes 2 and 3) with a concomitant increase in their mean associated fluorescence, for electric field values higher than a threshold value Ep, in agreement with equation ***(2)***.
5. Reversibility of the electropermeabilized state of the cell membranes can be analyzed by flow cytometry. Cells that have been electropermeabilized will progressively become unpermeable again. These cells belong initially to class 2. With time, class 2 population decreases to give an increase in class 1 population. The mean fluorescence in class 2 is decreasing. In the case of irreversible damages, a subpopulation of cells can move to class 3. These processes are affected by the temperature ***(10)***: at 37°C, the rate of the resealing process (transfer between class 2 and class 1) is high. Viability is also enhanced (lower rate of transfer between class 2 and class 3).

4. Notes

1. Storage conditions: Media for electropulsation and washing can be stored sterile for several months if kept at 4°C, and almost indefinitively if kept at –20°C.

2. Electropulsation: Some laboratories use other pulsing media (essentially phosphate buffer plus salts), but we have found that replacing salts with sucrose improves cell viability by decreasing Joule effect associated to electric pulses.
3. In these experiments, we used electrodes which were 0.5 cm apart. This distance can, however, be adjusted depending on the size of the cells to be electropermeabilized. Smaller cells require a larger electric field strength. Therefore decreasing the distance between the electrodes is a means of increasing the electric field strength for a given voltage ***(10)***.

   The electropulsator must be connected to an oscilloscope to monitor the actual shape of pulses. This is highly important because it is the only way to detect possible deformation of electric pulses that sometimes occurs at high currents. The major advantage of a square wave electropulsator (as used in this protocol) is that the voltage is kept constant during all the pulse duration, and that it does not depend on the electrical resistance of the sample. Thus, very accurate control of the experimental conditions results.
4. Cell density on Petri dishes can in fact vary by as much as one log depending on the size of the cells. The main fact is to have enough cells to analyze. These cells must be in the exponential stage of growth in order to preserve cell viability. Usually, concentration between 10^4 and 10^5 cells/cm^2 has to be used. The optimum delay between plating cells on dishes and electropulsation can vary from 4 to 20 h.
5. Concentrations of PI other than 100 μM can be used (from 10 μM to 1 mM). According to **Eq. 2**, higher concentrations of PI increase the values of cell associated fluorescence intensity.
6. After pulsation, cells can be directly analyzed by flow cytometry or kept on ice until being analyzed. Cells can last a few hours without any change in their fluorescence. It is however better not to wait for more than 4 h because cell death takes place, leading to the entrance of external PI into dead cells and then to artifactual results.
7. No difference among the three classes of cells are observed as far as light scattering is concerned.

References

1. Neumann, E. and Rosenheck, B. (1972) Permeability induced by electric impulses in vesicular membranes. *J. Membr. Biol.* **10,** 279–290.
2. Knight, D. E. and Scruton, M. C. (1986) Gaining access to the cytosol: the technique and some applications of electropermeabilization. *Biochem. J.* **234,** 497–506.
3. Förster, W. and Neumann, E. (1989) Gene transfer by electroporation: a practical guide, in *Electroporation and Electrofusion in Cell Biology* (Neumann, E., Sowers, A. E., and Jordan, C. A., eds.), Plenum, New York, pp. 298–378.
4. Mir, L. M., Orlowski, S., Belehradek, J., Jr., Teissié, J., Rols, M. P., Sersa, G., Miklavcic, D., Gilbert, R., and Heller, R. (1995) Biomedical applications of electric pulses with special emphasis on antitumor electrochemotherapy. *Bioelectrochem. Bioenerg.* **38,** 203–207.

5. Bernhardt, J. and Pauly, H. (1973) On the generation of potential differences across the membranes of ellipsoidal cells in an alternating electric field. *Biophysik* **10,** 89–98.
6. Teissié, J. and Tsong, T. Y. (1981) Electric field induces transient pores in phospholipid bilayer vesicles. *Biochemistry* **20,** 1548–1554.
7. Teissié, J. and Rols, M. P. (1993) An experimental evaluation of the critical potential difference inducing cell membrane electropermeabilization. *Biophys. J.* **65,** 409–413.
8. Rols, M. P. and Teissié, J. (1990) Electropermeabilization of mammalian cells. Quantitative analysis of the phenomenon. *Biophys. J.* **58,** 1089–1098.
9. Prausnitz, M. R., Lau, B. S., Milano, C. D., Conner, S., Langer, R., and Weaver, J. C. (1993) A quantitative study of electroporation showing a net plateau in net molecular transport. *Biophys. J.* **65,** 414–422.
10. Teissié, J. and Rols, M. P. (1988) Electropermeabilization and electrofusion of cells, in *Dynamics of Membrane Proteins and Cellular Energetics* (Latruffe, N., Gaudemer, Y., Vignais, P., and Azzi, A., eds.), Springer-Verlag, Berlin, pp. 249–268.

15

Flow Cytometric Detection and Quantitation of Cell–Cell Electrofusion Products

Mark J. Jaroszeski, Richard Gilbert, and Richard Heller

1. Introduction

Cell–cell electrofusion (CCE) is a process that involves forcing cells into close juxtaposition and then inducing fusion by delivering electric pulses to the cells. CCE has proven to have many practical applications. It has been used for monoclonal antibody (Mab) production *(4,5)*, hybridoma production *(1–3)*, and membrane surface marker transfer *(6)*. Many other applications are described in this volume. In addition, the study of membrane fusion mechanisms has been the focus of some researchers *(7–12)*. Electrofusion techniques seldom result in 100% yields between fusion partners. Therefore, a major aspect of all CCE applications is the ability to detect and quantitate fusion products.

Traditional methods for rating the success or failure of fusion are visual. Light and fluorescence microscopy have been implemented for this purpose. One popular light microscopic technique uses a polynucleation index *(12–14)*. Fluorescence methods include using a variety of different internal dyes *(15–17)*. Microscopic methods have been used extensively because they are simple and use a one-cell-at-a-time approach to analysis. However, they also have several disadvantages. First, they are very time consuming. Manual enumerations can take from 5 to 15 min per sample, depending on the application. Time becomes particularly important if large sets of fusion samples are analyzed. Second, it is only practical to determine fusion success for a small number of cells. Microscopic methods typically use visual determinations from several hundred cells as an analytic basis. Most fusion methods utilize several thousand to several million cells per fusion sample. Thus, manual determinations are normally based on a small fraction of the cells in a fusion sample. Third, microscopic methods are subject to human bias and error. Alternative assays for quantitat-

From: *Methods in Molecular Biology, Vol. 91: Flow Cytometry Protocols*
Edited by: M. J. Jaroszeski and R. Heller © Humana Press Inc., Totowa, NJ

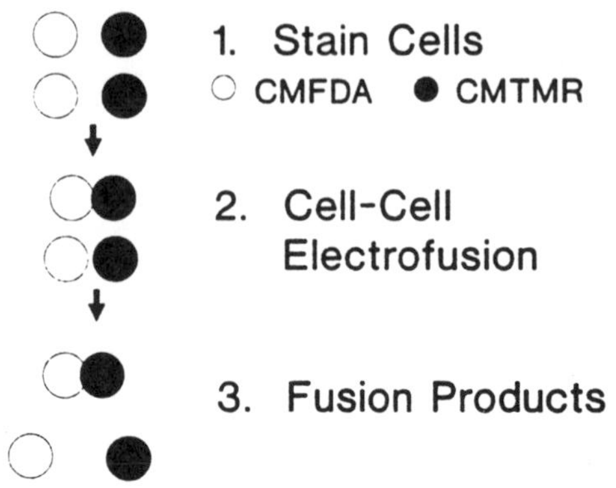

Fig. 1. The premise of cytometric detection and quantitation. 1. Opposing fusion partners are first stained with different fluorescent dyes, CMFDA, and CMTMR. 2. Cell–cell electrofusion is performed on the stained cells. 3. Fusion products will be composed of three different cell types. Fused cells will be dual fluorescing. Unfused cells will retain the single fluorescence of their respective stained fusion partners.

ing fusion include analyzing for the secretion of a specific product ***(4)***, flow cytometric methods ***(18,19)***, and spectrofluorometric methods ***(20)***.

Flow cytometry offers several advantages over microscopic methods while retaining a one-cell-at-a-time approach to analysis. The time required to cytometrically analyze a fusion sample is on the order of seconds. Therefore, cytometric analysis offers a time savings that is of practical significance. The number of cells analyzed is very large compared to microscopic methods. Common cell numbers for cytometric samples range from several thousand to >40,000. Another advantage of cytometry is that human bias and error are removed from the detection and quantitation method. Finally, many cytometers have sorting capability. Therefore, it is possible to separate hybrid cells from a sample.

The premise of the cytometric detection and quantitation method given in this chapter is dual fluorescence (**Fig. 1**). Opposing fusion partners are first stained with different fluorescent dyes. One cell type is loaded with 5–chloromethylfluorescein diacetate (CMFDA) and the other is loaded with 5-(and 6)-([(4–chloromethyl)benzoyl]amino) tetramethylrhodamine (CMTMR). After loading, CCE is performed using the stained cells. After the fusion process is complete, three different cell types will be present. Cells that exhibit the fluorescence of both dyes (dual fluorescence) are hybrid cells. The other two cell types will exhibit single fluorescence of the respective stains; these are unfused cells. Flow cytometry is utilized to detect and quantitate the number of hybrids and unfused cells based on fluorescence.

This chapter describes the method used to stain cells prior to fusion. It also addresses the cytometric protocol. The detection and quantitation scheme has been used with a variety of cell types. Some of these are human promyelocytic leukemia cells (ATCC CCL 240), murine lymphoma cells (ATCC TIB 53), rat Sertoli cells from primary culture, human fibroblasts from primary culture, human breast carcinoma cells (ATCC HTB 132), human bladder carcinoma cells (ATCC HTB 9 5637), rat Sertoli cells (ATCC CRL 1715), and hamster β cells (ATCC CRL 1777). The method detailed in **Subheading 3.** has yielded consistent results regardless of the cell types used for fusion. This indicates that it is applicable to many different cell–cell systems.

2. Materials

1. Two different cell types to be fused; alternatively, cells from primary sources can be used. Cells of the same type can also be used as opposing fusion partners.
2. Growth media used to maintain the cell type(s) used for fusion.
3. Stock solution of CMFDA (Molecular Probes, Eugene, OR): 5 m*M* solution (*see* **Note 1**) in dimethyl sulfoxide (DMSO; Sigma, St. Louis, MO).
4. Stock solution of CMTMR (Molecular Probes): 5 m*M* solution (*see* **Note 1**) in DMSO.
5. A flow cytometer that is capable of acquiring fluorescent data from two different band widths of light. Cytometers that are set up to detect cells that are labeled with fluorescein and rhodamine compounds should provide excellent results. For example, a Becton Dickinson (San Jose, CA) FACStar Plus is used to detect fusion products from a variety of different cell–cell systems. Excitation is provided by an 80 mW argon laser tuned to an wavelength of 488 nm. CMFDA emission is detected in the FL1 (green) channel; CMTMR emission is detected in the FL2 (red) channel. Detection wavelengths for FL1 and FL2 channels are 530 ± 15 nm and 585 ± 21 nm, respectively. Forward light scatter (FSC) and side angle light scatter (SSC) data are also collected. A Becton Dickinson FACScan instrument has also been used with equivalent results.
6. Dulbecco's phosphate-buffered saline solution (PBS; Mediatech, Washington, DC).
7. Conical-bottomed centrifuge tubes with a 15 or 45 mL capacity. Tube size will depend on the number of cells stained with the fluorescent dyes.

3. Methods

3.1. Cell Staining

1. Mix separate suspensions of each fusion partner in the media used for growth. Cell densities of approx 1.5×10^6 cells/mL yield excellent results. Mix the suspensions in conical-bottomed centrifuge tubes to facilitate subsequent washing. Note the total volume in each tube; this will be referred to as one staining volume. If adherent cells are under investigation, detach cells in order to make the appropriate suspension(s).

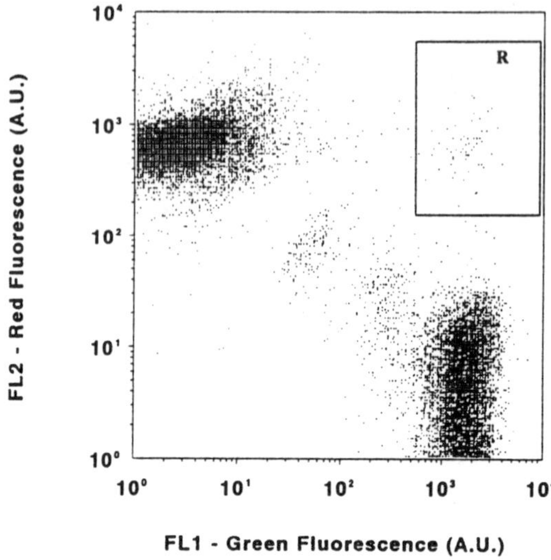

Fig. 2. A typical fluorescent dot plot from a sample that contained a mixture of two stained fusion partners. Cells stained with CMFDA appear as a population near the FL1 axis. Cells stained with CMTMR populate plots near the FL2 axis. Note that the stained fusion partners both appear as separate and distinct populations on the plot. This facilitates detection and quantitation of fusion products that will be present in region R after fusion. A background level of aggregated cells are typically present in region R for samples that have not been electrically treated. Fluorescence was measured in arbitrary units (A.U.). Reprinted with permission ***(19)***.

2. Stain one cell type by adding CMFDA stock solution to the appropriate suspension so that the dye concentration is 0.5 μ*M*. Stain the other cell type by adding CMTMR stock solution to make the final dye concentration 6.0 μ*M*.
3. Incubate both suspensions for 30 min at 37°C. Shield the mixtures of stain and cells from light during incubation.
4. Wash each suspension twice by centrifugation. Replace the liquid between washes with one staining volume of growth media. After the second wash, suspend the cells in one staining volume of growth media.
5. Incubate the stained cell suspensions for 60 min at 37°C. The suspensions should be shielded from light (*see* **Note 2**).
6. Wash each suspension three times by centrifugation. Use one staining volume of PBS to replace the liquid after each wash.
7. Cytometrically analyze a mixture of both stained fusion partners. The mixture should contain approximately equal fractions of CMFDA and CMTMR stained cells. A plot of FL1 (green fluorescence) vs FL2 (red fluorescence) should be made in order to determine if each fusion partner has a suitable fluorescent magnitude. The resulting plot should appear similar to **Fig. 2**. Note that the CMFDA- and CMTMR-stained cell populations in the figure are separate and distinct. This

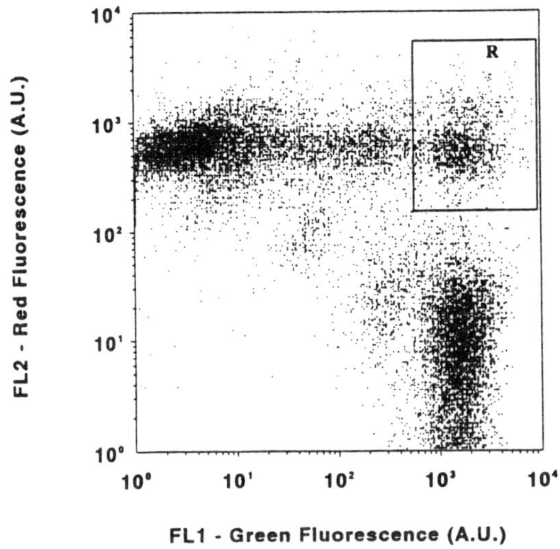

Fig. 3. A typical fluorescent dot plot from a fusion sample. The population within region R contains hybrid cells. The hybrids within the region represent 6.9% of the 20,000 cells represented by the plot. Unfused fusion partners retain their respective positions near the FL1 and FL2 axes. Cells located between the hybrid and unfused cell populations are typically damaged cells or debris that result from electrofusion. Fluorescence was measured in arbitrary units (A.U.). Reprinted with permission ***(19)***.

characteristic is critical to the detection and quantitation method. Fusion products will appear as a dual fluorescing population with FL1 and FL2 magnitudes that are approximately equal to those of the stained fusion partners. This region is labeled R in **Fig. 2**. If the populations are not separate and distinct, graphical resolution of the hybrid population will be impossible. Dye concentrations used for staining should be increased if the fusion partner populations can not be distinguished (*see* **Notes 3** and **4**).

8. Perform fusion after establishing that the staining procedure will yield a graphically resolvable hybrid population (*see* **Notes 5** and **6**).

3.2. Detection and Quantitation of Fusion Products

1. Cytometrically analyze cells that remain after fusion. Acquire FL1, FL2, FSC, and SSC data.
2. Examine the FL1 vs FL2 plot from each fusion sample. A typical dot plot from a fusion sample is given as **Fig. 3** (*see* **Notes 7** and **8**). Note that the plot shown contains three major populations. The population within the region labeled R contains fused cells. The populations near each of the axes are unfused CMFDA- and CMTMR-stained cells.
3. Quantitation of fusion products is conducted using flow cytometer software. A region similar to R in **Fig. 3** should be drawn around the hybrid population. Flow

cytometer software then tabulates the number of cells within the region. Most software packages express the population within any region as a percentage of the cells represented on the plot and as the number of cells within the region. If multiple samples are fused, the same region is applicable to all samples.

4. Notes

1. Stock solutions of CMFDA and CMTMR should be stored frozen (–10°C) and shielded from light. Consistent staining is achieved if the stock solutions are quickly defrosted prior to use and then refrozen as soon as possible. If used in this manner, the stock solutions will provide reproducible staining if used within several months.
2. According to the dye manufacturer (Molecular Probes) CMFDA and CMTMR pass freely through the membranes into the cytosol. Once they pass through the membranes, both dyes undergo what is thought to be a glutathione *S*-transferase-mediated reaction that renders the dyes membrane-impermeable. Therefore, the purpose of incubating cells in **Subheading 3.1., step 5** is to allow time for this reaction to occur.
3. The CMFDA and CMTMR staining concentrations that yield distinct and separate populations on FL1 vs FL2 cytometric plots varies with cell type. Fluorescent dye concentrations ranging from 0.25 to 0.8 μ*M* CMFDA have been used successfully to load a variety of cell types. Concentrations of 4.0–12.0 μ*M* CMTMR have been utilized for several different cell lines.
4. Quenching does not appear to be a complicating factor for CMFDA or CMTMR. Stained cell samples exposed to room light for several hours did not show decreased fluorescent magnitudes with time.
5. Several different cell lines containing CMFDA and CMTMR have been placed back into culture. Fluorescence is retained for several days with no marked cytotoxic affect.
6. The staining method yields cells and hybrids that can be visualized using fluorescence microscopy. Standard filters for viewing fluorescein isothyocyanate allow CMFDA-stained cells to be viewed, and filters normally used for phycoerythrin allow CMTMR-stained cells to be seen. We prefer to view emission from both dyes simultaneously using a Leitz Orthoplan 2 microscope (Leica, Wetzlar, Germany). Filters are designed for simultaneous excitation by 490 ± 20 and 575 ± 30 nm. In addition, the filters allow wavelengths of 525 ± 40 and 633 ± 20 nm to be simultaneously viewed.
7. Electrofusion can produce small cellular debris. This debris may hinder interpretation of FL1 vs FL2 plots. Examination of a FSC vs SSC plot is useful for detecting this type of debris. It is generally present as a particles that are smaller in FSC magnitude than the major population of FSC vs SSC plots. The small particles can be gated out of sample data prior to making FL1 vs FL2 plots.
8. In principle this method can be used with other dye combinations as long as both stains can be detected in the FL1 (530 ± 14 nm) and FL2 (535 ± 15 nm) cytometer channels. Many stain combinations were investigated during development of the

detection and quantitation method. The combination of CMFDA and CMTMR provides consistent and reproducible results. A common problem with fluorescent dye combinations other than CMFDA and CMTMR (i.e., combinations of fluorescein isothiocyanate (Sigma); 5(and -6) sulfofluorescein diacetate (Molecular Probes) hydroethidine (Polysciences, Warrington, PA); and calcium crimson (Molecular Probes) was that mixtures containing cells stained with both dyes migrated off their respective axes in time. Both stained cell populations moved toward the dual fluorescing region of FL1 vs FL2 plots over the course of minutes to hours. This complicates interpretation because mixtures of stained cells are present during and after fusion procedures. One possible explanation for this is that dye may leak out of a stained cell type and then diffuse into the other stained cell type. This problem can seriously hinder detection and quantitation because cytometric populations will not occupy the same positions on plots over a normal experimental time frame. Also, migration toward the dual fluorescing region would increase the likelihood of false-positive results. Mixtures of CMFDA- and CMTMR-stained cells showed no population migration in time; therefore, they were judged suitable for use. If the methods presented in this chapter are applied using different dyes, it should be verified that stained cell populations do not migrate in time.

References

1. Lo, M. M. S., Tsong, T. Y., Conrad, M. K., Strittmatter, S. M., Hester, L. D., and Snyder, S. (1984) Monoclonal antibody production by receptor-mediated electrically induced cell fusion. *Nature* **310,** 792–794.
2. Foung, S. K. H. and Perkins, S. (1989) Electric field-induced cell fusion and human monoclonal antibodies. *J. Immunol. Methods* **116,** 117–122.
3. Glassy, M. (1988) Creating hybridomas by electrofusion. *Nature* **333,** 579–580.
4. Hewish, D. R. and Werkmeister, J. A. (1989) The use of an electroporation apparatus for the production of murine hybridomas. *J. Immunol. Methods* **120,** 285–289.
5. Wojchowski, D. M. and Sytkowski, A. J. (1986) Hybridoma production by simplified avidin-mediated electrofusion. *J. Immunol. Methods* **90,** 173–177.
6. Grasso, R. J., Heller, R., Cooley, J. C., and Haller, E. M. (1989) Electrofusion of individual animal cells directly to intact corneal epithelial tissue. *Biochim. Biophys. Acta* **980,** 9–14.
7. Abidor, I. G. and Sowers, A. E. (1992) Kinetics and mechanism of cell membrane electrofusion. *Biophys. J.* **61,** 1557–1569.
8. Zimmerman, U., Pilwat, B., and Riemann, F. (1974) Dielectric breakdown of cell membranes. *Biophys. J.* **14,** 881–899.
9. Dimitrov, D. S., Apostolova, M. A., and Sowers, A. E. (1990) Attraction, deformation, and contact of membranes induced by low frequency electric fields. *Biochim. Biophys. Acta* **1023,** 389–397.
10. Sowers, A. E. (1989) The mechanism of electroporation and electrofusion in erythrocyte membranes, in *Electroporation and Electrofusion in Cell Biology* (Neumann, E., Sowers, A. E., and Jordan, C. A., eds.), Plenum, New York, pp. 229–256.

11. Rols, M. P. and Teissie, J. (1989) Ionic-strength modulation of electrically induced permeabilization and associated fusion of mamalian cells. *Eur. J. Biochem.* **179,** 109–115.
12. Teissie, J. and Blangero, C. (1984) Direct experimental evidence of the vectorial character of the interaction between electric pulses and cells in cell electrofusion. *Biochim. Biophys. Acta* **775,** 446–448.
13. Teissie, J. and Rols, M. P. (1986) Fusion of mammalian cells in culture is obtained by creating the contact between cells after their electropermeabilization. *Biochem. Biophys. Res. Comm.* **140,** 258–266.
14. Sukharev, S. I., Bandrina, I. N., Barbul, A. I., Fedorova, L. I., Abidor, I. G., and Zelenin, A. V. (1990) Electrofusion of fibroblasts on the porous membrane. *Biochim. Biophys. Acta* **1034,** 125–131.
15. Sowers, A. E. (1988) Fusion events and nonfusion contents mixing events induced in erythrocyte ghosts by an electric pulse. *Biophys. J.* **54,** 619–626.
16. Sowers, A. E. (1986) A long-lived fusogenic state is induced in erythrocyte ghosts by electric pulses. *J. Cell Biol.* **102,** 1358–1362.
17. Bakker Schut, T. C., Kraan, Y. M., Barlag, W., de Leij, L., de Grooth, B. G., and Greve, J. (1993) Selective electrofusion of conjugated cells in flow. *Biophys. J.* **65,** 568–572.
18. Shi, T., Eaton, A. M., and Ring, D. B. (1991) Selection of hybrid hybridomas by flow cytometry using a new combination of fluorescent vital stains. *J. Immunol. Methods* **141,** 165–175.
19. Jaroszeski, M. J., Gilbert, R., and Heller, R. (1994) Detection and quantitation of cell-cell electrofusion products by flow cytometry. *Anal. Biochem.* **216,** 271–275.
20. Heller, R. (1992) Spectrofluorometric assay for the quantitation of cell-tissue electrofusion. *Anal. Biochem.* **202,** 286–292.

16

Cell Cycle Analysis of Asynchronous Populations

Michael G. Ormerod

1. Introduction

Measurement of a DNA histogram can be achieved by fixing or permeabilizing cells and staining them with a DNA-binding dye, such as propidium iodide (PI). The histogram will yield the percentage of cells in the G1, S, and G2/M phases of the cell cycle *(1)*. Although some inferences about the movement of cells through the cycle may be drawn, the information gained is essentially static. For example, it is not known whether a cell with S-phase DNA content is actually synthesizing DNA; also the presence of subpopulations with different cycle times cannot be detected.

Dynamic information about cell cycle progression can be obtained by labeling cells with 5'-bromodeoxyuridine (BrdUrd) which is incorporated into DNA in place of thymidine. Detection of BrdUrd in the DNA allows the fraction of cells in S phase to be enumerated and, if samples are taken at different time intervals, also gives information about the cell cycle kinetics.

Three methods have been used. After the application of a pulse-label, monoclonal antibodies that react specifically with BrdUrd reveals those cells that are in S phase. Counter-staining with propidium iodide shows the cell cycle. If cells are harvested at times after the application of the pulse-label, the movement of labeled cells through the cell cycle can be followed. This method has been applied in vitro *(2)* and in vivo *(3,4)*. The other two methods exploit the observation of Latt that the fluorescence of *bis*-benzimidazoles (Hoechst 33342 and 33258) bound to DNA is quenched by BrdUrd *(5)*. One method, in which the cells are pulse-labeled and then stained with a combination of Hoechst 33258 and mithramycin, requires a dual laser flow cytometer and a complex analysis in which fluorescence signals are subtracted in real time *(6)*. In the method described in this chapter, cells are continuously labeled with BrdUrd,

From: *Methods in Molecular Biology, Vol. 91: Flow Cytometry Protocols*
Edited by: M. J. Jaroszeski and R. Heller © Humana Press Inc., Totowa, NJ

permeabilized, and stained with Hoechst 33258 and PI. The quenching of Hoechst/DNA fluorescence reveals whether the cells have incorporated BrdUrd; the PI/DNA fluorescence (unaffected by BrdUrd) gives the cell cycle phase. This method was originally applied to quiescent cells that had been stimulated into the cell cycle *(7)*. It can also be applied to asynchronous populations of cells *(8)*. Although the data analysis is more complex, a wealth of information can be derived. In particular, cell cycle-specific effects of toxic treatments can be observed without resorting to artificial cell synchronization *(8–11)*.

It should be noted that, in all three methods, the cells have to be fixed or permeabilized to allow access of reagents to the DNA. Analysis of viable cells is not possible. In addition, for the antibody method, the DNA of the cells has to be partially denatured to allow access of the antibody to the BrdUrd incorporated into the DNA.

In the method described here, BrdUrd is added and cells are harvested at fixed time intervals, typically, every 4 h for 36 h. Samples may be collected and the cells frozen prior to analysis. For analysis, cells are suspended in a buffer containing Hoechst 33258 and a detergent, which releases the nuclei, and PI is then added. Hoechst 33258 is exited in the UV and the analysis requires a flow cytometer equipped with a source of UV light.

2. Materials

1. 5'-bromodeoxyuridine (Sigma [St. Louis, MO], catalog no. B 5002): Make up a stock solution of 10 m*M*. Store frozen.
2. Hoechst 33258 (Sigma, catalog no. B2883 or Molecular Probes [Eugene, OR], catalog no. H-1398).
3. Propidium iodide (Sigma, catalog no. P 4170 or Molecular Probes, catalog no. P-1304): Make up a stock solution of 100 µg/mL in distilled water. Store in the dark at 4°C. Stable for at least 6 mo.
4. Ethidium bromide (Sigma, catalog no. E 8751 or Molecular Probes): Make up a stock solution of 100 µg/mL in distilled water. Store in the dark at 4°C. Stable for at least 6 mo.
5. Staining solution: 100 m*M* Tris-HCl, pH 7.4, 154 m*M* NaCl, 1 m*M* $CaCl_2$, 0.5 m*M* $MgCl_2$, 0.1% (v/v) Nonidet-P40, 0.2% (w/v) BSA, 1.2 µg/mL Hoechst 33258. Make up at 10X final strength and store in the dark at 4°C. The solution is stable for at least 6 mo. Batches of staining buffer can be prepared weekly from the 10X concentrated stock solution in distilled water.

3. Methods

3.1. Analysis of Cells

1. If desired, treat the cells (radiation, drug, heat, and so on).
2. Immediately after the treatment, add a suitable concentration of BrdUrd to the cell culture (*see* **Notes 1–3**).

3. At fixed time intervals (3–8 h, depending on the cell cycle time), harvest an aliquot of cells (*see* **Notes 4–6**).
4. Centrifuge the cells and resuspend in 500 µL ice-cold staining buffer. Briefly vortex mix. Stand on ice for 15 min.
5. Add 10 µL PI solution. Briefly vortex mix. Store on ice (*see* **Notes 7** and **8**).
6. Analyze on the flow cytometer recording red (PI/DNA) and blue (Hoechst/DNA) fluorescences. Adjust the flow rate to approx 500 particles/s. If possible, perform a pulse shape analysis on the red fluorescence signal and gate to exclude any clumped nuclei (see **ref. *1*** and **Fig. 1**). Display a cytogram of red vs blue fluorescence (*see* **Notes 9–12**).

3.2. Data Analysis

Figure 2 shows cytograms obtained from an untreated human embryonic fibroblast cell line (MRC5/34C). At time 0, G1, S, and G2/M phases of the cell cycle could be identified from both the red (PI) and blue (Hoechst) fluorescence. When the cells were incubated in 40 µ*M* BrdUrd, as the cells progressed through S phase, their red fluorescence increased but their blue fluorescence (which was partially quenched by BrdUrd) did not change. After 4 h in BrdUrd, there had been a small progression giving the S phase a slight bow shape on a plot of red vs blue fluorescence. At 8 h, the bow shape was more pronounced; also many of the cells originally in G2/M had divided and moved into G1 (unlabeled). At 16 h, all the cells in S phase at time 0 had reached G2 (labeled G2*) and a few had divided (G1*). Some of the cells which had been in G1 at time 0 h were now in S phase (Sf); some had progressed as far as G2/M (G2f). By 28 h, some cells that begun the experiment in G2 had completed a cell cycle and returned to G1 again (labeled G1').

Figures 3–5 illustrate typical effects of cell cycle perturbation. **Figure 3** shows MRC5/34 cells that had been given 5 gy γ radiation immediately prior to adding the BrdUrd. The cells suffered a G2 block. Cells that were irradiated in all phases of the cell cycle have become arrested in G2. (Compare **Fig. 3** to **Fig. 2**, 32-h time point). **Figure 4** shows W1L2 cells (human lymphoblastoid cell line) that had been incubated with cisplatin for 2 h before adding BrdUrd. Only cells treated with drug in G1 and early S phase are arrested in G2. The other cells have divided. **Figure 5** shows a human medulloblastoma cell line (D283) that underwent a G1 block after γ radiation. Only a proportion of the cells in G1 became blocked in G1, some of the G1 cells progressed through the cycle and divided (in compartment G1'). Presumably these were cells irradiated in late G1 phase.

4. Notes

1. The correct concentration of BrdUrd must be determined by a preliminary experiment. The concentration should be such that cells in G1 after one round of replication have about half the blue fluorescence of unlabeled cells in G1.

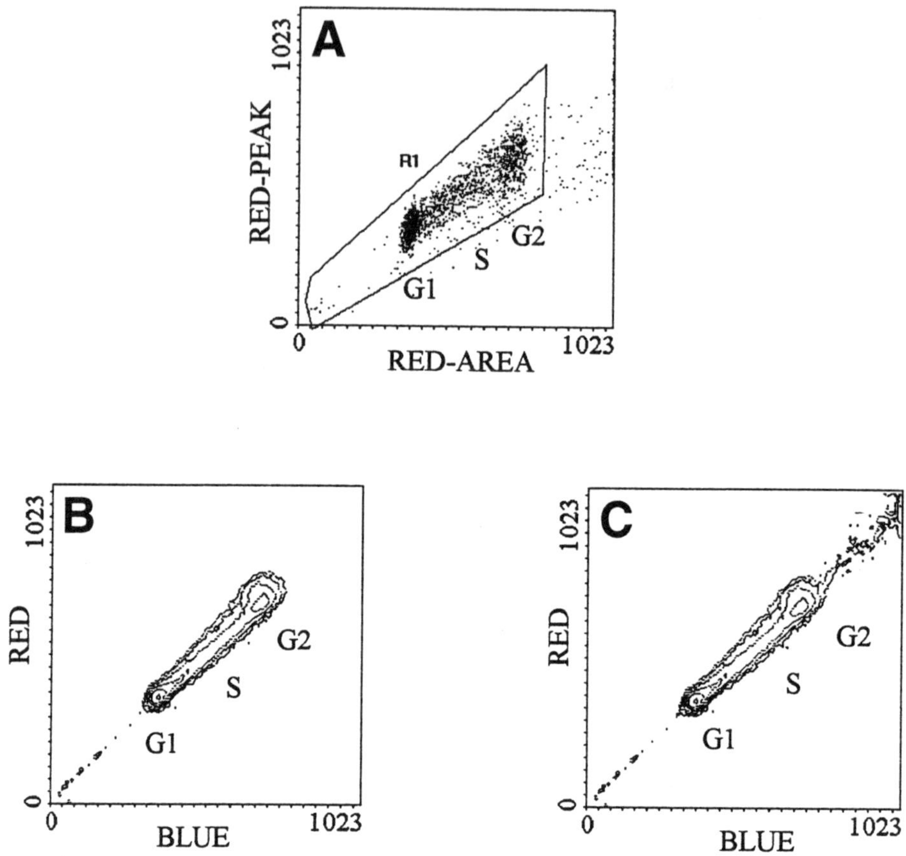

Fig. 1. Hoechst 33258-PI analysis of nuclei from a human embryonic fibroblast cell (MRC5/34) without incubation in BrdUrd. **(A)** PI/DNA fluorescence (red) showing a cytogram of the peak of the red fluorescence signal vs the integrated area. A gate has been set to include single nuclei and to exclude clumps. The cytogram is displayed as a dot plot. **(B,C)**. Hoechst/DNA (blue) vs PI/DNA (red) fluorescence (cytograms displayed as contour plots). The cytogram in **(C)** shows ungated data; that in **(B)** shows the effect of gating on region R1 in cytogram **(A)**. Cells prepared by David Gilligan and data recorded by Jenny Titley on a Coulter Elite ESP using a Spectra-Physics argon-ion laser tuned to produce 100 mW in the UV. Red (>630 nm) and blue (460 nm) fluorescence were measured. Data were acquired on an IBM-PC compatible computer and the figure was prepared using the WINMDI program supplied by Joe Trotter, Salk Institute.

2. Concentrations of BrdUrd vary from 10 to 100 μ*M*. If the concentration of BrdUrd is >20 μ*M*, the BrdUrd may become exhausted by 24 h. The remedy is to either work at a lower cell density or to replenish the BrdUrd every 12 h.
3. A trial experiment should be performed to check that the incorporation of BrdUrd into the DNA is not inhibiting the progression of the cells through the cycle.

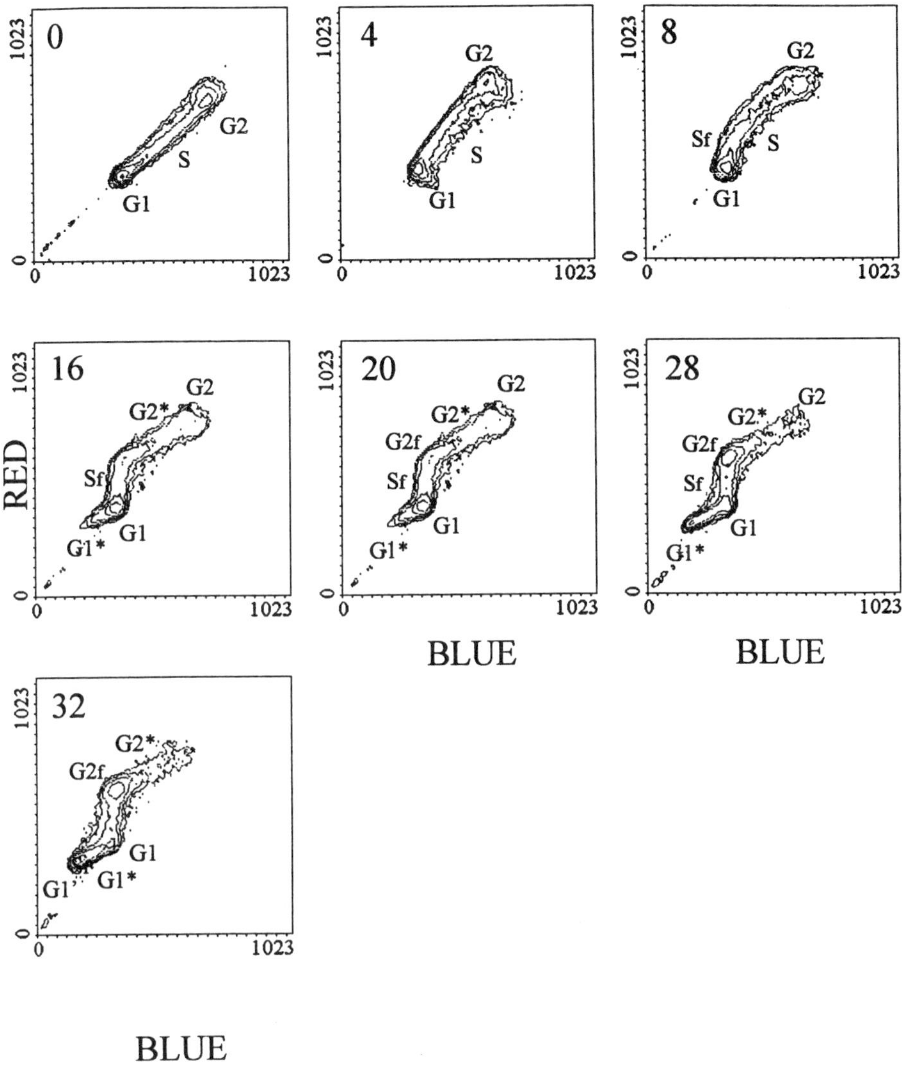

Fig. 2. Cytograms of PI-DNA (red) vs Hoechst 33258-DNA fluorescence of nuclei from MRC5/34 cells incubated in 40 μ*M* BrdUrd and 40 μ*M* deoxycytidine BrdUrd for the times shown on the panels. Other details as in **Fig. 1**. For a description of the data, *see* the main text.

After different times of incubation with BrdUrd, fix cells in 70% ethanol, centrifuge, and resuspend in phosphate-buffered saline containing 20 μg/mL PI and 0.1 mg/mL RNase. Incubate at 37°C for 1 h and record the DNA histogram. Addition of deoxycytidine (equimolar with the BrdUrd) can reduce any effect of BrdUrd.

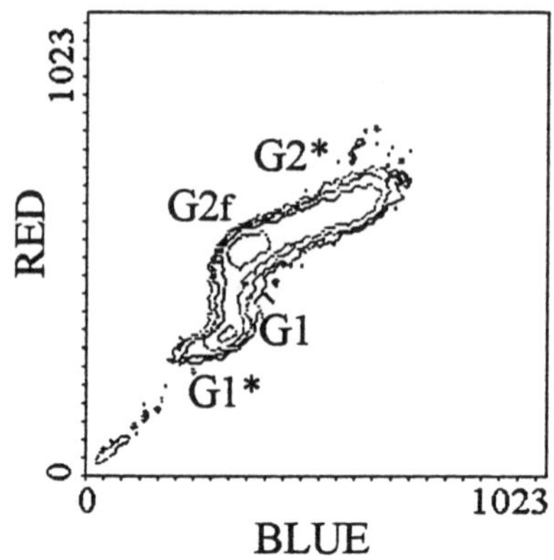

Fig. 3. A cytogram of PI-DNA (red) vs Hoechst 33258-DNA fluorescence of nuclei from MRC5/34 cells given 5 gy γ radiation and then incubated in 40 μM BrdUrd and 40 μM deoxycytidine BrdUrd for 32 h. Other details as in **Fig. 1**. For a description of the data, *see* the main text.

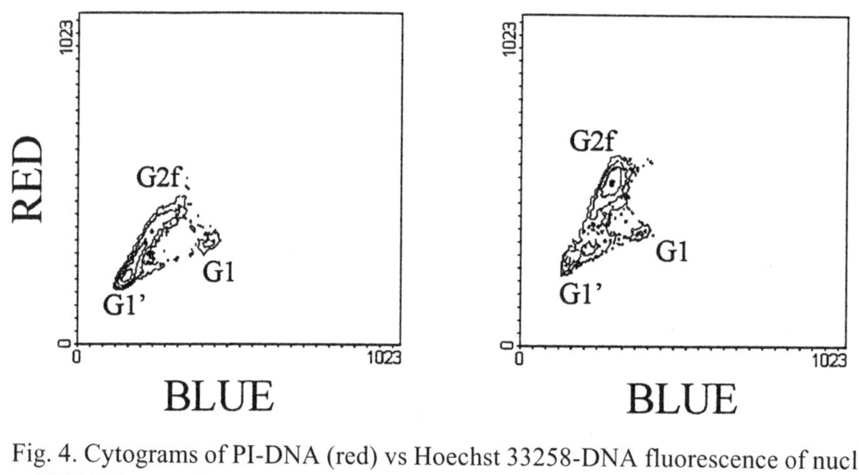

Fig. 4. Cytograms of PI-DNA (red) vs Hoechst 33258-DNA fluorescence of nuclei from W1L2 (human lymphoblastoid cell line) cells either untreated **(left)** or incubated for 2 h with 20 μM cisplatin **(right)** and then incubated in 40 μM BrdUrd for 24 h. Experiment run by the author. Other details as in **Fig. 1**.

4. For suspension cultures, shake to resuspend cells and remove 3 mL. For adherent cell cultures, use a separate flask for each time point. Harvest the cells by a short incubation with trypsin.
5. After harvesting, cells can be frozen and stored for later analysis.
6. The detergent in the staining buffer releases nuclei. At this stage, the samples are stable for several hours on ice.

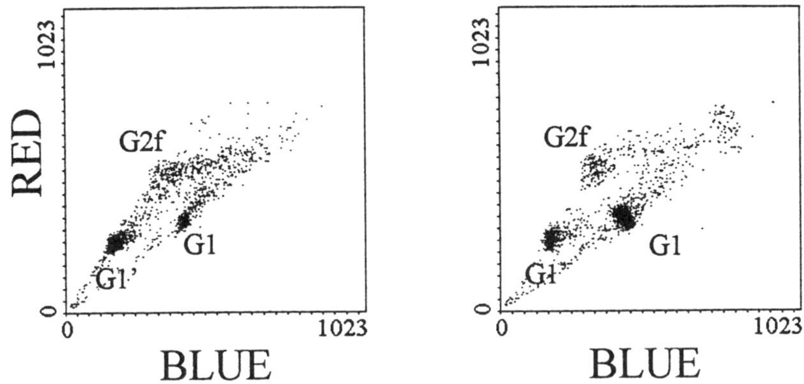

Fig. 5. Cytograms of PI-DNA (red) vs Hoechst 33258-DNA fluorescence of nuclei from a medulloblastoma cell line (D283) either untreated **(left)** or given 8 gy γ radiation **(right)** and then incubated in 40 μM BrdUrd and 40 μM deoxycytidine BrdUrd 32 h. Cells prepared by Cyd Bush. Other details as in **Fig. 1**.

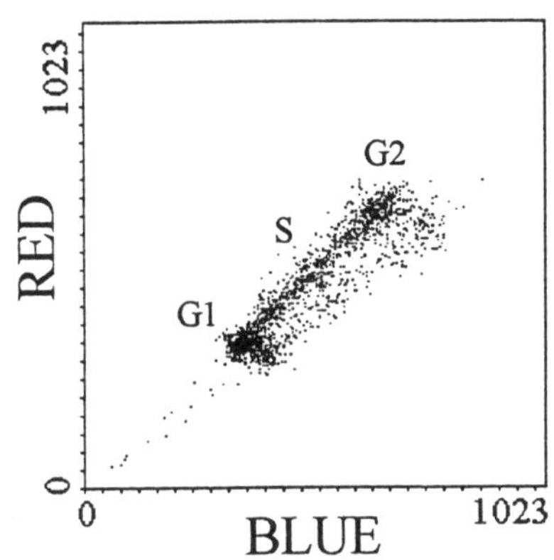

Fig. 6. A cytogram of PI-DNA (red) vs Hoechst 33258-DNA fluorescence of nuclei from MRC5/34C cells. The cells had not lysed completely. The nuclei and the partially lysed cells stain slightly differently creating a shadow in the cytogram. Experimental details as in **Fig. 1**.

7. The cell concentration is important and should be between 5×10^5 and 2×10^6. If the concentration is too high, the nuclei will be under-stained and may give distorted cytograms.
8. Either PI or ethidium bromide (EB) may be used as a counter-stain for DNA.
9. When setting a gate on a plot of DNA-peak signal vs DNA-area, be careful to include material of DNA content less than that of cells in G1.

10. If it is not possible to perform a pulse-shape analysis, clumps may usually be differentiated on a cytogram of forward light scatter vs red fluorescence.
11. A PI/DNA complex is excited by UV. It also absorbs blue light and will be excited by energy transfer from the Hoechst dye. When the Hoechst fluorescence is quenched, there will be a consequent reduction in PI fluorescence. If a He-Cd laser is used as a source of UV (325 nm) (rather than an argon-ion laser, 360–390 nm), direct absorption will predominate and the secondary quenching will be reduced. For this reason, it is preferable to use a He-Cd laser *(12)*.
12. If the pattern produced from the sample at 0 h has a shadow (*see* **Fig. 6**), the cells have not lysed properly. Incubate the samples at 37°C for 5 min and rerun.

Acknowledgments

I thank Jenny Titley, David Gilligan, and Cyd Bush, Institute of Cancer Research, Sutton, UK, for supplying the data used to illustrate this chapter. Their work was supported by the Cancer Research Campaign.

References

1. Ormerod, M. G. (1994) Analysis of DNA: general methods, in *Flow Cytometry: A Practical Approach,* 2nd ed. (Ormerod, M. G., ed.), IRL Press, Oxford, UK, pp. 118–135.
2. Dolbeare, F. A., Gratzner, H. G., Pallavicini, M. G., and Gray, J. W. (1983) Flow cytometric measurement of total DNA content and incorporated bromodeoxyuridine. *Proc. Natl. Acad. Sci. USA* **80,** 5573–5577.
3. Begg, A. C., McNally, N. J., Shrieve, D. C., and Karcher, H. (1985) A method to measure the duration of DNA synthesis and potential doubling time from a single sample. *Cytometry* **6,** 620–626.
4. Wilson, G. D. (1994) Analysis of DNA—measurement of cell kinetics by the bromodeoxyuridine/anti-bromodeoxyuridine method, in *Flow Cytometry: A Practical Approach,* 2nd ed. (Ormerod, M. G., ed.), IRL, Oxford, UK, pp. 137–156.
5. Latt, S. A. (1973) Microfluorometric detection of deoxyribonucleic acid replication in human metaphase chromosomes. *Proc. Natl. Acad. Sci. USA* **70,** 3395–3399.
6. Crissman, H. A. and Steinkamp, J. A. (1987) A new method for rapid and sensitive detection of bromodeoxyuridine in DNA replicating cells. *Exp. Cell Res.* **173,** 256–261.
7. Rabinovitch, P. S., Kubbies, M., Chen, Y. C., Schindler, D., and Hoehn, H. (1988) BrdUrd-Hoechst flow cytometry. A unique tool for quantitative cell cycle analysis. *Exp. Cell Res.* **74,** 309–318.
8. Ormerod, M. G. and Kubbies, M. (1992) Cell cycle analysis of asynchronous cells by flow cytometry using bromodeoxyuridine label and Hoechst-propidium iodide stain. *Cytometry* **13,** 678–685.
9. Ormerod, M. G., Imrie, P. R., Loverock, P., and Ter Haar, G. (1991) A flow cytometric study of the effect of heat on kinetics of cell proliferation of Chinese hamster V79 cells. *Cell Prolif.* **25,** 41–51.

10. Ormerod, M. G., Orr, R. M., and Peacock, J. H. (1994) The role of apoptosis in cell killing by cisplatin, a flow cytometric study. *Brit. J. Cancer* **69,** 93–100.
11. Poot, M. and Ormerod, M. G. (1994) Analysis of proliferation using the bromodeoxyuridine-Hoechst/ethidium bromide method, in *Flow Cytometry: A Practical Approach,* 2nd ed. (Ormerod, M. G., ed.), IRL Press, Oxford, UK, pp. 157–167.
12. Kubbies, M., Goller, B., and Van Bockstaele, D. R. (1992) Improved BrdUrd-Hoechst bivariate cell kinetic analysis by helium-cadmium single laser excitation. *Cytometry* **13,** 782–789.

17

Simultaneous DNA/RNA Analysis of Solid and Hematoreticular Malignancies

Kim Steck and Adel El-Naggar

1. Introduction

Acridine orange (AO) is the most notable member of the acridine fluorescent dye family. Because of its versatility and diverse biological applications, it has been widely used in basic and clinical research fields *(1–10)*. For flow cytometric application, AO offers distinct advantages over other fluorochromes by simultaneously binding to DNA and RNA in individual cells. RNA quantitation augments conventional DNA analysis by indirectly assessing the biological activity of tumor populations as it relates to the transcriptional activity, cell proliferation, and differentiation status of cells *(11)*. However, flow cytometric application of this dye remains limited, because of concerns regarding procedure tediousness and its notoriety for contaminating tubing of flow cytometers. These difficulties can be overcome and successful results can be obtained by adherence to strict staining protocols and careful maintenance of instrumentation.

AO, [3,6 (dimethylamino) acridine], interacts with double-stranded nucleic acids regardless of base composition or ligand type *(12)*. AO intercalates into the double-stranded conformation of nucleic acids, and on excitation, fluoresces approx 530 nm (green) and electrostatically aggregates on single-stranded nucleic acids, emitting a red fluorescence >600 nm, thus allowing for the simultaneous analysis of DNA and RNA. However, because of the positive charge of AO, electrostatic binding to cellular proteins may also occur. AO may also bind by intercalation to the dsRNA that is typically present in variable proportions in different cells, leading to spurious elevation of green fluorescence. To circumvent these reactions, staining techniques must include reagents that ensure specific interaction of the dye with the respective nucleic

From: *Methods in Molecular Biology, Vol. 91: Flow Cytometry Protocols*
Edited by: M. J. Jaroszeski and R. Heller © Humana Press Inc., Totowa, NJ

acids. This has been successfully achieved through a two-step staining method that enhances the differential staining of DNA and RNA *(13)*.

2. Materials

2.1. General Guidelines

Stock solutions should be made prior to specimen preparation, with the exception of the AO working solution, which is prepared only at the time of sample staining. Glassware used for reagent preparation must be RNase free. Rigorous observation of molarity and pH conditions should also be adhered to for optimal staining.

Peripheral blood lymphocytes and bone marrow aspirates are harvested by ficoll-hypaque gradient centrifugation. Solid-tumor and bone-marrow biopsies should be carefully minced and agitated to release individual cells. Fluids and fine needle aspirates are either washed directly in buffer or separated by gradient centrifugation depending on the extent of red blood cell contamination. Specimens should subsequently be washed in phosphate-buffered saline (PBS) with $MgCl_2$ and cell concentrations adjusted to 1×10^6 cells/mL. If additional analyses, such as surface membrane markers are to be performed, PBS with $MgCl_2$ (*see* **Subheading 2.2.4.**) should be eliminated from specimen processing. A separate aliquot of the cell suspension can be rinsed in PBS with $MgCl_2$ prior to acridine orange staining.

2.2. Stock Solutions

1. 1 mg/mL AO: (AO is a mutagen. Handle with care.) Measure 50 mg of AO powder (Polysciences, Warrington, PA; *see* **Note 1**) into a clean glass beaker. Add 50 mL distilled H_2O and stir until dissolved, protected from light. Filter solution through a #1 Whatman filter paper. Store protected from light at 4°C. Stable for 1 yr (*see* **Note 2**).
2. 0.2 *M* Citric acid: Add 19.21 g of citric acid (Sigma, St. Louis, MO) to 500 mL distilled H_2O. Stir until dissolved and store at 4°C. Stable for 2 mo.
3. 10 m*M* Ethylenediaminetetra-acetic acid (EDTA): Add 2.17 g of EDTA (Gibco-BRL, Grand Island, NY) to 500 mL distilled H_2O, stir until dissolved, and store at 4°C. Stable for 2 mo.
4. PBS with 2 m*M* $MgCl_2$: Dilute 10X PBS without calcium and magnesium salts (Irvine Scientific, Santa Ana, CA) to a 1X solution. Add 0.408 mL of 4.9 *M* $MgCl_2$ for each liter of PBS. Adjust pH to 7.2–7.4. Bulk preparations may be stored at room temperature up to 3 mo. Store working aliquots at 4°C.
5. 1 *M* Sodium chloride (NaCl): Add 29.0 g of NaCl (Fisher, Houston, TX) to 500 mL distilled H_2O and stir until dissolved. Store at 4°C. Stable for 3 mo.
6. 0.4 *M* Sodium phosphate dibasic (Na_2HPO_4): Add 28.39 g of Na_2HPO_4 (Fisher) to 500 mL distilled H_2O and stir until dissolved. Store at room temperature, 22°C. Stable for 2 mo.
7. Triton 10X: Add 10 mL Triton X-100 (CMS, Houston, TX) to 90 mL distilled water. Store at 4°C. Stable for 3 mo.

2.3. Working Solutions

Working solutions are prepared from the stock solutions described above.

1. Solution A: Mix 1 mL Triton 10X, 8 mL 1 *N* HCl, 15 mL 1 *M* NaCl, and 76 mL distilled H_2O. Adjust to pH 1.2 with 1 *N* NaOH or 1 *N* HCl. Store at 4°C. Stable for 2 wk.
2. Solution B: Add 50 mL of 10 m*M* EDTA, 75 mL 1 *M* NaCl solution, 157.5 mL of 0.4 *M* Na_2HPO_4, 92.5 mL of 0.2 *M* citric acid to 120 mL distilled water. Adjust to pH 6.0 with 1 *N* NaOH or 1 *N* HCl. Store at 4°C. Stable for 1 mo.
3. AO working solution: Add 0.1 mL AO stock solution to 9.9 mL solution B. Prepare daily and keep on ice. Because of hazards associated with AO, contact local safety officials for proper disposal guidelines.

3. Method

3.1. Staining

All steps are performed at 0–4°C in disposable glass tubes.

1. Prepare a single cell suspension at a concentration of 1.0×10^6 cells/mL in PBS with 2 m*M* $MgCl_2$.
2. Place solution A and AO working solution on ice.
3. Aliquot 0.2 mL of freshly disaggregated cell suspension into a 12 × 75 mm glass disposable test tube.
4. Add 0.4 mL of solution A (*see* **Note 3**). Incubate for 45 s at 0–4°C.
5. Add 1.2 mL of AO working solution and analyze immediately.
6. Repeat **steps 3–5** for each specimen.
7. After all samples have been collected, clean cytometer with bleach and rinse thoroughly with distilled H_2O (*see* **Note 4**).

3.2. Analysis

3.2.1. Instrumentation and Acquisition

Excitation of AO occurs between 455 and 490 nm. Since the emission spectra of AO DNA binding overlaps with the red fluorescence from RNA, filter combinations should be chosen to reduce such overlap. We suggest a 550LP dichroic with a 525BP for DNA fluorescence and a 630LP for RNA fluorescence. However, compensation may be necessary to further reduce overlapping emissions (*see* **Note 5**).

Instrument performance should be monitored according to manufacturer guidelines, with emphasis on obtaining consistent resolution and mean peak channels for calibration particles. Test protocols should be structured to acquire light scatter signals, dual parameter DNA/RNA histograms, along with single parameter fluorescence distributions of each fluorescent signal. Doublet discrimination should be performed if possible. A possible acquisition protocol is depicted in **Fig. 1**.

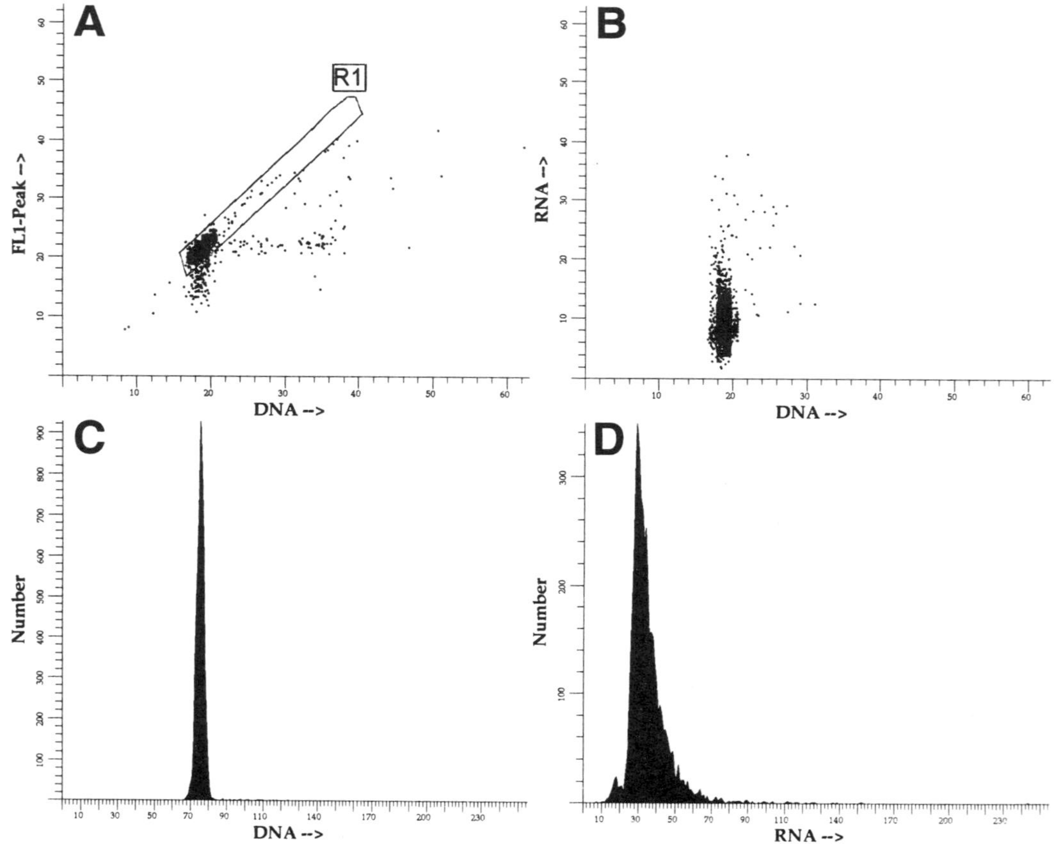

Fig. 1. A typical acquisition protocol displaying a doublet discrimination histogram (peak vs integral fluorescence) (**A**), a two-parameter DNA/RNA display (**B**), and one-parameter DNA and RNA histograms **(C** and **D**, respectively).

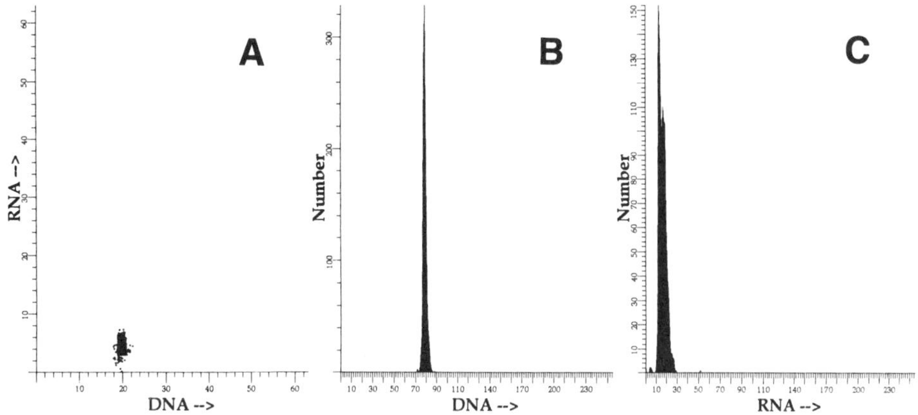

Fig. 2. Two-parameter DNA/RNA (**A**) and one-parameter DNA and RNA histograms (**B** and **C**, respectively) of NPBL stained with AO.

3.2.2. Controls

Normal peripheral blood lymphocytes (NPBL) are used as a biological control to assess the quality of AO staining and the consistency of the diploid G0/1 mean channels (**Fig. 2**). NPBL may also be used to establish an acceptable range of RNA mean channel values (*see* **Note 6**). Lymphocytes should be harvested after ficoll-hypaque gradient separation and cell count adjusted to 1×10^6 cells/mL. The lymphocyte cell suspension should be stained as outlined above and run concurrently with each set of specimens analyzed.

NPBL can also be used to determine DNA ploidy, especially in tumors with near-diploid peaks (*see* **Notes 7** and **8**). Equal parts (i.e., 100 µL each) of NPBL and tumor cell suspension are mixed together and stained with acridine orange (**Fig. 3**).

3.2.3. Computations

DNA indices and cell cycle data can be obtained using commercial software programs or manually. For manual analysis, cursors may be placed around G0/G1 peaks and S (+G2M) areas as illustrated in **Fig. 4**. DNA indices are calculated by dividing the mean channel value (X-axis) for the G0/1 aneuploid peak by the mean channel value for the intrinsic diploid G0/1 peak. By definition, diploid DNA index is equal to 1.00 ***(14)***. Cell cycle percentages may be calculated using the actual number of cells included within the cursor boundaries (area). RNA indices are calculated by dividing the mean G0/1 RNA fluorescence of the test sample by the mean G0/1 RNA fluorescence for NPBL.

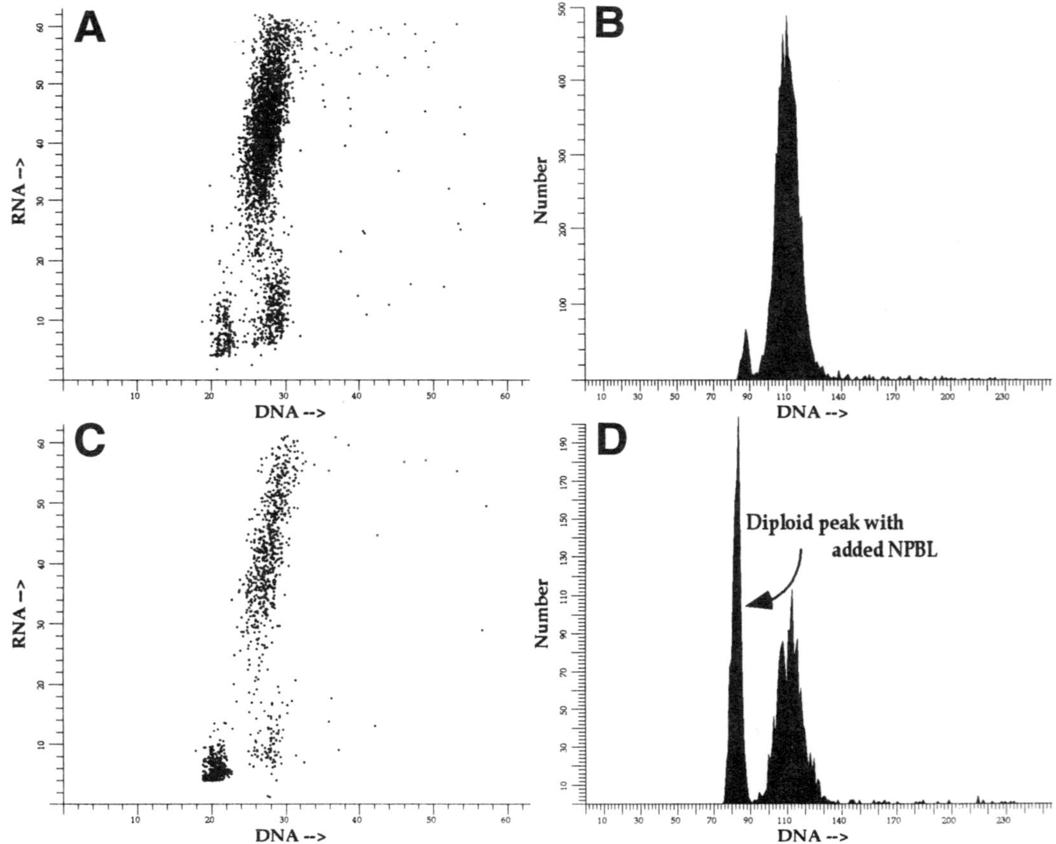

Fig. 3. Two parameter (**A**) DNA/RNA and one-parameter (**B**) DNA histograms of a tumor specimen before mixing with NPBL. (**C**, **D**) AO analysis after mixing equal portions of the tumor cell suspension with NPBL. Note the increase in cell number for the diploid G0/G1 peak in (D).

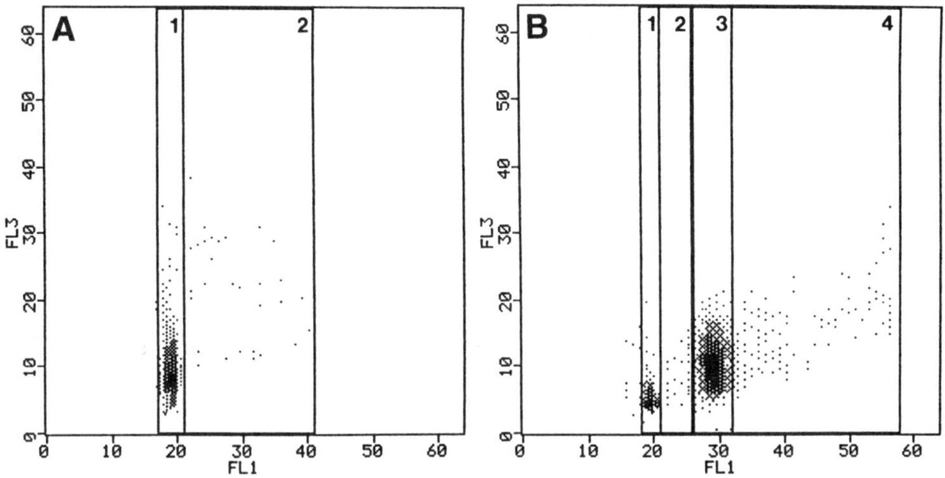

Fig. 4. DNA diploid histogram (**A**) with cursors placed around the diploid G0/1 peak (1) and the S + G2M (2). Panel (**B**) shows cursor placement for a DNA aneuploid tumor with diploid and aneuploid G0/1 peaks (1 and 3, respectively) and the diploid S (2) and aneuploid S + G2M (4).

3.2.4. Interpretation

Interpretation of simultaneous DNA/RNA measurements depends on the quality of specimen preparation and staining conditions (*see* **Note 9**). Assuming that technical aspects and quality control issues are well executed, consistent, and reproducible high-quality histograms are readily achieved.

Various applications of this technique have included: differentiation of leukemic subtypes ***(15–19)***, definition of cell cycle subsets ***(20,21)***, and the prognostic assessment of hematologic malignancies and solid tumors ***(22–31)***.

3.2.4.1. Lymphoreticular Malignancies

Cells from leukemias and the leukemic phase of lymphoma are typically uniform and do not require disaggregation, thus retaining an intact cell membrane. Therefore, RNA values can reliably be determined. Analysis of bone marrow specimens, however, varies depending on the extent of neoplastic involvement and residual normal marrow elements. RNA values and the histogram profile of normal bone marrow aspirates should be used to guide interpretation of data generated from these specimens. Examples of AO staining in different hematologic malignancies are displayed in **Figs. 5** and **6**.

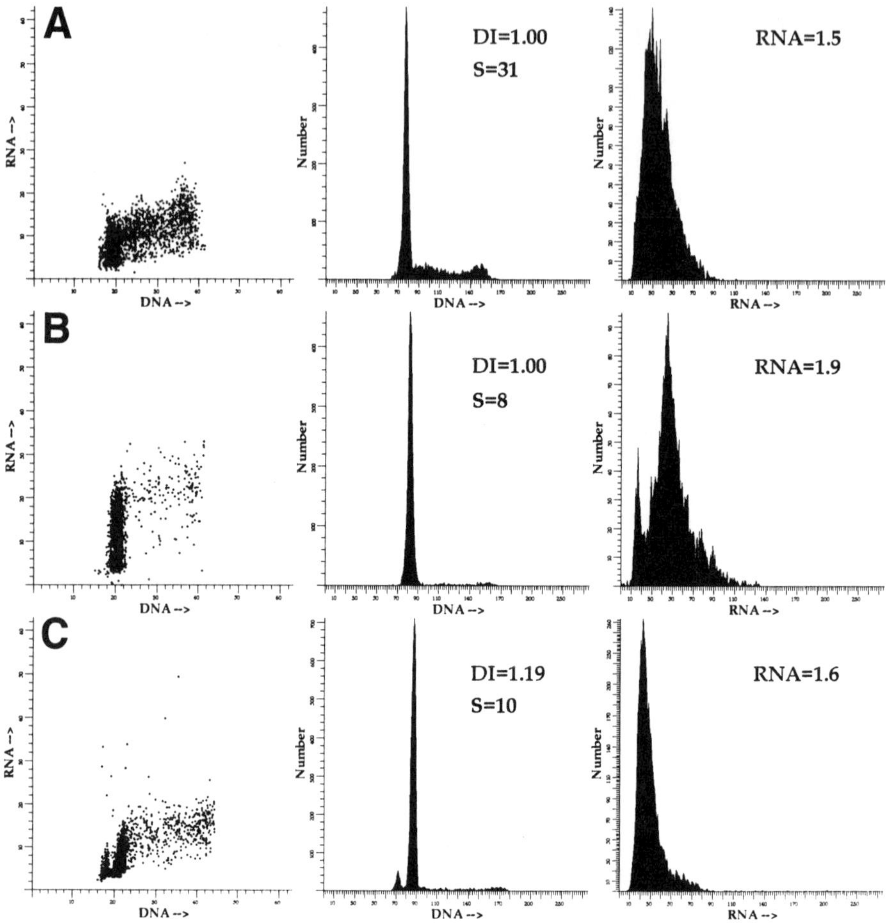

Fig. 5. (**A–C**) Typical acute lymphocytic leukemia patterns (left: DNA/RNA; middle: one-parameter DNA histogram; right: one-parameter RNA distribution). The specimen in (A) shows a high proliferative fraction and low RNA content. (B) shows an increasing RNA content with lower proliferative activity; (C) illustrates a classic DNA aneuploid ALL pattern.

3.2.4.2. Solid Tumors

AO analysis of solid neoplasms are more complicated because of tumor cell heterogeneity, cell membrane disruption, and contamination by host normal cells. Evaluation of Wright-Giemsa stained cytospin preparations is of vital importance in specimen qualification and interpreting these results (*see* **Note 10**). If sufficient intact tumor cells are present, histogram interpretation can be

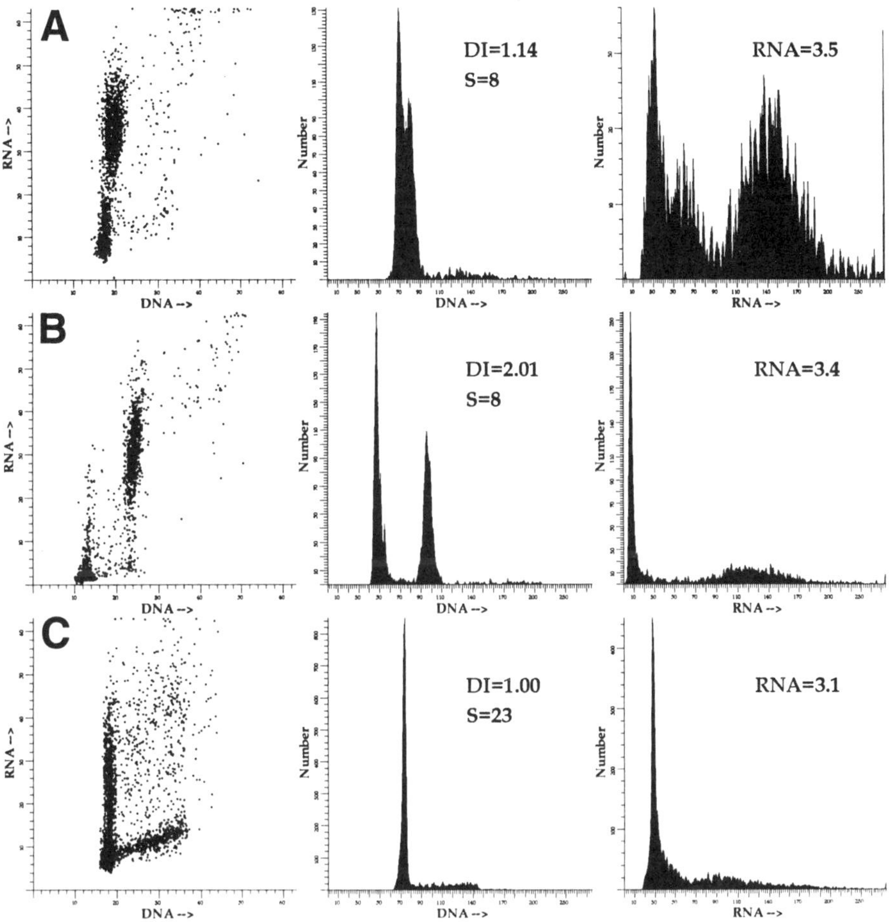

Fig. 6. (**A**, **B**) Two multiple myeloma distributions with high RNA levels and different DNA indices. (**C**) Typical AML pattern, with high-trailing RNA and high proliferation (left: DNA/RNA; middle: one-parameter DNA histogram; right: one-parameter RNA distribution).

reliably analyzed (**Fig. 7**). In general, at least 70% of the cells in a given specimen should contain neoplastic elements and, of these, 60% should be intact. RNA indices should be reported for the diploid stemline in DNA diploid histograms and the aneuploid stemline for DNA aneuploid histograms.

4. Notes

1. The purity of AO is critical to the successful implementation of this procedure. We recommend AO from Polysciences (Warrington, PA).

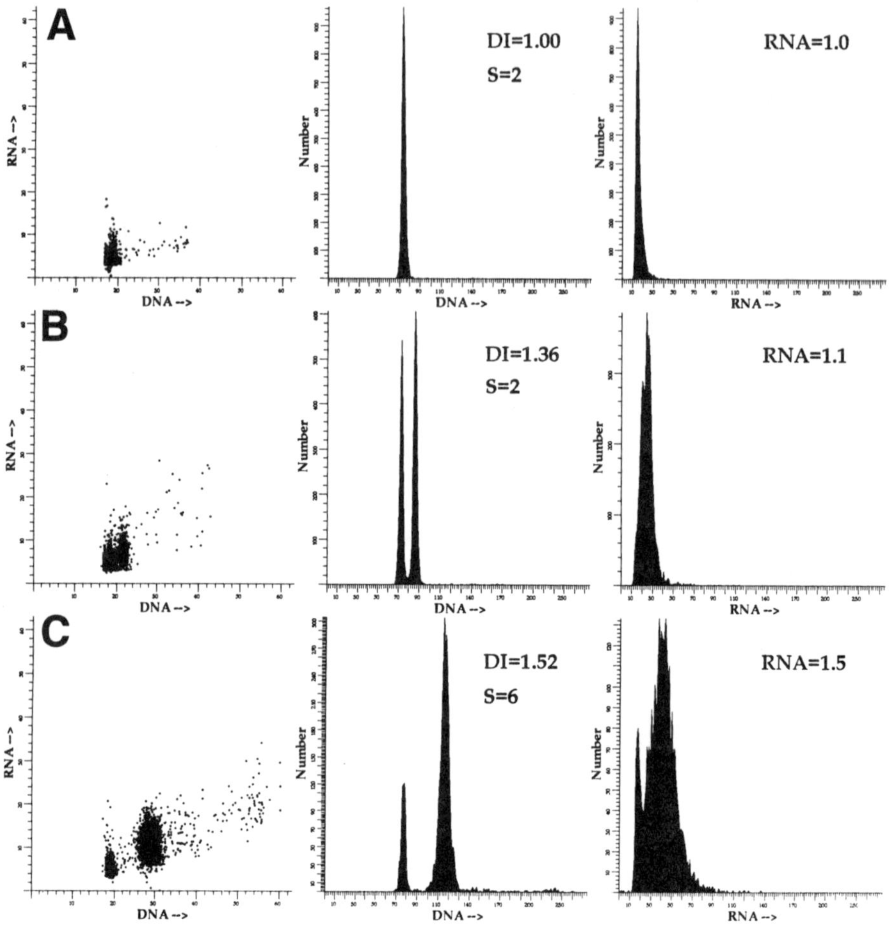

Fig. 7. (**A–C**) AO analysis of solid tumor specimens. (A) shows a lymph node biopsy with low proliferation and RNA content. In (B), a fine-needle aspirate of a retroperitoneal tumor with low degree DNA aneuploidy and relatively low RNA content is shown. (C) shows a DNA aneuploid lung carcinoma (left: DNA/RNA; middle: one-parameter DNA histogram; right: one-parameter RNA distribution).

2. It may be necessary to adjust the AO concentration to achieve optimum stainability for different instruments in a given setting ***(1)***. During the initial adaptation of this procedure, trials should be performed to determine the optimal concentration for the instrument in use.
3. Solution A is used to permeabilize cells and remove histones and acid-soluble proteins. Incubate cell suspensions in this solution at 0–4°C (**Subheading 3.1., step 4.**). Vortex gently, as vigorous agitation of cell suspension in solution A may disrupt cell membranes.

4. Cleaning flow cytometers is essential after using AO. This is further stressed if other applications are to follow. Regular cleansing with bleach followed by water should alleviate any AO carry over. If problems persist after cleaning, tubing can be changed between protocols.
5. Paired samples treated with and without RNase and DNase should be stained with AO ***(1)***. The purpose is twofold; to assess the specificity of staining and determine the amount of spectral overlap. Acquire DNase-treated preparations to ensure AO specificity for RNA (red fluorescence). Likewise, RNase-treated preparations should be acquired to collect DNA (green) fluorescence. Acquire consecutive, independent samples and lymphocyte preparations with and without RNase. These data are important in determining the extent of spectral overlap to adjust compensation accordingly.
6. A series of lymphocyte preparations should be used to determine an acceptable range of RNA values only after spectral overlap has been accounted for (**Note 5**). Acquire NPBL with each group of specimens and record the mean RNA fluorescence.
7. Cell concentration is critical, particularly for lymphocyte and tumor mixtures. Equal portions of each are necessary to accurately classify near-diploid populations as hypodiploid (DNA index <1.00) or hyperdiploid (DNA index >1.00).
8. Chromatin condensation differences in solid tumor cells and NPBL may lead to incorrect histogram classification. Epithelial-derived tumor cells may allow more accessibility to the dye, therefore fluorescing brighter than NPBL.
9. Debris or RBC contamination appear on the left of the diploid population on histograms and may interfere with the detection of a hypodiploid peak and/or cause inaccurate S-phase determination. Cell clumping may result in an artificial elevation of S and G2/M phase values, but the use of commercial programs with debris subtraction and aggregate modeling may overcome these problems.
10. Specimens with more than 30% nonneoplastic elements should be eliminated. If lymphocytes are predominant, gate on light scatter and reacquire DNA/RNA fluorescence. Should more than 40% bare nuclei be present, the specimen should be deemed inappropriate for RNA evaluation. Specimens should be used only for DNA ploidy and cell-cycle analysis.

References

1. Darzynkiewicz, Z. (1990) Differential staining of DNA and RNA in intact cells and isolated cell nuclei with acridine orange, in *Methods in Cell Biology,* vol. 33 (Darzynkiewicz, Z. and Crissman, H. A., eds.), Academic, San Diego, CA, pp. 285–298.
2. Myc, A., Traganos, F., Lara, J., Melamed, M. R., and Darzynkiewicz, Z. (1992) DNA stainability in aneuploid breast tumors: comparison of four DNA fluorochromes differing in binding properties. *Cytometry* **13,** 389–394.
3. Traganos, F. and Darzynkiewicz, Z. (1994) Lysosomal proton pump activity: supravital cell staining with acridine orange differentiates leukocyte subpopulations, in *Methods in Cell Biology,* vol. 41 (Darzynkiewicz, Z., Robinson, J. P., and Crissman, H. A., eds.), Academic, San Diego, CA, pp. 185–194.

4. Darzynkiewicz, Z. (1994) Simultaneous analysis of cellular DNA and RNA content, in *Methods in Cell Biology,* vol. 41 (Darzynkiewicz, Z., Robinson, J. P., and Crissman, H. A., eds.), Academic, San Diego, CA, pp. 185–194.
5. Larson, A. M., Dougherty, M. J., Nowowielski, D. J., Welch, D. F., Matar, G. M., Swaminathan, B., and Coyie, M. B. (1994) Detection of *Bartonella (Rochalimaea) quintana* by routine acridine orange staining of broth blood cultures. *J. Clin. Microbiol.* **32,** 1492–1496.
6. Preisler, H. D., Raza, A., Gopal, V., Banavali, S. D., Bokhari, J., and Lampkin, B. (1994) The study of acute leukemia cells by means of acridine orange staining and flow cytometry. *Leukemia & Lymphoma* **13,** 61–73.
7. Frey, T. (1995) Nucleic acid dyes for detection of apoptosis in live cells. *Cytometry* **21,** 265–274.
8. Smithwick, R. W., Bigbie, M. R., Ferguson, R. B., Kartix, M. A., and Wallis, C. K. (1995) Phenolic acridine orange fluorescent stain for mycobacteria. *J. Clin. Microbiol.* **33,** 2763, 2764.
9. Lopez-Roman, A. and Armengol, J. A. (1995) A fast and easy fluorescent counterstaining method for neuroanatomical studies by using acridine orange. *J. Neurosci. Methods* **60,** 39–42.
10. Gonzalez, K., McVey, S., Cunnick, J., Udovichenko, I. P., and Takemoto, D. J. (1995) Acridine orange differential staining of total DNA and RNA in normal and galactosemic lens epithelial cells in culture using flow cytometry. *Curr. Eye Res.* **14,** 269–273.
11. Grunwald, D. (1993) Flow cytometry and RNA studies. *Biol. Cell* **78,** 27–30.
12. Darzynkiewicz, Z. and Kapuscinski, J. (1990) Acridine orange: a versatile probe of nucleic acids and other cell constituents, in *Flow Cytometry and Sorting,* 2nd ed. (Melamed, M. R., Lindmo, T., and Mendelsohn, M. L., eds.), Wiley-Liss, New York, pp. 291–314.
13. Traganos, F., Darzynkiewicz, Z., Sharpless, T., and Melamed, M. R. (1977) Simultaneous staining of ribonucleic and deoxyribonucleic acids in unfixed cells using acridine orange in a flow cytofluorometric system. *J. Histochem. Cytochem.* **25,** 46–56.
14. Hiddemann, W., Schumann, J., Andreeff, M., Bariogie, B., Herman, C. J., Leif, R. C., Mayall, B. H., Murphy, R. F., and Sandberg, A. A. (1984) Convention on nomenclature for DNA cytometry. *Cytometry* **5,** 445, 446.
15. Andreef, M., Darzynkiewicz, Z., Sharpless, T. K., Clarkson, B. D., and Melamed, M. R. (1980) Discrimination of human leukemic subtypes by flow cytometric analysis of cellular DNA and RNA. *Blood* **55,** 282–293.
16. Barlogie, B., Maddox, A. M., Johnston, D. A., Raber, M. N., Drewinko, B., Keating, M. J., and Freireich, E. J. (1983) Quantitative cytology in leukemia research. *Blood Cells* **9,** 35–55.
17. Barlogie, B., McLaughlin, P., and Alexanian, R. (1987) Characterization of hematologic malignancies by flow cytometry. *Anal. Quant. Cytol. Histol.* **9,** 147–155.
18. Andreeff, M. (1990) Flow cytometry of leukemia, in *Flow Cytometry and Sorting,* 2nd ed. (Melamed, M. R., Lindmo, T., and Mendelsohn, M. L., eds.), Wiley-Liss, New York, pp. 697–724.

19. Andreef, M. (1990) Flow cytometry of lymphoma, in *Flow Cytometry and Sorting,* 2nd ed. (Melamed, M. R., Lindmo, T., and Mendelsohn, M. L., eds.), Wiley-Liss, New York, pp. 725–743.
20. Darzynkiewicz, Z., Sharpless, T., Stalano-Coico, L., and Melamed, M. R. (1980) Subcompartments of the G1 phase of the cell cycle detected by flow cytometry. *Proc. Natl. Acad. Sci. USA* **77,** 6696–6700.
21. Darzynkiewicz, Z., Traganos, F., and Melamed, M. R. (1980) New cell cycle compartments identified by multiparameter flow cytometry. *Cytometry* **1,** 98–108.
22. Tyrer, H. W., Golden, J. F., Vansickel, M. H., Echols, C. K., Frost, J. K., West, S. S., Pressman, N. J., Allbright, C. D., Adams, L. A., and Gill, G. W. (1979) Automatic cell identification and enrichment in lung cancer. II. Acridine orange for cell sorting of sputum. *J. Histochem. Cytochem.* **27,** 552–556.
23. Collste, L. G., Darzynkiewicz, Z., Traganos, F., Sharpless, T., Sogani, P., Grabstald, H., Whitmore, W. F., and Melamed, M. R. (1980) Flow cytometry in cancer detection and evaluation using acridine orange metachromatic nucleic acid staining of irrigation cytology specimens. *J. Urol.* **123,** 478–485.
24. Barlogie, B., Alexanian, R., Gehan, E. A., Smallwood, L., Smith, T., and Drewinko, B. (1983) Marrow cytometry and prognosis in myeloma. *J. Clin. Invest.* **72,** 853–861.
25. Barlogie, B., Alexanian, R., Dixon, D., Smith, L., Smallwood, L., Delasalle, K. (1985) Prognostic implications of tumor cell DNA and RNA content in multiple myeloma. *Blood* **66,** 338–341.
26. Srigiey, J., Barlogie, B., Butler, J. J., Osborne, B., Thick, M., Johnston, D., Kantarlian, H., Reuben, J., Batsakis, J., and Freireich, E. (1985) Heterogeneity of non-Hodgkin's lymphoma probed by nucleic acid cytometry. *Blood* **65,** 1090–1096.
27. Andreef, M., Hansen, H., Cirrincione, C., Filippa, D., and Thaler, H. (1986) Prognostic value of DNA/RNA flow cytometry on B-cell non-Hodgkin's lymphoma: development of laboratory model and correlation with four taxonomic systems. *Ann. NY Acad. Sci.* **486,** 368–386.
28. El-Naggar, A. K., Batsakis, J. G., Teague, K., Giacco, G., Guinee, V. F., and Swanson, D. (1990) Acridine orange flow cytometric analysis of renal cell carcinoma. *Am. J. Pathol.* **137,** 275–280.
29. Enker, W. E ., Kimmel, M., Cibas, E. S., Cranor M. L., and Melamed, M. R. (1991) DNA/RNA content and proliferative fractions of colorectal carcinomas: a five year prospective study relating flow cytometry to survival. *J. Natl. Cancer Inst.* **83,** 701–707.
30. El-Naggar, A. K., Barlogie, B., McCabe, K., Teague, K., Ensign, L. G., and Pollock, R. E. (1994) Acridine orange DNA/RNA content analysis of soft-tissue tumors: correlation with clinicopathologic factors and biological behavior. *C.M.B.* **1,** 237–247.
31. El-Naggar, A. K., Kemp, B. L., Sneige, N., Hurr, K. G., Steck, K., Tu, Z. N., Fritsche, H. A., Singietary, S. E., and Balch, C. M. (1996) Bivariate RNA and DNA content analysis in breast carcinoma: biological significance of RNA content. *Clin. Cancer Res.* **12,** 419–426.

18

Solid Tumor DNA Ploidy Analysis

Kim Steck and Adel El-Naggar

1. Introduction

Flow cytometric DNA content analysis of human neoplasia provides quantitative information on DNA ploidy, clonal DNA heterogeneity, and proliferative activity. These submicroscopic features are important for the biological and clinical evaluation of tumors as demonstrated in clinical studies of a wide spectrum of human lymphoreticular and solid malignancies *(1–24)*. Although valuable, such information should not be used in isolation from other conventional diagnostic and biological parameters in the risk assessment of cancer patients.

Flow cytometry is an excellent technique in the analysis of DNA content and other cellular characteristics for clinical application. This, however, depends to a great extent on sample quality, careful preparation, instrument precision and accuracy, as well as technical expertise. Fresh and archived tissues, fine needle aspirations, and cells from body fluids can be used as a source of cells for the analysis. Current flow cytometers are highly sophisticated and can be equipped with user-friendly software for analysis. Sample integrity, tissue processing, and quality of cell preparation are central to the interpretation of results. A key to successful and accurate DNA flow cytometric analysis is sample representation of target cells. This can be routinely accomplished through the light microscopic evaluation of a cytospin of the cell suspension prior to the analysis. Criteria for acceptance and rejection of samples should be clearly defined and adhered to rigorously. If these issues are systematically observed, reasonably consistent results can be achieved.

From: *Methods in Molecular Biology, Vol. 91: Flow Cytometry Protocols*
Edited by: M. J. Jaroszeski and R. Heller © Humana Press Inc., Totowa, NJ

2. Materials

2.1. Fresh Solid Tumor Preparation

1. 0.5 mg/mL Propidium iodide (PI): PI is a potential carcinogen and should be handled carefully. Measure 50 mg of PI (Sigma, St. Louis, MO) into a clean glass beaker. Add 100 mL of distilled H_2O and stir until dissolved (approx 2 h). Filter through #1 Whatman filter paper. Store protected from light at 4°C. Stable for 1 yr.
2. 1 mg/mL Ribonuclease A (RNase): Thaw a 100-mg vial of RNase-A (Worthington Biochemical, Warrington, PA). Bring to a final volume of 100 mL by adding approx 94.63 mL of phosphate-buffered saline (PBS). Mix well. Aliquot 5 mL portions into 20 tubes and freeze at –20°C until needed. Aliquots are stable for 6 mo at –20°C. Thawed aliquots may be kept at 4°C for up to 2 wk. Concentrates may be stored at –20°C for 3 yr.
3. PBS: Dilute 10X PBS without calcium and magnesium salts (Irvine Scientific, Santa Ana, CA) to a 1X solution. Adjust to pH 7.2–7.4. Bulk preparations may be stored at room temperature up to 3 mo. Store working aliquots at 4°C.
4. 0.25% Trypsin: Dilute 1 mL stock trypsin (2.5%) (Sigma) in 9 mL of distilled H_2O. Store at 4°C for 1 wk.

2.2. Paraffin-Embedded Tissue Preparation

These solutions should be prepared in addition to the solutions listed above (with the exception of trypsin).

1. 0.5% Pepsin: Add 0.5 g pepsin (Sigma) to 95 mL of distilled H_2O. Adjust to pH 1.5 using 1 *N* HCl. Prepare fresh for each batch of specimens.
2. 0.05 mg/mL Pepstatin A: Empty a 5-mg vial of pepstatin A (97% pure) (Sigma) into a 150-mL beaker. Rinse the vial with 5 mL of 100% ethanol and add to the pepstatin A. Place several pieces of gauze on top of a heated stir plate and place beaker (with stir bar) on top. Cover beaker with aluminum foil to prevent evaporation. Adjust the mixer speed to a moderate setting and turn heat to the lowest setting. Do not allow solution to become warmer than tepid. Stir until completely dissolved and add 95 mL of distilled water. Stable at 4°C for up to 1 mo.
3. Histo-Solv X or another xylene substitute: (Curtin Matheson Scientific, Houston, TX) Store as directed on label.

3. Methods

3.1. Sample Preparation: Fresh Solid Tumors

1. Transfer the tissue into a 60 × 15 mm Petri dish. Add 2 mL of fresh cell culture media or PBS.
2. Mince the tissue into fine sections using a no. 21 surgical blade. Sieve the tissue and solution through a 500 μm stainless steel mesh (*see* **Note 1**).
3. Transfer the cell suspension to a new tube by filtering solution through a 35 μm nylon mesh (Small Parts, Miami Lakes, FL).

4. Add 3 mL of PBS and pipe to mix.
5. Spin cells in a table-top centrifuge at 500*g* for 5 min. Decant the supernatant.
6. Resuspend cells with 3 mL of PBS and mix well. Repeat **step 5**.
7. Resuspend cells in 1 mL of PBS and filter through a 35 μm nylon mesh (*see* **Note 2**).
8. Perform a cell count. Use 3×10^5 cells to prepare a cytospin.
9. Spin the remainder of the cell suspension for 5 min at 500*g*. Decant the supernatant.
10. Resuspend the cells gently with 1 mL of 0.9% saline.
11. Add 2.5 mL of cold 95% ethanol while vortexing at slow speed. Place samples at 0–4°C for at least 1 h or until ready to perform flow cytometric analysis.
12. Prepare two tubes for each specimen as follows:
    Tube 1: Patient cells.
    Tube 2: 1:1 mixture of patient cells and normal lymphocytes (*see* **Note 3**).
13. Aliquot a total of 1×10^6 70% ethanol-fixed cells into appropriate tubes.
14. Add 3 mL PBS, spin at 500*g* for 5 min. Decant supernatant. Repeat once.
15. Resuspend cells in 800 μL of PBS.
16. Add 70 μL RNase and 50 μL propidium iodide. Incubate in a 37°C water bath for 20 min.
17. Filter samples through 35 μm nylon mesh and place on ice.

3.2. Sample Preparation: Paraffin-Embedded Tissue

3.2.1. Initial Preparation

1. Label a 16 × 100 mm borosilicate culture tube and two 12 × 75 mm plastic culture tubes per block. Tape identification label to the upper circumference of tubes.
2. Cut a minimum of three 50 μm sections from each paraffin block. Blocks should be at room temperature (*see* **Note 5**).
3. Lift sections from the edge of the microtome knife with iris forceps and place into the bottom of the 16 × 100 mm tube. Handle sections carefully to avoid shattering
4. Chill the paraffin blocks on ice and cut 3 μm sections for H and E.
5. Tubes may be sealed and stored at 22°C until processing.

3.2.2. Day 1

Fill tubes with enough of the following solutions to cover tissue sections and incubate as directed (*see* **Notes 6** and **7**). At the end of each incubation period, remove as much solution as possible by vacuum aspiration, leaving tissue sections intact, at the bottom of the tube.

1. Histo-Solv X or another xylene substitute, 3 times for 15 min each.
2. Absolute ethyl alcohol, 2 times for 10 min each.
3. 95% EtOH, 2 times for 10 min each.
4. 80% EtOH, 1 time for 10 min.
5. 70% EtOH, 1 time for 10 min.
6. Add 50% EtOH, cover tubes, and leave overnight.

3.2.3. Day 2

1. Prepare pepsin as outlined in **Subheading 2.2, step 1** (*see* **Note 8**). Place in a 37°C water bath 30 min prior to use.
2. Remove 50% EtOH as above.
3. Rehydrate tissue section by resuspending in PBS for 10 min. Centrifuge at 500*g* for 5 min. Repeat once.
4. Remove PBS and add 2.5–3 mL prewarmed pepsin.
5. Place tubes in a 37°C water bath for 20–45 min (*see* **Note 9**).
6. Using 18-g needles and 3-cc syringes, aspirate the tissue sections with pepsin and dispel forcefully 1–2 times (*see* **Note 10**).
7. Add 175 µL of pepstatin A to the tube and agitate to mix. Place tubes on ice.
8. After all sections have been disaggregated spin at 500*g* for 5 min.
9. Resuspend in 2–3 mL PBS and spin. Remove supernatant and repeat.
10. Filter through a 35 µm nylon mesh into a clean 12 × 75-mm plastic tube.
11. Perform a cell count and adjust cell concentration to 1 × 10^6 cells/mL.
12. Transfer 900 µL of the cell suspension to the second 12 × 75-mm plastic tube, and add 100 µL of prewarmed RNase. Incubate for 30 min in a 37°C water bath. Remaining cell suspension may be placed at –20°C for storage (*see* **Note 11**).
13. Add 50 µL of PI. Incubate for 30 min in refrigerator or on ice.

3.3. Analysis

3.3.1. Instrumentation and Acquisition

PI is excited at 488 nm with emission >610 nm. Instrument performance should be monitored according to manufacturer guidelines, with emphasis on obtaining consistent resolution and mean peak channels for calibration particles on a daily basis. Linearity of photomultiplier tubes (PMTs) should be tested monthly. Test protocols should be structured to acquire forward scatter and DNA peak and integral fluorescence signals. **Fig. 1** illustrates a typical acquisition protocol (*see* **Notes 4** and **12**).

3.3.2. Controls

For fresh solid tumor analysis, normal peripheral blood lymphocytes (NPBL) are used as a biological control to assess the quality of PI staining and the consistency of the diploid G0/1 mean channels. Also, by mixing equal parts of tumor cell suspensions and NPBL, ploidy can be assessed for tumor specimens (**Figs. 2** and **3**; *see* **Note 3**). Lymphocytes should be harvested after gradient separation and prepared as outlined in **Subheading 3.1., steps 4–11**.

3.3.3. Computations

DNA indices and cell cycle data may be derived using commercial software programs or manually. For manual analysis, cursors may be placed around G0/G1 peaks and S (+G2/M) areas. DNA indices are calculated by dividing the mean channel value (X-axis) for the G0/1 aneuploid peak by the mean

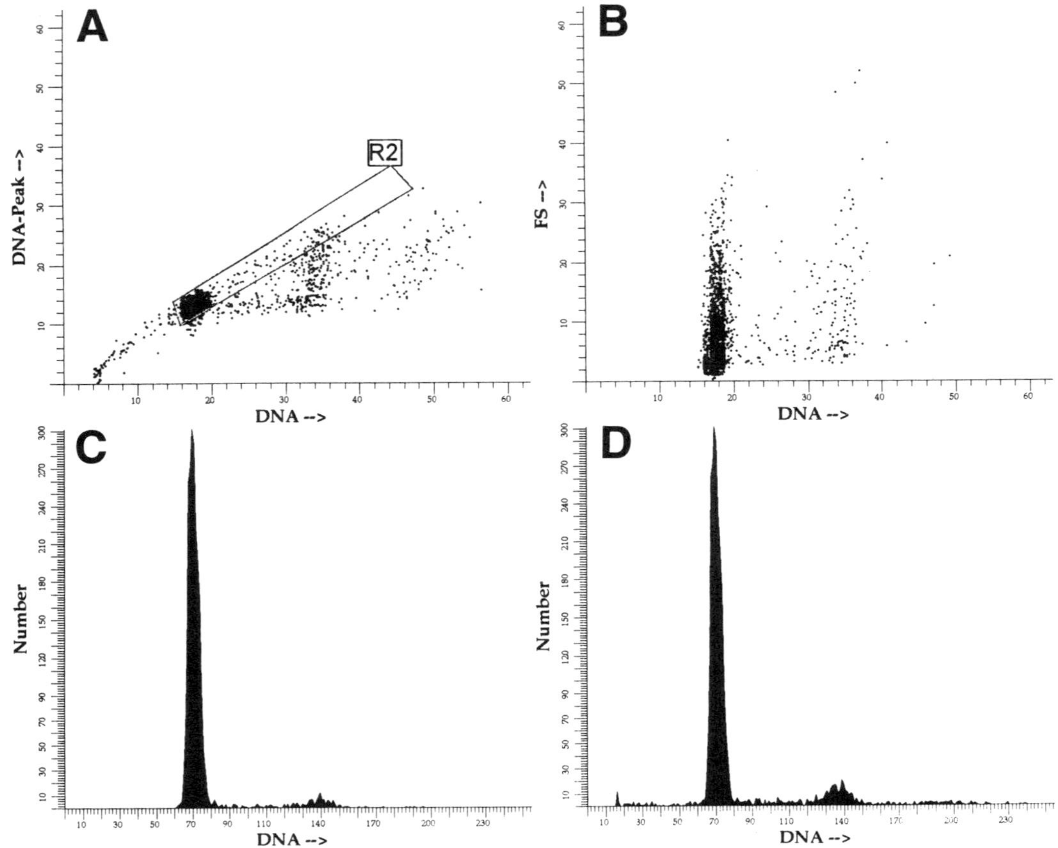

Fig. 1. A typical acquisition protocol with doublet discrimination (**A**), DNA vs forward light scatter (**B**), and gated and ungated DNA histograms (**C** and **D**, respectively).

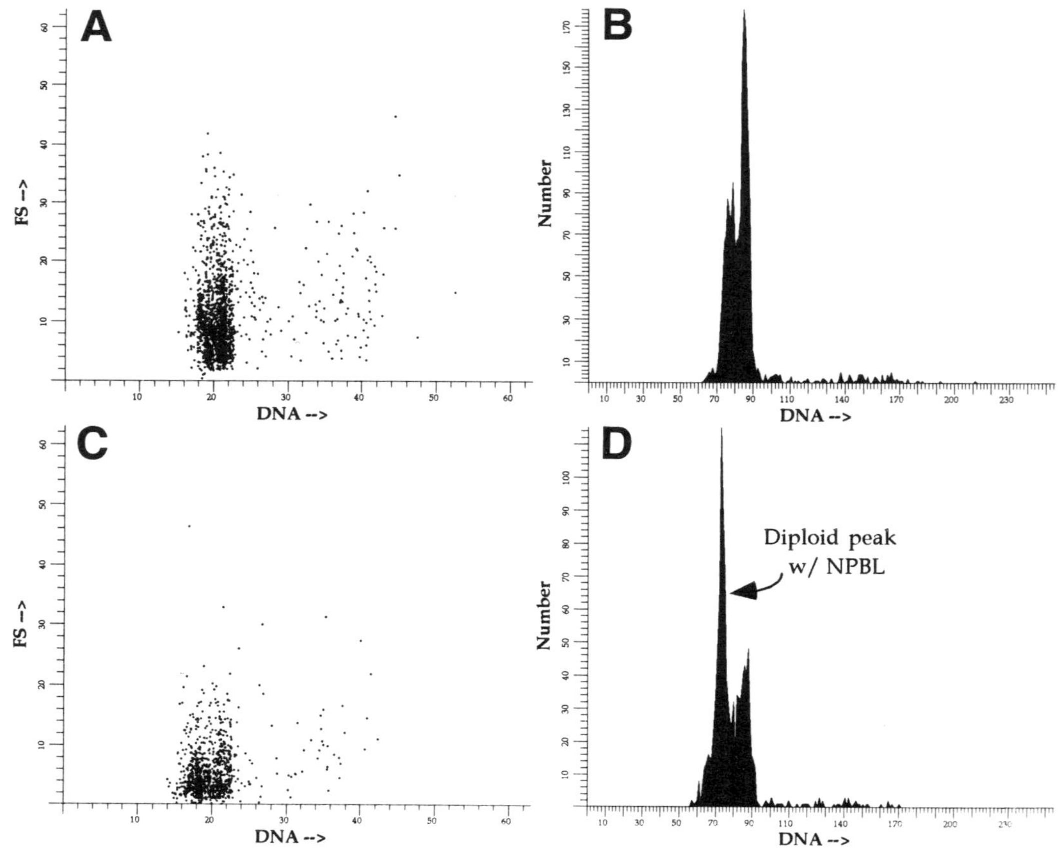

Fig. 2. (**A,B**) DNA distributions of a tumor with a near-diploid aneuploid peak. (**C,D**) DNA histograms of the tumor specimen mixed with an equal portion of NPBL. Note the increase in cell number for the diploid G0/G1 peak in (D).

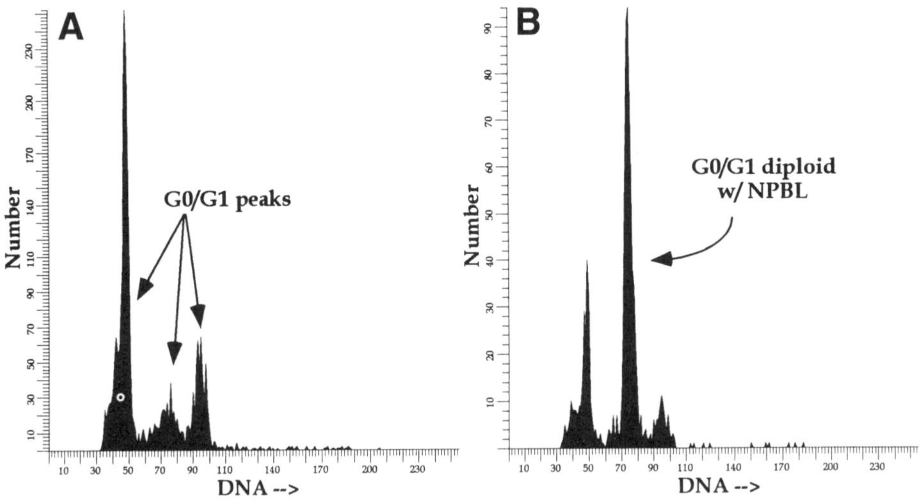

Fig. 3. Example of a DNA histogram with multiple G0/G1 peaks (**A**) and the NPBL mixture (**B**), that reveals DNA hypodiploid and DNA hyperdiploid peaks.

channel value for the intrinsic diploid G0/1 peak. By definition, diploid DNA index is equal to 1.00 *(3)*. Cell cycle percentages may be calculated using the actual number of cells (area) for the respective cell cycle compartment divided by the total number of cells analyzed.

3.3.4. Interpretation

3.3.4.1. DNA Ploidy

DNA ploidy analysis may be used along with other clinicopathologic characteristics to aid in the interpretation of borderline lesions and reactive vs neoplastic tissues. DNA aneuploidy should not be used as a sole criteria for malignancy, especially in fine needle aspiration materials. This is especially relevant to the interpretation of thyroid and other neuroendocrine tumors (*see* **Note 13**). Ploidy analysis of histologically malignant tumors may provide prognostic information as a guide in stratification of patients for management protocols.

Current flow cytometers can resolve a >5% DNA content difference between tumor cell populations. Therefore, minor numerical chromosomal changes may not be detected. Assuming proper instrument performance and a stable biological control (normal lymphocytes or normal tissue from the same individual) are obtained, histogram interpretation can be reproducibly achieved (**Fig. 4**). Generally, DNA ploidy is defined as DNA diploid, a single G0/G1 peak corresponding to the same channel of the biological control, and DNA aneuploid, a

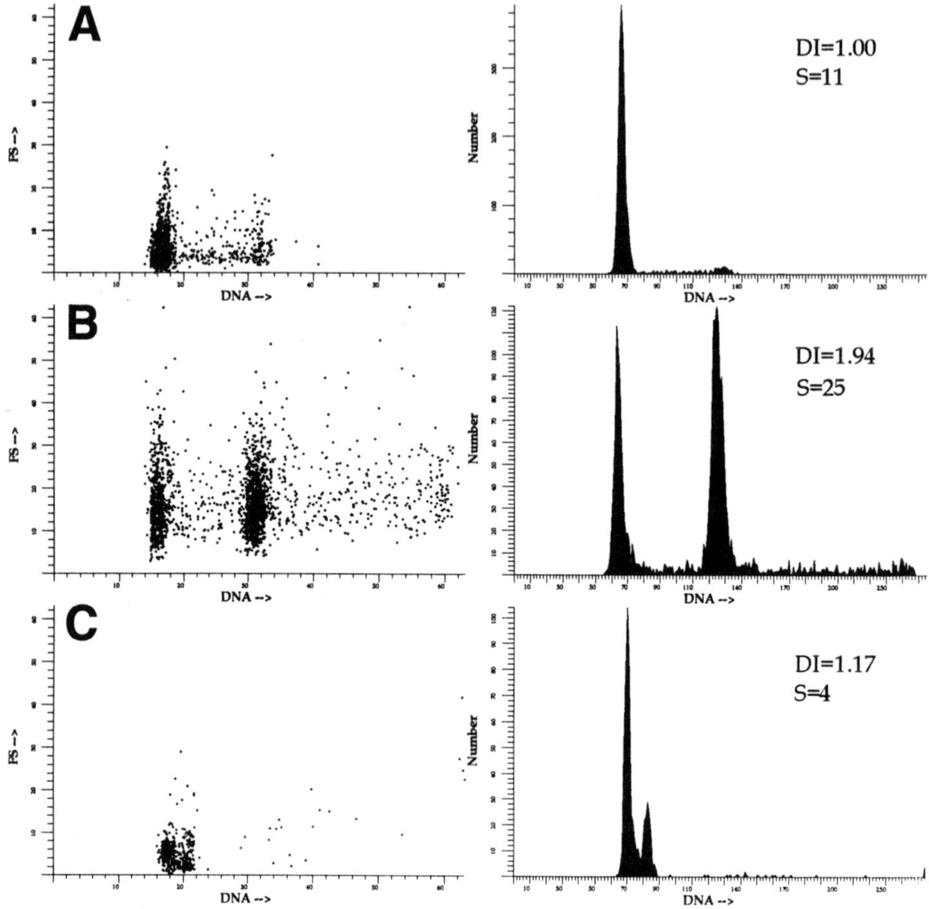

Fig. 4. DNA analyses of solid tumors. Panels show a DNA diploid tumor with intermediate proliferative activity (**A**), a DNA tetraploid tumor with high proliferative activity (**B**), and a DNA hyperdiploid tumor with low proliferative activity (**C**).

separate distinct G0/G1 peak from the diploid G0/G1 *(3)*. DNA aneuploidy may be further characterized as hypodiploid (DNA index <1.00) and hyperdiploid (DNA index >1.00). The tetraploid (DNA index = 2.00) category should be strictly defined. A narrow, well-defined peak of >10% of cells analyzed can be defined as a tetraploid peak. On the other hand, a peak with a Gaussian distribution, up to 20% of the total population, should be considered as G2/M (**Fig. 5**). Although defining the ploidy patterns of most histograms is readily achievable, classifying near-diploid aneuploidy into hypodiploid or hyperdiploid may be difficult. Generally, this is possible in the analysis of fresh tissues, where a mixture of normal lymphocytes and the sample can allow for the

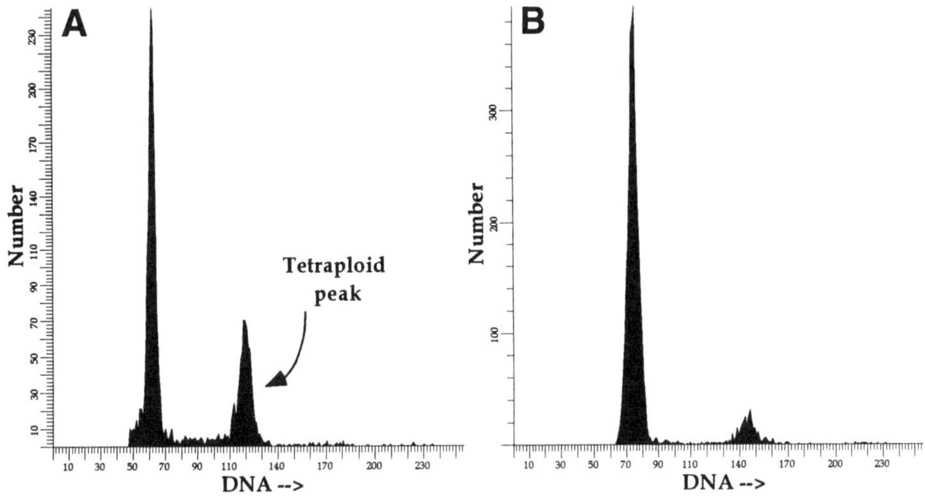

Fig. 5. DNA tetraploid (**A**) and diploid (**B**) histograms. Note the appearance of the G0/G1 tetraploid peak vs the diploid G2/M distribution.

definitive determination of DNA indices in the majority of cases (*see* **Note 14**). A perfect superimposition of the normal biologic control channel number and that of the sample is necessary for such analysis. This, however, is difficult to accomplish using paraffin-embedded tissue *(25–27)*.

A definable limit of acceptable CVs may be restrictive. Establishing a cut-off may lead to the exclusion of otherwise reasonable histograms that reflect the innate properties of different cell populations in certain solid tumors, such as cellular polymorphism. Interpretation should be based on the features of each individual histogram and the knowledge of the cellular composition after reviewing cytospins of the disaggregated sample (*see* **Notes 15–17**). Histograms with extensive debris, wide CV, and skewed ascending and descending limbs of the G0/G1 peaks should be considered unfit for analysis.

3.3.4.2. S-Phase (Proliferative Fraction)

Measurement of the proliferative fraction is assumed to reflect tumor growth rate and aggressiveness. The assessment of this parameter depends on the quality and the simplicity of the histogram analyzed. Diploid DNA and single-peak, aneuploid DNA histograms with minimal debris contamination and good CVs are suitable for S-phase analysis. The analysis of this parameter in multiploid, uneven G0/G1 (skewed) peaks, and large amounts of debris background may not be suitable for such analysis. It must be realized that cells within the S-phase boundaries may not represent cycling cells, but possibly include apoptotic, noncycling, or DNA aneuploid cells.

4. Notes

4.1. Technical Considerations for Fresh Solid Tissue Preparation

1. To improve recovery of intact cells, tumors may be minced without pressing through the stainless steel sieve. In general, this will also decrease the amount of cellular debris.
2. If the tumor sample is clumping after mechanical disaggregation, pellet cells, decant, and add 1 mL of diluted trypsin (0.25%). Incubate at room temperature for 15 min. Gently resuspend the cells with a Pasteur pipets to achieve a good suspension. Wash with PBS, perform a cell count and adjust to 1×10^6 cells/mL. Proceed with cell fixation as described in **Subheading 3.1., steps 9–11**.
3. Cell concentration is critical, particularly for lymphocyte and tumor mixtures. Equal portions of each are necessary to accurately determine ploidy for near-diploid populations.
4. PMT voltages may be adjusted during acquisition to accommodate the following possibilities. Specimen preparations and mixtures with normal peripheral blood lymphocyes (NPBL) may be acquired at higher PMT voltages to separate near-diploid G0/G1 peaks (**Fig. 6**). For acquisition of tumors with DNA hyperdiploid peaks, PMT voltages should be decreased to include all cycling cells on the histogram display (**Fig. 7**).

4.2. Technical Considerations for Paraffin-Embedded Tissue Preparation

5. For paraffin-embedded small core biopsies and tissues measuring <1 cm, more sections may be required to yield adequate nuclei for staining. Blocks should be sectioned at room temperature to prevent shattering *(28)*.
6. Overexposure to absolute alcohol will increase the likelihood of cell shattering. Strict adherence to incubation times is recommended.
7. Agitate specimens during the solvent and hydration steps to aid in unrolling tissue sections. Do not fill culture tubes more than necessary. Proper agitation of sections may be inhibited when tubes are more than half full of fluid.
8. Prepare only enough pepsin to treat the number of specimens being processed (2.5–3.0 mL/specimen).
9. Monitor pepsin incubation carefully to avoid overincubation. Tubes may be agitated by thumping with forefinger to aid in cell dispersal. Suspension will begin to cloud as cells are released. Tubes should be removed from the water bath at this point, syringed, and treated with pepstatin A to halt enzymatic action.
10. Expelling liquid forcefully from the syringe is usually sufficient to release cells. Minimize the shearing of nuclei by limiting the number of times sections are aspirated and dispelled. Observe the turbidity of the solution to monitor release of cells.
11. If cell yield permits, do a cytospin on each sample from diluted, filtered cells which have not been exposed to RNase.
12. Observe mean channel position of the diploid G0/1 peak during the initial phases of specimen acquisition. Restart the acquisition frequently until a stable mean channel is achieved.

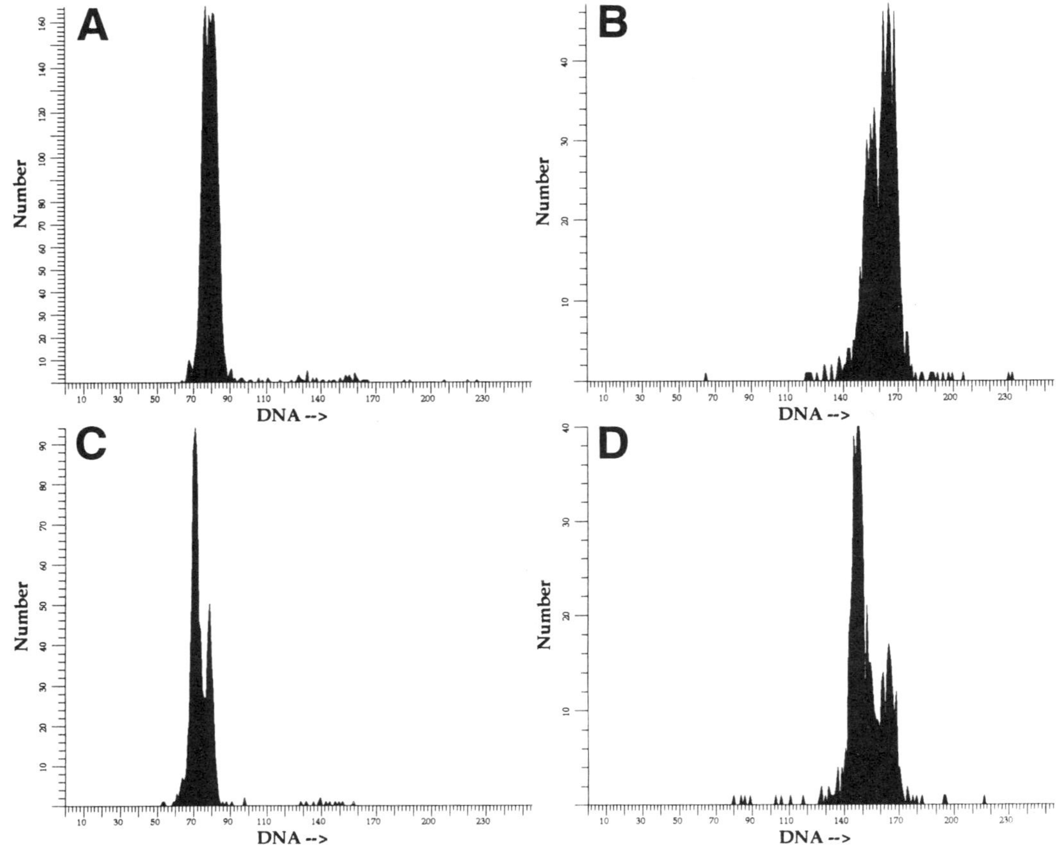

Fig. 6. Example of a high PMT voltage acquisition to separate near-diploid peaks. Histograms of the tumor specimen at the normal voltage (**A**), at higher voltage (**B**), and the mixture of tumor specimen and NPBL at normal voltage (**C**), and higher voltage (**D**). This combination of analyses reveals a hyperdiploid peak.

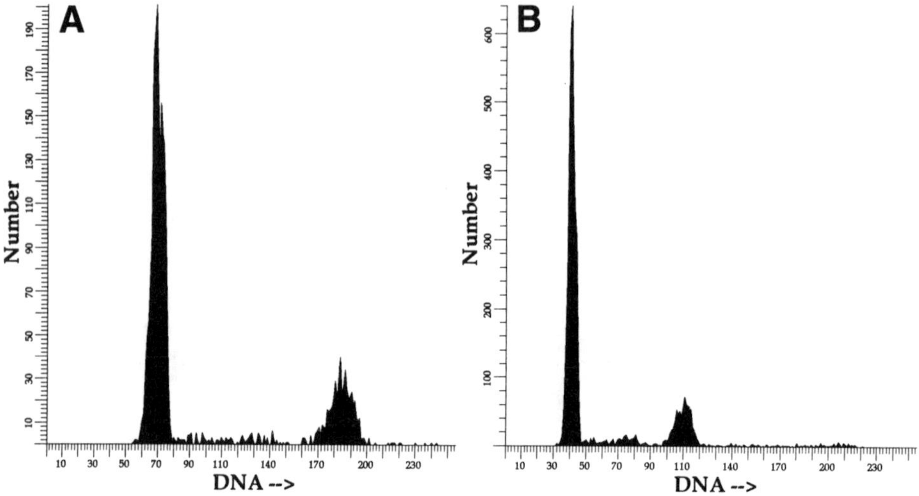

Fig. 7. Initial acquisition of a tumor specimen with a hypertetraploid G0/G1 peak (**A**). The tumor was reanalyzed at a lower PMT voltage (**B**) to include all of the DNA aneuploid cycling cells.

4.3. Interpretive Considerations

13. Occasionally, benign and reactive tumors have shown aneuploidy, by flow cytometry and *in situ* hybridization. DNA aneuploidy in these tumor types should not be interpreted as diagnostic of malignancy. Correlation with histologic evaluation is necessary.
14. Consider chromatin condensation when interpreting mixtures with NPBL. Epithelial-derived tumor cells may allow more accessibility to the dye, therefore fluorescing brighter than NPBL, and leading to classification of the histogram as aneuploid. Histograms should only be classified as aneuploid when two separate distinct peaks are discernable for the tumor sample (i.e., without lymphocytes added) ***(3)***.
15. Processing procedures require solid tumors to be dispersed by mechanical methods, therefore selective loss of neoplastic elements may occur in tumors that do not dissociate easily. Prepare cytospin preparations of all disaggregated solid tumors and paraffin-embedded tissues to assess cellular integrity and normal contamination. Cytospins may be stained with Wright/Giemsa.
16. Host cell contamination may markedly affect tumor cell representation ***(29, 30)***. Specimens with more than 30% normal cells should be deemed uninterpretable if a diploid histogram is obtained. This may be circumvented by acquiring DNA fluorescence of a tumor-enriched population, through cytokeratin gating or lymphocyte depletion (LCA staining) ***(31–35)***.
17. Cellular debris or RBC contamination may interfere with the detection of aneuploidy and/or cause inaccurate S-phase determination. Additionally, cell clumping may result in an artificial elevation of S- and G2/M-phase values. Commercial software programs for debris subtraction and aggregate modeling should alleviate this problem (**Fig. 8**).

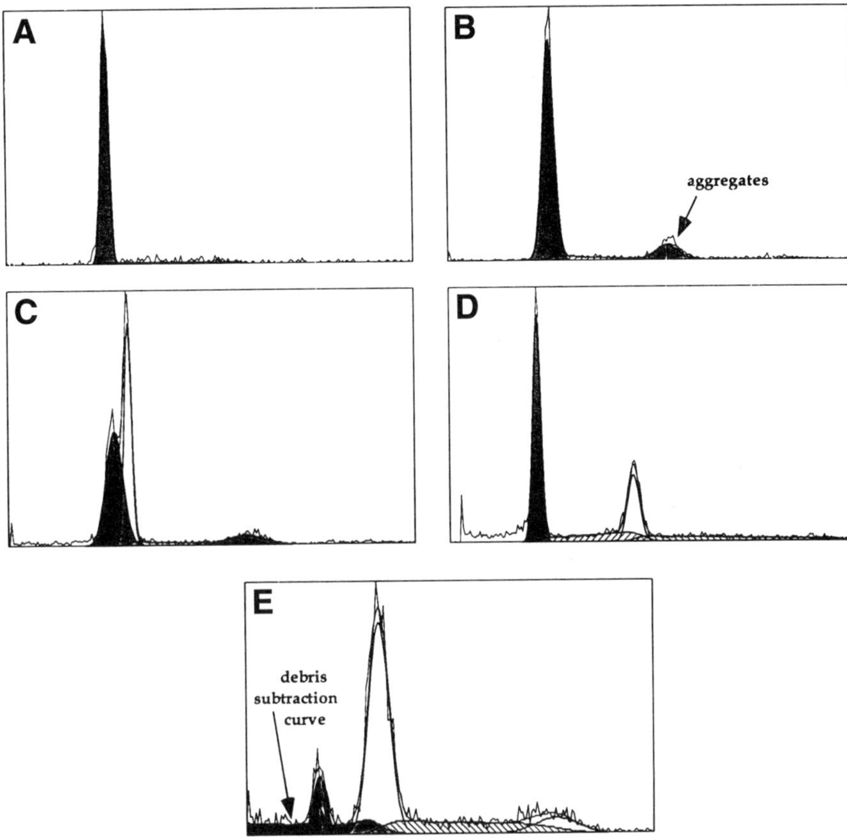

Fig. 8. Commercial software DNA distributions showing a DNA diploid tumor with low proliferative fraction (**A**), a DNA diploid tumor with aggregates (**B**), a DNA hyper-diploid tumor (**C**), a DNA tetraploid tumor (**D**), and a debris subtraction model for a DNA aneuploid tumor (**E**).

References

1. Barlogie, B., Göhde, W., Johnston, D. A., Smallwood, L., Schumann, J., Drewinko, B., and Freireich, E. J. (1978) Determination of ploidy and proliferative characteristics of human solid tumors by pulse cytophotometry. *Cancer Res.* **38,** 3333–3339.
2. Barlogie, B., Drewinko, B., Schumann, J., Göhde, W., Dosik, G., Latreille, J., Johnston, D. A., and Freireich, E. J. (1980) Cellular DNA content as a marker of neoplasia in man. *Am. J. Med.* **42,** 195–203.
3. Hiddemann, W., Schumann, J., Andreeff, M., Barlogie, B., Herman, C. J., Leif, R. C., Mayall, B. H., Murphy, R. F., and Sandberg, A. A. (1984) Convention on nomenclature for DNA cytometry. *Cytometry* **5,** 445, 446.
4. Merkel, D. E., Dressler, L. G., and McGuire, W. L. (1987) Flow cytometry, cellular DNA content, and prognosis in human malignancy. *J. Clin. Oncol.* **5,** 1690–1703.

5. Koss, L. G., Czerniak, B., Herz, F., and Wersto, R. P. (1989) Flow cytometric measurements of DNA and other cell components in human tumors: a critical appraisal. *Hum. Pathol.* **20,** 528–548.
6. Seckinger, D., Sugarbaker, E., and Frankfurt, O. (1989) DNA content in human cancer. *Arch. Pathol. Lab. Med.* **113,** 619–626.
7. Shapiro, H. M. (1989) Flow cytometry of DNA content and other indicators of proliferative activity. *Arch. Pathol. Lab. Med.* **113,** 691–597.
8. Vohn, M. (1989) Prognostic implications of DNA aneuploidy and proliferative activity in solid human tumors. *Tumor Diagn Ther* **10,** 229–232.
9. Raber, M. N. and Barlogie, B. (1990) DNA flow cytometry of human solid tumors, in *Flow Cytometry and Cell Sorting,* 2nd ed. (Melamed, M. R., Lindmo, T., and Mendelsohn, M. L., eds.), Wiley-Liss, New York, pp. 745–754.
10. Williams, N. N. and Daly, J. M. (1990) Flow cytometry and prognostic implications in patients with solid tumors. *Surgery* **171,** 257–265.
11. Wersto, R. P., Liblit, R. L., and Koss, L. G. (1991) Flow cytometric DNA analysis of human solid tumors: a review of the interpretation of DNA histograms. *Hum. Pathol.* **22,** 1085–1098.
12. Camplejohn, R. S. (1992) Critical summaries, flow cytometry. *J. Pathol.* **166,** 323–326.
13. El-Naggar, A. K. (1992) Clinical relevance of the flow cytometric analysis in cancer. *Cancer J.* **5,** 321–327.
14. Sasaki, K. and Murakami, T. (1992) Clinical application of flow cytometry for DNA analysis of solid tumors. *Acta. Pathologica. Japonica.* **42,** 1–14.
15. Barlogie, B., McLaughlin, P., and Alexanian, R. (1987) Characterization of hematologic malignancies by flow cytometry. *Anal. Quant. Cytol. Histol.* **9,** 147–155.
16. Andreeff, M. (1990) Flow cytometry of leukemia, in *Flow Cytometry and Cell Sorting,* 2nd ed. (Melamed, M. R., Lindmo, T., and Mendelsohn, M. L., eds.), Wiley-Liss, New York, pp. 697–724.
17. Andreef, M. (1990) Flow cytometry of lymphoma, in *Flow Cytometry and Cell Sorting,* 2nd ed. (Melamed, M. R., Lindmo, T., and Mendelsohn, M. L., eds.), Wiley-Liss, New York, pp. 725–743.
18. Hedley, D. W., Shankey, T. V., and Wheeless, L. L. (1993) DNA cytometry consensus conference. *Cytometry* **14,** 471.
19. Shankey, T. V., Rabinovitch, P. S., Bagwell, B., Bauer, K. D., Duque, R. E., Hedley, D. W., Mayall, B. H., and Wheeless, L. (1993) Guidelines for the implementation of clinical DNA cytometry. *Cytometry* **14,** 472–477.
20. Wheeless, L. L., Badalament, R. A., de Vere White, R. W., Fradet, Y., and Tribukait, B. (1993) Consensus review of the clinical utility of DNA cytometry in bladder cancer. *Cytometry* **14,** 478–481.
21. Hedley, D. W., Clark, G. M., Cornelisse, C. J., Killander, D., Kute, T., and Merkel, D. (1993) Consensus review of the clinical utility of DNA cytometry in carcinoma of the breast. *Cytometry* **14,** 482–485.
22. Bauer, K. D., Bagwell, C. B., Giaretti, W., Melamed, M., Zarbo, R. J., Witzig, T. E., and Rabinovitch, P. S. (1993) Consensus review of the clinical utility of DNA flow cytometry in colorectal cancer. *Cytometry* **14,** 486–491.

23. Duque, R. E., Andreeff, M., Braylan, R. C., Diamond, L. W., and Peiper, S. C. (1993) Consensus review of the clinical utility of DNA flow cytometry in neoplastic hematopathology. *Cytometry* **14,** 492–496.
24. Shankey, T. V., Kallioniemi, O.-P., Koslowski, J. M., Lieber, M. L., Mayall, B. H., Miller, G., and Smith, G. J. (1993) Consensus review of the clinical utility of DNA content cytometry in prostate cancer. *Cytometry* **14,** 497–500).
25. Kute, T. E., Gregory, B., Galleshaw, J., Hopkins, M., Buss, D., and Case, D. (1988) How reproducible are flow cytometry data from paraffin-embedded blocks? *Cytometry* **9,** 494–498.
26. Hedley, D. W. (1989) Flow cytometry using paraffin-embedded tissue: five years on. *Cytometry* **10,** 229–241.
27. Joensuu, H. and Kallioniemi, O-P. (1989) Different opinions on classification of DNA histograms produced from paraffin-embedded tissue. *Cytometry* **10,** 711–717.
28. McLemore, D. D., El-Naggar, A., Stephens, L. C., and Jardine, J. H. (1990) Modified methodology to improve flow cytometric DNA histograms from paraffin-embedded material. *Stain. Technol.* **65,** 279–291.
29. Nervi, C., Badaracco, G., Maisto, A., Mauro, F., Tirindelli-Danesi, D., and Starace, G. (1982) Cytometric evidence of cytogenetic and proliferative heterogeneity in human solid tumors. *Cytometry* **2,** 303–308.
30. Kallioniemi, O. P. (1988) Comparison of fresh and paraffin-embedded tissue as starting material for DNA flow cytometry and evaluation of intratumor heterogeneity. *Cytometry* **9,** 164–169.
31. van der Linden, J. C., Herman, C. J., Boenders, J. G., van de Sandt, M. M., and Lindeman, J. (1992) Flow cytometric DNA content of fresh tumor specimens using keratin-antibody as second stain for two-parameter analysis. *Cytometry* **13,** 163–168.
32. Frei, J. V. and Martinez, V. J. (1993) DNA flow cytometry of fresh and paraffin-embedded tissue using cytokeratin staining. *Modern Pathol.* **6,** 599–605.
33. Park, C. H. and Kimler, B. F. (1994) Tumor cell-selective flow cytometric analysis for DNA content and cytokeratin expression of clinical tumor specimens by "cross-gating". *Anticancer Res.* **14,** 29–36.
34. Kenyon, N. S., Schmittiing, R. J., Siiman, O., Burshteyn, A., and Bolton, W. E. (1994) Enhanced assessment of DNA/proliferative index by depletion of tumor infiltrating leukocytes prior to monoclonal antibody gated analysis of tumor cell DNA. *Cytometry* **16,** 175–183.
35. Zarbo, R. J., Brown, R. D., Linden, M. D., Torres, F. X., Nakhleh, R. E., and Schultz, D. S. (1994) Rapid (one-shot) staining method for two-color multiparametric DNA flow cytometric analysis of carcinomas using staining for cytokeratin and leukocyte common antigen. *Am. J. Clin. Pathol.* **101,** 638–642.

19

FISHing for Cytokines

Methodology Combining Flow Cytometry and In Situ *Hybridization*

Kenneth J. Pennline

1. Introduction

Changes in steady-state mRNA levels may represent early molecular events in immune cells that can ultimately predict and define cellular function, activation, differentiation, and transformation *(1)*. The ability to detect and accurately quantitate these changes in specific cells in a rapid fashion would provide an extremely sensitive measure of the regulatory process in cells leading to immunological activity (i.e., inflammation, infectious disease, autoimmunity, and so forth). Cytokine specific mRNA has been measured in tissue and cells by a variety of methods, including Northern blot analysis *(2)*, polymerase chain reaction (PCR) *(3)*, and *in situ* hybridization *(4)*. The sensitivity of blot analysis, which requires radioactive labeled probes, is low, requiring many cells (or sufficient tissue) with high expression of mRNA for detection *(5)*. A more important limitation of this technique is that the cellular source of the specific RNA species can not be identified. PCR probably represents the most sensitive method to detect mRNA. However, there are inherent problems with quantitation *(3)* and, as stated for Northern analysis, there is no indication of the cellular source of the message in a heterogeneous population of cells. This information is vital for determining whether changes in cytokine mRNA levels reflect changes in the frequency of cells expressing a specific message, alterations in the level of cytokine gene expression, or both *(6)*. On the other hand, methods of *in situ* hybridization have been described for detecting specific mRNA in cells and tissue using radioactive probes *(6, 7)*, such as ^{35}S, as well as

From: *Methods in Molecular Biology, Vol. 91: Flow Cytometry Protocols*
Edited by: M. J. Jaroszeski and R. Heller © Humana Press Inc., Totowa, NJ

nonradioactive probes that have been labeled with fluorochromes ***(8)***, immunocytochemically detectable enzymes ***(7)***, or biotin ***(9)***. Analysis of tissue or cells prepared by cytospin can easily be examined microscopically to obtain detailed information about the location, spatial arrangement (tissue), and cellular source of mRNA. However, a disadvantage of this technique is based on the fact that in analysis of histologically prepared cells or tissue, clear access to the target mRNA can be limited. This becomes a critical issue in instances where the cytokine mRNA expression level is low since it may be difficult to quantitate the total hybridization signal for a particular cell type.

Analysis of gene expression in suspended cells by a methodology that combines fluorescent *in situ* hybridization (FISH) and flow cytometry (FC) readily permits the accurate quantitation of mRNA in specific cells ***(1,10–12)***. The initial results using FC-FISH were reported by Bauman et al. ***(11)*** and demonstrated the quantitation of ribosomal RNA in a murine leukemia cell line with a biotinylated antisense RNA probe that was detected with fluorescein-conjugated avidin. An additional study employing biotinylated probes and FC-FISH indicated β-globin mRNA could be detected in specific populations of murine bone marrow and spleen cell preparations ***(12)***. In both of these studies FC-FISH was used to detect a message that was constitutively produced and present at a high expression level. Our laboratory advanced the technology by slightly modifying the methodology and demonstrating that cytokine (IL-1) mRNA could be induced in vivo and then detected in specific cells by flow cytometry ***(10)***. These results represented the first indication that message detection and immunophenotyping could be combined in the FC-FISH analysis. We have continued to progress the technology by incorporating new reagents, standardizing a microplate methodology, analyzing several more cytokine mRNAs, and performing the procedure on flow-sorted cells.

2. Materials

2.1. Animals

When cells from animals were used they were obtained from Balb/c mice that were purchased from Taconic Farms (Germantown, NY). These animals were maintained and fed under pathogen-free conditions.

2.2. Cell Suspensions

The techniques used to obtain splenic or peritoneal exudate (PEC) cell suspensions have been described in detail ***(10)***. Briefly, PEC were collected into conical centrifuge tubes by gentle peritoneal lavage with 5-mL aliquots of calcium- and magnesium-free phosphate-buffered saline (cmfree-PBS). Spleen cells were obtained by carefully teasing the tissue with forceps in 5–

10 mL of PBS and then filtering the suspension through several layers of gauze. The cell preparations were washed in cmfree-PBS and resuspended to a desired concentration.

2.3. Cell Line

The TH1 T-cell helper clone, HDK, was maintained in culture with IL-2 supplement. This cell line, when stimulated with concanavalin A (ConA), will express and produce cytokines characteristic of the TH1 helper cell population, including IL-2, IFN-γ, and TNF. Because it requires IL-2 for growth, there is a cyclic positive feedback mechanism that causes these cells to express IL-2 mRNA. The messages for other cytokines is induced only after stimulation.

2.4. Cell Stimulation

Cytokine message was induced in mouse leukocytes after in vivo or ex vivo stimulation. IL-1 mRNA was induced in PEC by the IP injection of 20 μg lipopolysaccharide (LPS), a commercially available bacterial endotoxin. Four hours after injection, PECs were harvested and prepared for FC-FISH. In contrast, IL-2 and IFN-γ gene expression was induced in murine splenic T-cells by culturing the cells in vitro for 4 h with ConA, a potent T-cell stimulator.

2.5. Biotinylated Riboprobes

IL-1α, IL-2, and IFN-γ cDNAs were cloned into the appropriate sites of the polylinker vector pGEM3 (Promega, Madison, WI). Prior to transcription, purified plasmid template DNAs were linearized with the appropriate restriction enzymes to generate either sense (SP6) or antisense (T7) RNA probes (**Fig. 1**). Linearized template (2 μg) was transcribed in RNase-free conditions at 37°C for 60 min in a total volume of 100 μL containing 1X transcription buffer: 10 m*M* DTT, 2.4 U/μL RNasin, 0.4 m*M* each of ATP, CTP, and GTP (Promega), 1 m*M* biotinylated (bio)-11-UTP (Enzo Diagnostics, New York, NY), and 1 U/μL T7 or SP6 RNA polymerase (Promega). DNase (Promega) was then added at a concentration of 0.04 U/μL for 15 min to digest the DNA template. The reaction mixture was ethyl alcohol (ETOH)-precipitated, resuspended in a volume of 55 μL, and a 5 μL sample was then electrophoresed in a 5% polyacrylamide gel in 1X TBE to check the quality and size of the transcripts. In general, the biotinylated riboprobes (*see* **Note 3**) appeared to be approximately twofold larger on gels when compared to nonbiotinylated probes (probably resulting from changes in electrophoretic mobility attributable to the biotin molecule). Biotinylated antisense (~700 bases) and sense (~600 bases) probes were reduced in size by alkaline treatment to ~100 bases and then reprecipitated in ETOH prior to use in FISH (*see* **Note 2**).

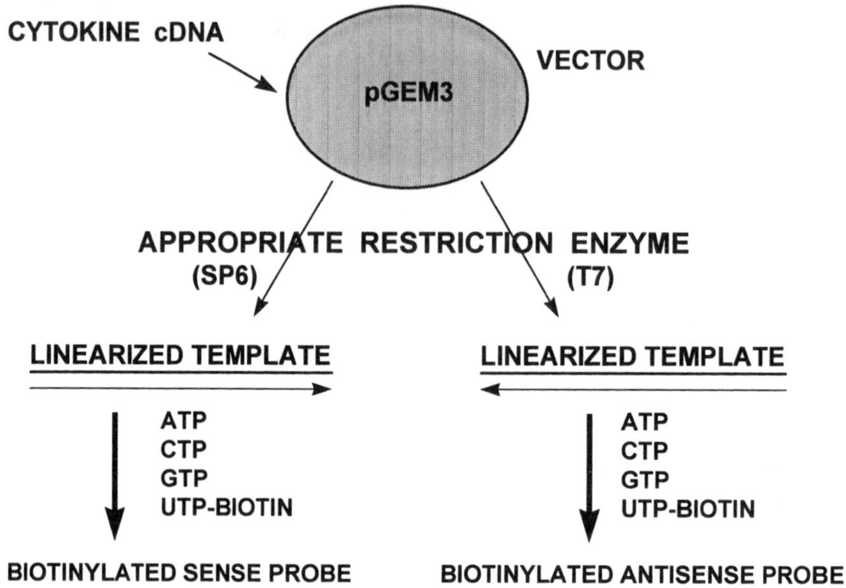

Fig. 1. Diagrammatic representation showing construction of sense and antisense probes. Restriction enzymes SP6 and T7 are used to "cut" the DNA strand into linearized templates that will transcribe in an appropriate direction to yield sense and antisense probes respectively.

3. Methods

3.1. Cell Surface Staining

Murine cells obtained from stimulated animals or in vitro cultures were washed in staining buffer (PBS, 2% fetal bovine serum [FBS], 0.1% sodium azide), resuspended at 2×10^7 cells/mL, and phenotyped for MAC, IA, or Thy1.2 (CD14-, HLADR-, and CD3-equivalent in human) as previously described ***(10)***, prior to performing FISH. Following a 5-min incubation with Fc block (Pharmingen, San Diego, CA), the cells were stained by sequential 30-min incubations (on ice) with purified monoclonal anti-MAC, -IA, or -Thy1.2 (Pharmingen) and goat-antirat IgG-PE (phycoerythrin). The indirect method was chosen here in order to augment the intensity of the cell surface staining, which may be highly susceptible to the harsh effects of the FISH procedure (*see* **Note 3**). Stained cells were washed and resuspended in preparation for fixation, permeabilization, and hybridization.

3.2. FISH

The results presented in this chapter were obtained using our original and current method of FISH (*see* **Note 4**). The latter is a modification of the original technique that is described in detail in a previous publication *(10)*. In the initial technique, the entire hybridization procedure (17 h) was performed at 37°C in microcentrifuge tubes and the method of permeabilization involved storage and/or incubation in cold 70% ETOH. In the modified/improved technique, cells were hybridized (4 h) at 45°C in 96-well plates and fixation and permeabilization were achieved using a single reagent, Ortho Permeafix (Ortho Diagnostics, Raritan, NJ) (*see* **Note 1**). The use of this perm/fix reagent was responsible for greatly improved cellular integrity (light scatter) and significant reduction in autofluorescence and cell attrition during the hybridization procedure.

3.2.1. FISH Protocol in V-Bottom Plates

1. Prepare single-cell suspensions of the cells to be hybridized and stimulate 4 h to overnight (1×10^6 cells/mL in complete medium) in culture at 37°C (Con A, phytohemaglutinin [PHA]), and LPS are the most common stimulants used in these studies).
2. Wash the cells thoroughly in Hank's BSS and aliquot 2×10^6 cells/well/100 μL of staining buffer (PBS, 2% FBS); then incubate with specific monoclonal antibody (MAb) for surface stain as described in **Subheading 3.1.**
3. Wash twice in PBS and resuspend in 250 μL/well/Permeafix (Ortho Diagnostics, Raritan, NJ). Incubate for 40 min at room temperature; wash once in staining buffer and then twice in PBS. (It is important to wash out all perm-fix reagent and all serum from other buffers.) At this point, further lysis of red blood cells (RBCs) can be done if necessary.
4. Centrifuge plates, aspirate fluid from the cell pellet (you can also "flick" the plate to expel the fluid), and then resuspend the cell pellets in hybridization buffer (Hyb-B) alone or with Hyb-B containing sense or antisense riboprobes. Mix well, seal the lid to the plate with parafilm, cover with foil, and then hybridize for 4–16 h at 45°C in a dry-air oven (*see* **Note 3**).
5. Add 200 μL of 2X SSC (saturated sodium chloride solution; NEN) to each well, mix well to break up any clumps, and then centrifuge the mixture at 500*g* for 2–5 min. Aspirate the liquid off the pellet carefully, add 100 μL of RNase A (1 mg/mL in 2X SSC), mix well, and incubate for 15 min at 37°C.
6. Add 100 μL of 1X SSC to each well and centrifuge as above. Aspirate the liquid carefully and then wash the cells sequentially with 200 μL of 1X SSC, 0.5X SSC, and HBSS.
7. Add 50 μL of HBSS to the cell pellets to resuspend them and then add 20 μL of streptavidin–allophycocyanin (SA-APC) to control, sense, and antisense wells (*see* **Note 5**). Incubate cells at room temperature for 15–30 min, wash twice in 200 μL of HBSS, and then fix in 1% paraformaldehyde prior to flow cytometric analysis.

3.2.2. Riboprobe Preparation

1. Mix ethanol-precipitated riboprobes well, remove 2 μL for each well to be hybridized and add to 200 μL of 70% ETOH in an RNase-free microcentrifuge tube (*see* **Note 2**).

2. Spin probes in a microcentrifuge at full speed (12,000*g*) for 30 min at 4°C, decant the supernatant, and dry completely in a lyophilizer or let air-dry in a tissue culture hood.
3. Resuspend dried probe in Hyb-B (50 µL/2 µL of probe), heat at 85°C for 3 min, cool briefly on ice, add 50 µL aliquots to appropriate wells immediately, mix well, and hybridize as detailed above.

3.3. Flow Cytometric Analysis: Instrument Settings

The data illustrated in this chapter represent information that was obtained on two types of Becton Dickinson (San Jose, CA) fluorescent activated cell sorters; one was the FACS 440 and the other was the FACS Vantage. The latter is the newer and more sensitive of the two instruments and is currently used for all data analysis.

3.3.1. FACS 440

This system has been described in detail in a previous publication ***(10)***. Briefly, the FACS 440 is a dual laser flow cytometer equipped with five decade logrithmic amplifiers. A coherent Innova-70 2 W laser was used to excite FITC and PE at 488 n*M* to emit at 520 and 590 n*M*, respectively. The second laser, a Helium Neon laser, was used to excite APC to emit at 650 n*M*. Dual fluorescence analysis was performed on all samples. MAC- and IA-positive cells were identified by PE fluorescence, whereas hybridized biotinylated probes were detected using APC fluorescence (*see* **Note 5**). The histograms are representative of the fluorescence of bound RNA probes that were analyzed by logrithmic amplification where 200 fluorescent channels equal one log. The mean fluorescent intensity (MFI) of hybridized cells is defined as the mean fluorescent log channel of the cells that were analyzed.

3.3.2. FACS Vantage

This system was described in detail in previous publications ***(13,14)***. Briefly, dual fluorescence analysis was performed on all samples where T-lymphocytes were phenotyped with Thy1.2-PE (excited at 488 nm and collected with a 465/30 bandpass filter) and hybridized with biotinylated riboprobes/SA-APC (excited with a 647 nm krypton line and fluorescence emission collected with a 670/14 bandpass filter) (*see* **Note 5**). For each sample, 30,000 events were acquired and analysis was performed on listmode data using LYSYS II and CellQuest (BDIS).

3.4. IL-1 Expression After In Vivo Stimulation

As stated in **Subheading 2.**, the endotoxin LPS, a powerful inducer of proinflammatory cytokines, such as IL-1, was injected ip into mice and specific mRNA was detected in spleen cells and PEC. The data presented in this section were obtained from analysis on a FACS 440 flow cytometer using the initial hybridization procedure that was developed. Dual fluorescent analysis was performed in which IA(+) or MAC(+)cells (PE) were isolated (or gated) and the APC fluorescence message for detection was displayed in histograms and contour maps (*see* **Notes 3** and **5**).

3.4.1. IL-1 Gene Expression in Spleen Cells

Figure 2 shows two histograms in which the fluorescence is exhibited for cells hybridized with no probe, sense IL-1 probe, or antisense IL-1 probe. The top histogram illustrates the fluorescent profiles for IA(+) cells obtained from control animals that did not receive an LPS injection. As can be seen, all three profiles are overlapping with very little change in the antisense signal. Any shift shown to be significant in the control cells could possibly represent constitutive expression of IL-1. In contrast, the bottom histogram shows the hybridization patterns exhibited by IA(+) cells from LPS-stimulated animals. The data clearly show that the hybridization signal with antisense probe was significantly increased above the overlapping fluorescent signals of no probe or sense probe hybridizations.

3.4.2. IL-1 Gene Expression in PEC

In **Fig. 3** the fluorescent hybridization signals for sense IL-1 and antisense IL-1 probes on stimulated PEC are displayed using contour maps. Again, dual fluorescent analysis was performed in which cells were surface-stained for MAC-PE (quadrant I; to identify the macrophages) and then dual stained with avidin-APC to detect the biotinylated riboprobes (quadrant II) (*see* **Notes 3** and **5**). The top plate shows the fluorescent profile for the sense probe and, as can be seen, there is a low percentage of cells (8%) in quadrant II. This display would represent any background fluorescence attributable to the control probe and hybridization procedure. On the other hand, the bottom plate illustrates the fluorescent profile of MAC(+) cells hybridized with the antisense probe. Contour mapping of these fluorescent signals demonstrated that 32% of the MAC(+) cells exhibited specific IL-1 mRNA expression.

3.5. mRNA Expression in In Vitro Stimulated Cells

In these experiments, murine T-cells or an established T-cell clone (HDK; Th1 T-cell clone) were stimulated in vitro with ConA and then probed for the expression of IL-2 or interferon (IFN) γ mRNA. Dual fluorescence analysis of the murine T-cells included surface staining for THY-1.2-PE (CD90) and the intracellular detection of the hybridized biotinylated IL-2 probe (APC). The HDK T-cell clone was not surface-stained but analyzed for fluorescent hybridization signals (APC) for the expression of IL-2 and IFN γ mRNA (*see* **Notes 3** and **5**).

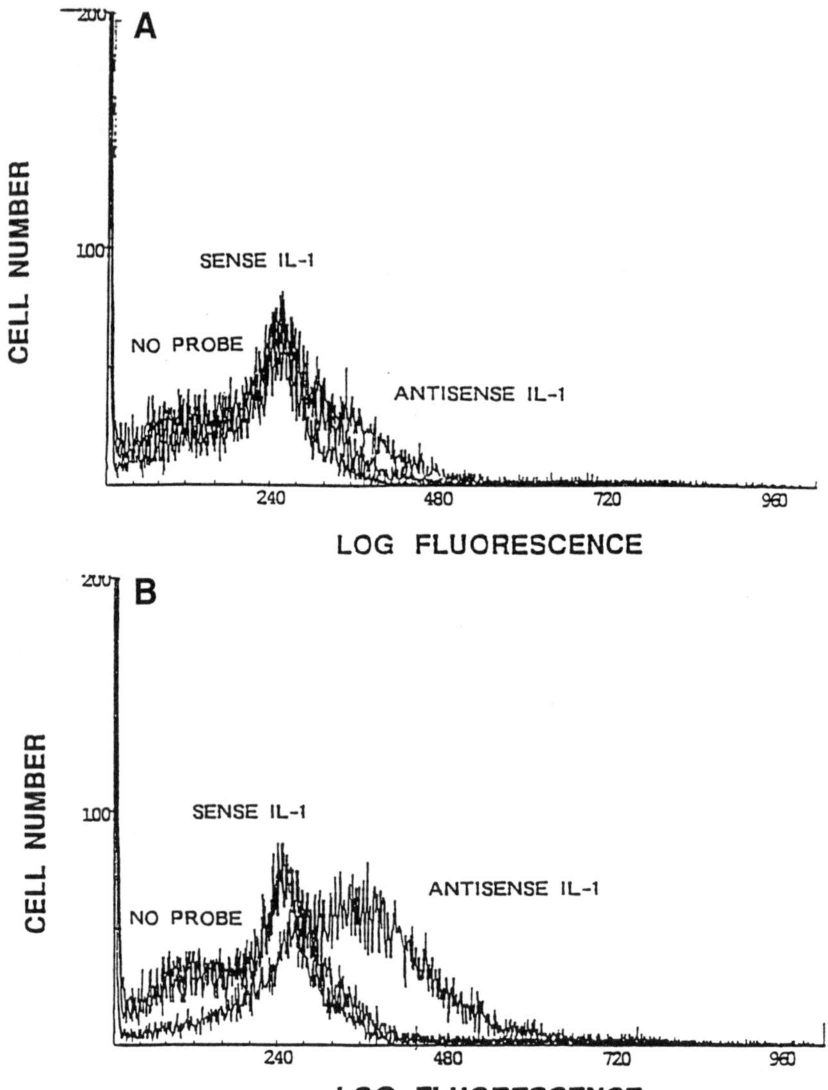

Fig. 2. Histograms showing the log-fluorescent signals of hybridized IA(+) spleen cells obtained from control mice (unstimulated, **A**) and animals that were injected ip with 20 µg of LPS (stimulated, **B**). (A) Data from control cells and illustrates the lack of IL-1 mRNA expression in unstimulated cells. In this histogram the signals from no probe, sense probe, and antisense probe hybridizations overlap and show no difference in fluorescence. (B) Demonstration of a dramatic shift in the fluorescent signal of the antisense probe when compared to control (no probe and sense probe) hybridizations of in vivo stimulated IA(+) spleen cells.

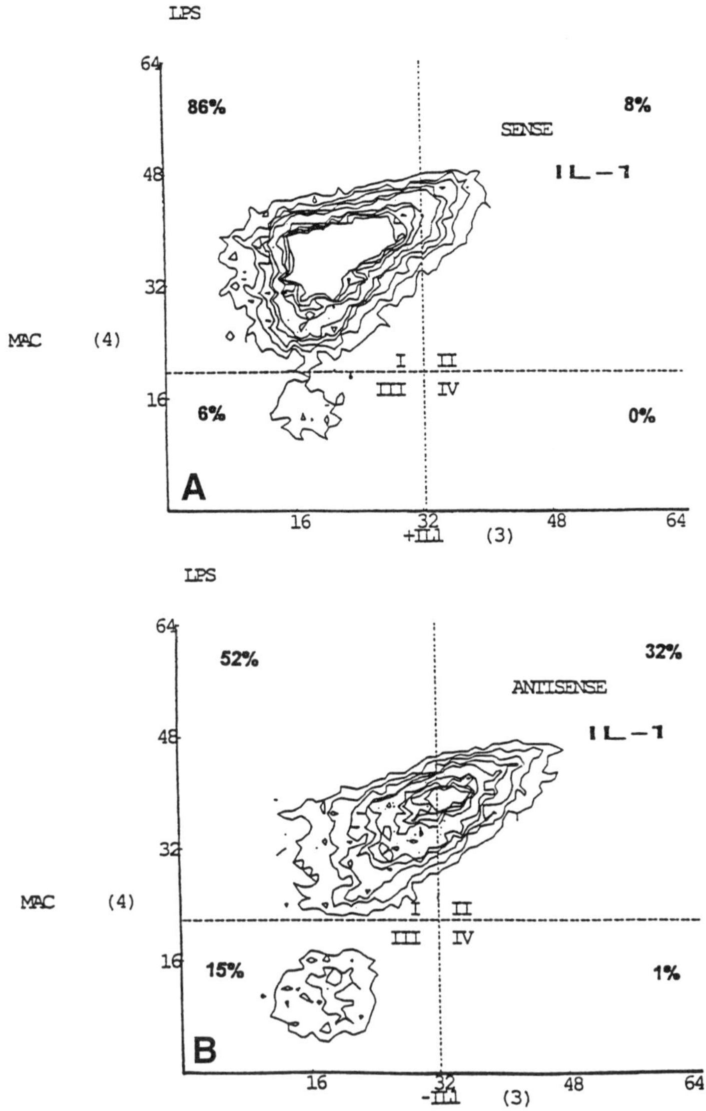

Fig. 3. Dual fluorescent contour plots illustrating the hybridization profiles of in vivo LPS-stimulated, MAC(+) peritoneal exudate cells (PECS). **(A)** The profile of the sense probe exhibiting 8% positive cells (nonspecific binding/background), whereas hybridization with the antisense probe **(B)** resulted in a strong positive signal and dramatic increase (32%) over that of the sense probe (8%). This illustration indicates that positive hybridization signals can be graphically represented as histograms or contour plots.

3.5.1. Analysis of Murine T-Cells

Maintaining cellular integrity and minimizing cell attrition throughout the procedure are critical factors associated with the ability to phenotype and hybridize cells simultaneously. As stated previously, one of the most important modifications of the initial technique was replacing the 70% ETOH with Ortho permeafix for fixation and permeabilization of the cells. When ETOH was used, hybridized samples exhibited dispersed light scatter patterns and significant cell loss. In contrast, the use of the permeafix reagent resulted in little or no cell loss and complete preservation of cellular integrity, which is illustrated by the data shown in **Fig. 4**. This figure shows the light scatter patterns of control and ConA stimulated splenocytes after hybridization using the permeafix reagent. As can be seen, the control cells exhibit a very "tight" lymphocyte profile within the leukocyte gate (top scattergram). The lymphocyte profile exhibited by the ConA-activated cells is more dispersed, which is to be expected because many of the T-cells have gone into blast formation, which causes increases in side- and forward-angle light scatter (cell granularity and cell size, respectively). These results are consistent with light-scatter patterns of cells that have not been permeabilized, fixed, or hybridized and clearly shows the efficiency of this reagent to maintain and preserve cellular integrity. Further analysis of these cells was performed by gating on CD90-PE(+) T-cells and then determining the APC-fluorescent profiles of cells hybridized with no probe (SA-APC), sense IL-2 probe (+IL-2), or antisense IL-2 probe (-IL-2). **Figure 5A–C** show the APC fluorescent patterns of hybridized ConA-activated cells throughout the course of probe titration (*see* **Note 1**). The results clearly show that the peak fluorescent signal (PFS) of the antisense IL-2 probe at 10, 5, and 2.5 μL is completely resolved from PFS of the two controls (SA-APC and +IL-2 probe). Hybridization of cells with 2.5 μL of the antisense IL-2 riboprobe exhibited optimized specific fluorescence and total reduction of background sense probe fluorescence. The remaining histogram (D) demonstrates that activated cells can be flow-sorted, based on a specific phenotypic marker (in this case THY1.2-PE), and then hybridized to determine expression of cytokine-specific mRNA. For this analysis, hybridization was performed using 2.5 μL of probe and, as can be seen, the fluorescent patterns are very similar to those illustrated in histogram C using the same amount of probe. As shown previously for IL-1 mRNA, the hybridization signal for IL-2 message can be displayed by contour mapping. **Figure 6** illustrates three contour maps for flow-sorted cells hybridized with no probe, sense probe, or antisense probe. Clearly the data shows that cells from the two control hybridizations map well within the first decade of fluorescence (*see* **Note 5**), whereas the majority of cells hybridized with antisense IL-2 map within the second decade of APC fluorescence.

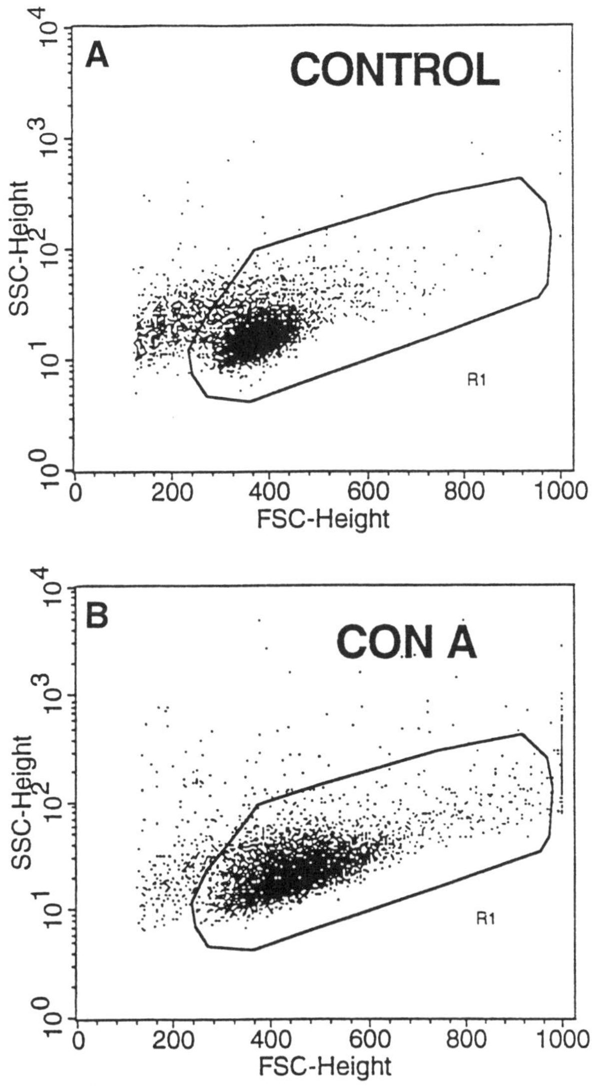

Fig. 4. Dot plots illustrating light scatter patterns of control (unstimulated, **A**) and in vitro-activated (ConA, **B**) murine spleen cells after hybridization. The purpose of this figure is to show that the permeafix reagent preserves cellular integrity of hybridized cells. The lymphocyte profile gate is compact and stable, which indicates continuity in cell size and granularity with very little cellular debris. The scatter is a little more dispersed for the activated cells, which is to be expected because these cells will undergo blast transformation and become larger and more granular.

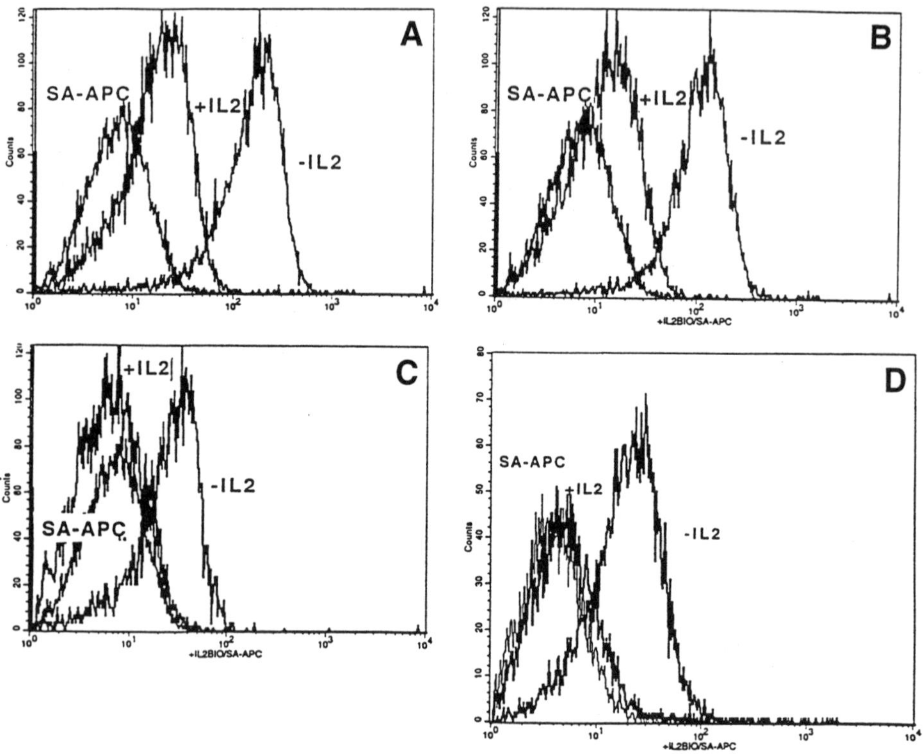

Fig. 5. Histograms of gated **(A–C)** and flow-sorted **(D)** murine splenic T-cells (THY1.2+) that have been stimulated in vitro with ConA and hybridized for IL-2 mRNA. Plots **(A–C)** illustrate data obtained from a probe titration study. As is shown in these three histograms, as the probe concentration is lowered the signals for both the sense (+)IL-2 and antisense (–)IL-2 probes are reduced. The lowest probe concentration used in (C) (2.5 μL) resulted in abrogation of background fluorescence (sense probe) while maintaining a strong signal from the antisense probe that was clearly resolved from the peak fluorescence of the controls. In (D) it was demonstrated that specifically phenotyped cells could be flow-sorted and then hybridized for detection of IL-2 mRNA. This analysis was performed on cells hybridized with 2.5 μL of probe (as shown in C), and shows a dramatic shift in the fluorescent profile of antisense IL-2 that is resolved from the signals exhibited by SA-APC alone and the sense IL-2 probe. Probe concentrations: (A) 10 μL; (B) 5 μL; (C) 2.5 μL.

3.5.2. Analysis of the HDK Th1 T-Cell Clone

The availability of defined T-cell clones has prompted their use in experimentation to study immune function and regulation of cytokine production. The Th1 T-cell clone HDK was used to illustrate specific mRNA expression

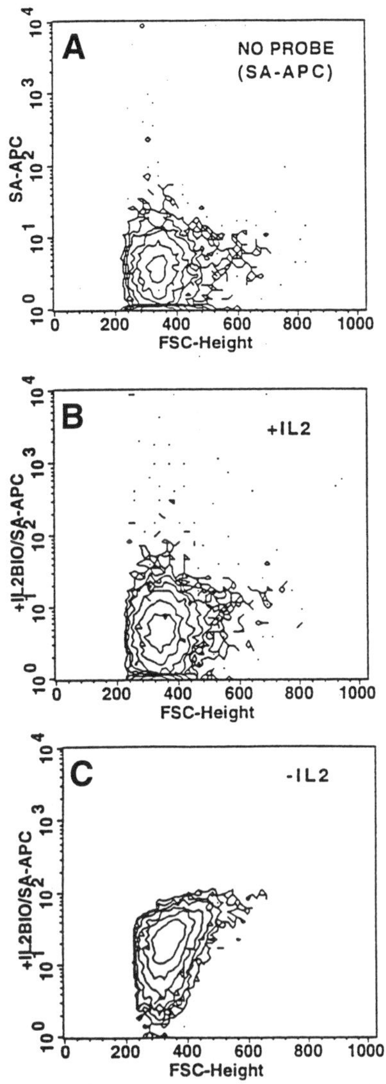

Fig. 6. Contour plots representing the histogram data shown in **Fig. 5(D)**. As shown for the histograms, only hybridization with antisense probe (-IL-2) **(C)** resulted in a shift upward into the second decade of fluorescence. The contours depicting profiles of hybridizations with no probe **(A)** and sense probe **(B)** were confined mostly to the first decade of fluorescence.

for two cytokines by the same cell type. This particular clone is maintained on cell feeder layers in the presence of IL-2 with periodic stimulation ConA. It is known to have a high constitutive level of IL2 and on stimulation will pro-

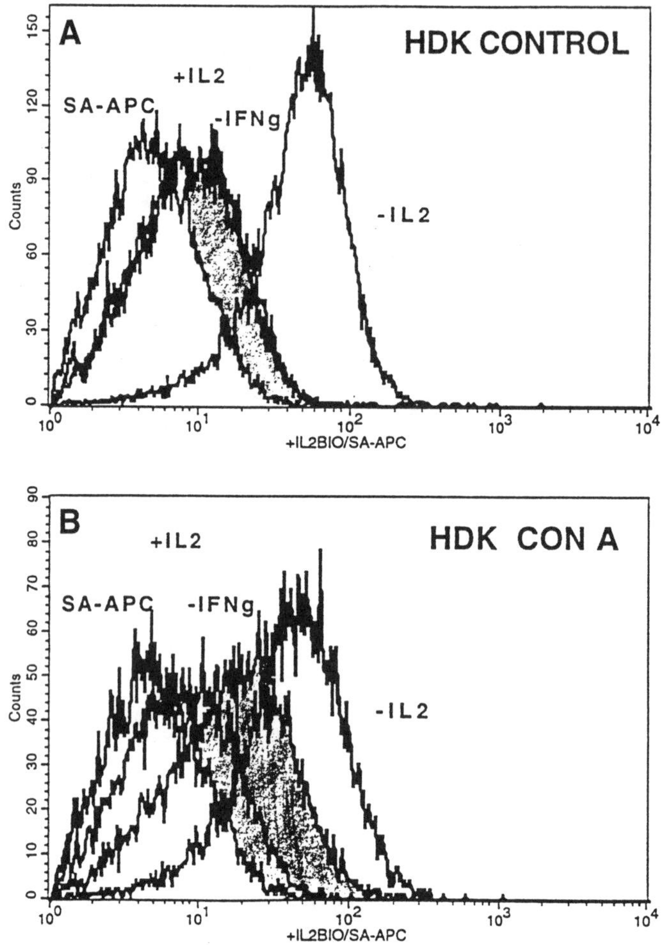

Fig. 7. Histograms demonstrating hybridization signals for IL-2 and IFN-γ antisense probes in control **(A)** and ConA-stimulated **(B)** HDK cells (TH1 T-cell helper clone). This cell line is maintained in IL-2 with periodic stimulation. The fluorescent signal of antisense IL-2 in unstimulated and stimulated cells, which is clearly resolved from the control sense probe, indicates that the HDK cell line has constitutive expression of IL-2 mRNA. On the other hand, hybridization with antisense IFN-γ demonstrates a defined increase in expression of IFN mRNA when fluorescent signals from unstimulated and stimulated cells are compared (shaded area). Profiles of hybridizations with no probe or sense probe exhibited low background fluorescence.

duce other cytokines. The histograms in **Fig. 7** represent the data obtained from the hybridization of unstimulated and stimulated (4 h after ConA) HDK cells. The data are presented as separate histograms that were overlayed to show

Table 1
mRNA Expression in HDK-1 Cells Hybridized with IL-2 and IFN-γ Riboprobes After In Vitro Simulation with ConA

	Mean fluorescent channel[a]	
Hybridization	(–) Con A	(+) Con A
No probe/unstained	4.3	4.5
No probe/SA-APC[b]	5.6	7.4
(+)[c]IL-2/SA-APC	10.3	11.6
(–)[d]IL-2/SA-APC	55.3	60.4
(–) IFN-γ/SA-APC	11.4	25.9

[a]Peak channel of fluorescence for hybridized cells.
[b]Streptavidin–allophycocyanin.
[c]Sense probe.
[d]Antisense probe.

differences in expression of specific IL-2 and IFN γ mRNA. The results do not represent simultaneous message detection for both cytokines in the same cell, but rather specific mRNA detection for IL-2 and IFN γ in separate aliquots of cells from the same cell population. In the top histogram the PFS of the sense (+IL-2) and antisense (-IL-2) probes in control cells indicates that a high level of constitutive IL-2 message is present. This expression does not change much after stimulation and may indicate that the message is maximized for that particular cytokine. In contrast, the expression of IFN-specific mRNA (shaded area) is relatively low in control cells (MFI = 11.4) and is increased significantly (MFI = 25.9) after stimulation with ConA (**Fig. 7** and **Table 1**). This analysis shows quite nicely that constitutive as well as induced mRNA can be determined in cells using this methodology (*see* **Note 4**).

This methodology will undoubtedly make a significant impact in the scientific and clinical arenas. There are several areas in which the application of this technology would be important. One involves situations in which monitoring the expression levels of immunomodulating cytokines or other proteins can determine the course of diseases. The other involves tracking the expression of viral particles in specific cell types. It is obvious in such diseases as AIDS, hepatitis, and cytomegalovirus (CMV) that this technology would be critical in determining the severity of the infection and quantitating cellular viral load. The technology is advancing continuously and future developments, such as "in cell" PCR, will certainly gain credibility and acceptance as clinical and diagnostic tools.

4. Notes

1. Hybridization temperature/time. In general, higher temperatures (85–90°C) result in more avid and efficient hybridizations. However, exposing cells, even fixed cells, to temperatures that high for any length of time will result in cell degradation. The key ingredient here is formamide, which is contained in the hybridization buffer. The presence of 50% formamide in the buffer allows for efficient hybridizations at 45°C over 4–5 h. In some cases, hybridizations can be performed at 37°C if the incubation time is extended to 17–24 h (the avidity of these hybridization may be less that). Modifications of the technique are usually dictated by the probe. What may work consistently with one probe may not work as well for the next since the probes themselves may be as heterogeneous as the proteins they represent. The data shown here were obtained by RNA–RNA hybridization of control and stimulated leukocytes using single-stranded riboprobes. With relation to temperature and time, the major advantage for using single-stranded riboprobes is that they are known to give higher hybridization efficiencies because of their lack of in-solution renaturization ***(5)*** as well as their ability to generate more stable complexes than the RNA–DNA or DNA–DNA hybrids. Furthermore, in comparative studies with RNA probes, LPS-stimulated cells hybridized with biotinylated IL-1 cDNA probes did not exhibit fluorescent signals that were greater than those of controls (data not shown). In addition, hybridization time and temperature were also related to other issues, such as high autofluorescence and high background binding by sense probes. Further developments and modifications in the technique resolved these problems by the addition of stringency washes (using 0.1–0.5X SSC) and RNase treatments to the procedures.
2. Probe size. The size of the riboprobes used in the studies shown here ranged between 100 (IL-2 and IFN) and 300 (IL-1) bases. The initial IL-1 probes were synthesized to 300 bases, whereas the 100-base IL-2 and IFN probes were obtained by controlled alkaline treatment of larger sense (~600 bases) and antisense products. The size of the probe is a critical issue that may determine binding specificity and binding efficiency. Since the makeup of each individual mRNA probe will vary to some degree depending on the protein (cytokine) it was specifically made for, it is necessary to optimize the size as well as the amount of probe used in hybridization protocols. Using a very small probe may allow for easier entry into permeabilized cells but it may also contribute to a degree of nonspecific binding, if indeed a probe was required to be larger in order to impart specificity. Equally important is optimizing the amount of probe used in the hybridization procedure. As was shown in **Fig. 4**, it may be necessary to sacrifice some of the specific antisense signal in order to reduce the background fluorescence that is attributable to the sense probe. Again, this my be entirely probe specific and may also depend on the level of gene expression within a give cell type or cell population.
3. Surface staining. The issue concerning surface staining is that the antibody and fluorochrome must have the avidity and durability to withstand the entire hybrid-

ization procedure. There are many standard fluorochromes available to phenotype cells prior to FISH. In choosing one it is important that it have good resolution from the fluorochrome used to identify the bound probe. Based on our experiences there are several points of consideration concerning phenotyping and FISH. The first is a problem we encountered when we were standardizing the protocol and establishing the technique. In our initial attempts performing FC-FISH we were not capable of maintaining surface staining regardless of the method used (direct staining, biotin–avidin, secondary antibody). After a process of elimination we discovered that one of the standard RNase inhibitors, diethylpyrocarbinate (DEPC), that is routinely used in all hybridization buffers was in some way removing the fluorochrome and/or antibody from the cell surface. When this chemical was removed from the procedure the detection of the surface stain was restored. Secondly, we discovered that the resolution and brightness of fluorochromes used for surface staining were always reduced, sometimes significantly, after performing the hybridization procedure. We have been trying several maneuvers to counteract this problem. One approach was to use a directly labeled antibody, such as CD90-PE, and then restain with the appropriate secondary antibody that is also conjugated to PE. This has been moderately successful. The other approach was to use primary antibodies conjugated to fluorospheres as described in Chapter 22. The surface fluorescent signal was more efficiently preserved using the latter approach (data not shown).

4. Advantages of FC-FISH. There are major advantages of this methodology over techniques previously used to detect gene expression. First, the combination of flow cytometry and *in situ* hybridization affords the investigator the capability of quantifying mRNA levels by rapid analysis of large numbers of cells (at rate of approx 2000 cells/s). Second, employing this technique, large heterogenous populations of cells can be probed for changes in levels of mRNA expression in specific cells identified by immunophenotyping. Here flow technology provides the distinct advantage over other methods in the ability to "gate on" or flow-sort specific cells in which gene expression is to be determined. This becomes a critical issue in cases where the targeted cell type is present in low numbers. Third, with the use of flow cytometry one can exploit the sensitivity and high resolution of fluorescent dyes (and tandem dyes). This is extremely important when you consider the detection of mRNA species that may be present at very low levels. In addition, with future developments, applications using combinations of fluorescent dyes can permit the simultaneous quantitation of multiple hybridization signals within a given cell or cell population. Fourth, utilizing this combined methodology allows investigators to accumulate data rapidly.
5. Probe fluorescence. There are many avenues to explore when it comes to selection of the method to detect hybridized probes. In this study, biotinylated probes specific for induced cytokine mRNA were detected using avidin–APC. Biotin–avidin systems are usually very good for use in detection assays because they are considered a means to amplify a specific signal without increasing background noise. In this case there are two considerations. The first is probe-specific and is

totally dependent on the makeup of the probe and how many UTP-biotin bases are included in the construct (**Fig. 1**). It is difficult to quantitate the exact number of UTP–biotin bases present and, as can be expected, this number will vary from probe to probe. The second consideration is the avidin–fluorochrome conjugate that is used. In the results presented here avidin–APC was used to detect bound biotinylated probes. There are a number of conjugates available to use, including -FITC and -PE. The most important consideration here is to choose a fluorochrome that has high resolution from other signals as well as one that will not be so large that steric hindrance becomes a problem. There are alternatives to the biotin–avidin system. One is to use probes that are directly conjugated to fluorochromes. On the one hand, using this kind of system would reduce the number of steps in the procedure and possibly reduce background fluorescence that may occur in two-step staining methods. However, the intensity of the signal would depend on the number of fluorochrome molecules bound to the probe and the level of expression for the target mRNA. Without an amplification system low-level gene expression may be difficult to quantitate with direct conjugates unless the fluorochrome used is intensely bright. Another alternative to biotin–avidin systems is the use of digoxigenin (DIG)-labeled probes ***(1,15)***. In this case DIG is incorporated into the construct of the probe (DIG-UTP) and then after hybridization the bound probes are detected using fluorochrome-conjugated anti-DIG antibodies. This latter method represents an amplification system that, like the biotin–avidin system, is dependent on the fluorescent characteristics of the dye that is used, i.e., FTIC, PE, or APC. Considering all three methods it may be favorable to use either of the amplification systems since in these methods the detecting dye does not have to withstand the harshness of the hybridization procedure. There are, however, new fluorochromes to try, such as the cyanin dyes, including Cy3, Cy5, and CY7. These are small molecules with bright fluorescence and, if incorporated correctly into FC-FISH methods, may prove to be ideal molecules for detecting low-level gene expression.

Acknowledgments

I would like to acknowledge the efforts of two coworkers: Francis Pellerito-Bessette, who struggled through the initial methodology and was responsible for setting up the parameters for *in situ* hybridization using flow cytometry; and Andrew J. Beavis, who clearly was instrumental in fine-tuning the methodology by introducing the Permeafix reagent and standardizing the technique in microtiter wells. I am grateful to have worked with such dedicated individuals.

References

1. Morvan, P. Y., Picot, C., Deyour, R., Gillot, E., Genete, B., and Genetet, N. (1994) *In situ* hybridization and cytofluorometric analysis of cytokine mRNA during *in vitro* activation of human T cells. *Eur. Cytokine Netw.* **5,** 469–480.

2. Grabstein, K., Dower, S., Gillis, S., Urdal, D., and Larsen, A. (1986) Expression of interleukin-2, interferon-γ and the IL-2 receptor by human peripheral blood lymphocytes. *J. Immunol.* **136,** 4503.
3. Montgomery, R. A. and Dallman, M. J. (1991) Analysis of cytokine gene expression during fetal thymic ontogeny using the polymerase chain reaction. *J. Immunol.* **147,** 554.
4. Tron, V. A., Harley, C. B., Caussy, D., and Sauder, N. (1988) *In situ* detection of interleukin-1 mRNA in human monocytes. *Mol. Immunol.* **25,** 439.
5. Honjo, T., Shimizu, A., Tsuda, M., Natori, S., Kataoka, T., Dohomoto, C., and Mono, Y. (1977) Accumulation of immunoglobulin messenger ribonucleic acid in immunized mouse spleen. *Biochemistry* **16,** 5764.
6. Fong, T. C. and Makinodin, T. (1989) *In situ* hybridization analysis of the age-associated decline in IL-2 mRNA expressing murine T-cells. *Cell. Immunol.* **118,** 199.
7. Samoszuk, M. and Nansen, L. (1990) Detection of interleukin-5 messenger RNA in Reed-Sternberg cells of Hodgkins disease with eosinophilia. *Blood* **75,** 13.
8. Serke, S. and Pachmann, K. (1988) An immunocytochemical method for the detection of fluorochrome-labelled DNA probes hybridized *in situ* with cellular RNA. *J. Immunol. Methods* **112,** 207.
9. Langer, P. R., Waldrop, A. A., and Ward, D. C. (1981) Enzymatic synthesis of biotin-labelled polynucleotides: novel nucleic acid affinity probes. *Proc. Natl. Acad. Sci. USA* **78,** 6633.
10. Pennline, K. J., Pellerito-Bessette, F., Umland, S. P., Siegel, M. I., and Smith, S. R. (1992) Detection of *in vivo* induced IL-1 mRNA in murine cells by flow cytometry (FC) and fluorescent *in situ* hybridization (FISH). *Lymphokine Cytokin Res.* **11,** 65–71.
11. Bauman, J. G. J. and Bentvelen, P. (1988) Flow cytometric detection of ribosomal RNA in suspended cells by fluorescent *in situ* hybridization. *Cytometry* **9,** 517.
12. Bayer, J. A. and Bauman, J. G. J. (1990) Flow cytometric detection of β-globin mRNA in murine haemopoietic tissues using fluorescent *in situ* hybridization. *Cytometry* **11,** 132.
13. Beavis, A. J. and Pennline, K. J. (1994) Simultaneous measurement of five cell surface antigens by five-color immunofluorescence. *Cytometry* **15,** 371–376.
14. Beavis, A. J. and Pennline K. J. (1994) Tracking of murine spleen cells in vivo: detection of PKH26-labelled cells in the pancreas of non-obese diabetic (NOD) mice. *J. Immunol. Methods* **170,** 57–65.
15. Lalli, E., Gibellini, D., Santi, S., and Facchini, A. (1992) *In situ* hybridization in suspension and flow cytometry as a tool for the study of gene expression. *Analyt. Biochem.* **207,** 298–303.

20

Analysis of Apoptosis by Flow Cytometry

**Wojciech Gorczyca, Myron R. Melamed,
and Zbigniew Darzynkiewicz**

1. Introduction

Based on morphological, biochemical, and molecular criteria two distinct modes of cell death can be recognized: necrosis and apoptosis *(1–5)*. Necrosis generally results from enzymatic digestion of the cells and denaturation of its proteins as the response to gross injury. An early event of necrosis is swelling of cell mitochondria, followed by rupture of the plasma membrane and release of the cytoplasmic contents. The necrotic cells is eosinophilic, and its nucleus becomes either pycnotic or karyorrhectic, or it may disappear in the process of karyolysis. Fully developed myocardial infarct, encephalomalacia, or caseous changes in tuberculosis are examples of necrosis. The process triggers an inflammatory reaction in the surrounding tissues, which often results in scar formation.

In contrast to necrosis, which is a passive and degenerative process, apoptosis requires active participation of affected cells in their demise. Apoptosis (apo'pto'sis, or less popular ap'o-to'sis), which means in Greek "falling away of leaves from the tree," is often called "cell suicide," "programmed cell death," "accidental cell death," or "physiological cell death." The apoptotic cell is smaller because of loss of water, and the intracellular organelles are tightly packed but are not swollen, as in the case of necrosis. The most characteristic feature of apoptotic cells is the chromatin condensation. The condensed chromatin aggregates peripherally, under the nuclear membrane, and then it forms well-demarcated masses of various size and shape often separated from each other. A distinct hyperchromicity and homogeneity characterize DNA stainability in the condensed chromatin of apoptotic cells. The apoptotic cells show extensive surface blebbing, which is followed by formation of characteristic

From: *Methods in Molecular Biology, Vol. 91: Flow Cytometry Protocols*
Edited by: M. J. Jaroszeski and R. Heller © Humana Press Inc., Totowa, NJ

apoptotic bodies, containing both nuclear fragments and constituents of cytoplasm (including intact organelles). The apoptotic bodies detach from the dying cell and, when apoptosis occurs in vivo, are phagocytized by neighboring cells without invoking an inflammatory response ***(2,3,5–7)***. The apoptotic mode of cell death occurs in isolated cells; in contrast to necrosis, which affects many adjoining cells often within a macroscopically defined area within the tissue, apoptotic mode of cell death occurs in isolated individual cells.

At the biochemical level, the most characteristic feature of apoptosis is nuclear DNA degradation. DNA is initially cleaved to 50–300 kb size sections and subsequently (albeit not in all cell types) at the internucleosomal (linker) sections ***(2,8–10)***. The later step produces nucleosomal and oligonucleosomal fragments of low-mol-wt DNA approx 180 bp (and multiples of it) in size, which can be detected by gel electrophoresis by virtue of their characteristic "ladder" pattern. Since DNA laddering does not occur in all cases of apoptotic cell death its absence should not be used as the sole criterion to exclude the apoptotic mode of cell death ***(11,12)***.

The boundaries between apoptosis and necrosis are not always unequivocal. For numerous examples of cell death where the pattern of morphological and/or biochemical changes resembles neither typical apoptosis nor necrosis ***(11–13)***. Moreover, some injuries that typically induce necrosis may lead to apoptosis as well (e.g., liver ischemia). Very often in malignant tumors both types of cell death can be observed, either spontaneously or as a result of treatment with antineoplastic agents. Majno and Joris proposed in their interesting overview of cell death ***(14)*** yet another term: oncosis. They proposed the renaming of accidental cell death (presently defined as "necrosis") as oncosis (derived from *oncos*, meaning swelling), and reserving the term necrosis for the late, postmortem stage, when the changes subsequent to cell death by any mechanisms (apoptosis or oncosis), e.g., loss of the plasma membrane integrity, are observed ***(14)***.

Much knowledge about the regulation of apoptosis stems from the studies of the nematode *Caenorhabditis elegans*. In that organism, the *ced-3* and *ced-4* genes are required for the cells to die, and in mutants lacking *ced-3* and *ced-4* the process of programmed cell death does not occur. The action of those genes is antagonized by expression of *ced-9* gene, which prevents the cells from dying. In humans and other mammals, the *bcl-2* gene acts as an equivalent of *ced-9* ***(15–18)***. At least two distinct checkpoints regulate the cell's propensity to respond to various stimuli by apoptosis: one controlled by the bcl-2/bax family of proteins ***(16,18,19)***, another by the cysteine and serine proteases. Several other genes are involved in the regulation of apoptosis (e.g., *p53*, c-*myc*) ***(16,18,20–22)***, but a description of their role in apoptosis is beyond the scope of this chapter. Regulatory mechanisms associated with apoptosis are the subject of recent reviews ***(7,23)***.

Because apoptosis plays an important role in physiological and pathological processes during development, as well as adult organisms, the methods of detection of apoptosis have become of wide interest in many diverse fields of biology and medicine. Thus, tissue remodeling during embryogenesis and adulthood occurs by genetically programmed apoptosis ***(6,24–28)***. Death of cells exposed to chemical mutagens or ionizing radiation, which eliminates potentially malignant cells before tumor development, also occurs by apoptosis. Recent studies provide ample evidence that not only the change in rate of cell proliferation but also their loss of the ability to die on schedule is associated with tumor development and progression ***(5,25,29–31)***. Analysis of spontaneous and treatment-induced apoptosis in tumors appears to have wide prognostic and therapeutic implications. Because apoptosis is an essential event in clonal selection of T-cells and is implicated in many other normal and pathological reactions of the immune system (32-35) the methods of detection of apoptosis become common in immunology. Their role in acquired immune deficiency syndrome (AIDS) is now emerging by virtue of the possibility they offer in following the disease progression via monitoring apoptosis of peripheral blood T-lymphocytes ***(36–38)***.

Numerous methods can be used to analyze apoptosis, but a universal biochemical, cytometric or antigenic marker for apoptosis is still lacking. A variety of new procedures is being introduced at rapid pace. Although many have a solid experimental foundation and have been successfully tested on a variety of cell systems, some methods or commercially available "kits" were proposed after being tested on a single cell model. Their wider applicability has yet to be documented. Detailed analysis of difficulties or pitfalls in flow cytometric analysis of apoptosis is described in our recent review ***(39)***. The methods presented in this chapter are based on a variety of markers of apoptosis and necrosis and have been tested in numerous laboratories worldwide.

1.1. Cell Morphology

Light or ultraviolet (UV) light microscopy remains the gold standard in analysis of apoptosis. All the morphological features described above (shrinkage of the cell, chromatin condensation, DNA hyperchromicity, nuclear fragmentation, and formation of apoptotic bodies) are easily recognizable by inspection of cells by microscopy (**Fig. 1**) and should always be sought out when analyzing apoptosis by other means (flow cytometry or gel electrophoresis), especially when any ambiguity arises regarding the mechanism of cell death. Conventional staining (hematoxylin and eosin) or fluorescent DNA dyes, such as 4',6'-diamidino-2-phenylindole (DAPI), combined with sulforhodamine 101, acridine orange (AO), propidium iodide (PI), or

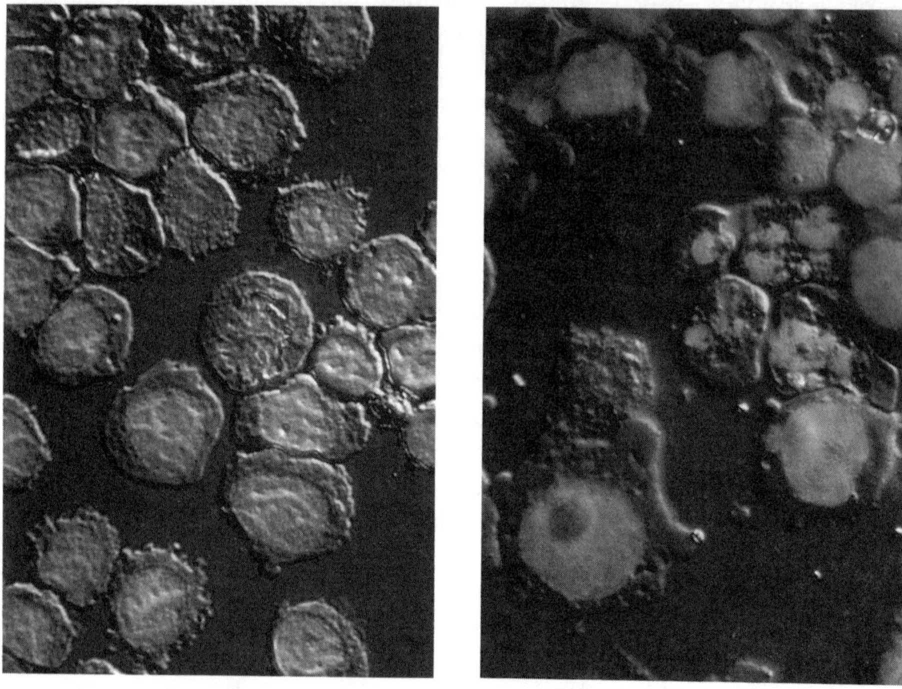

Fig. 1. Morphologic features of apoptotic cells. HL-60 cells were untreated (**left**) or treated with 0.15 μ*M* camptothecin for 4 h to induce apoptosis of S-phase cells (**right**). The cytospin preparations were stained with DNA fluorochrome DAPI (blue fluorescence). The cells were viewed under mixed illumination of incident UV light excitation and transmitted interference contrast (Nomarski). Note typical loss of nuclear structure, chromatin condensation, DNA hyperchromicity, and fragmentation of nuclei in apoptotic cells. Nikon Microphot–FXA, Objective ×40.

7-amino-actinomycin D (7-AAD) can be used with the same good results. More important than staining is the preparation of the specimen. The best results are obtained with cells cytospinned onto slides and then fixed. Cells initially fixed and stained in suspension and then transferred on slides are much less optimal for light microscopic inspection. They can be analyzed, however, by confocal microscopy.

1.2. Gel Electrophoresis

As already mentioned, activation of endonuclease(s), which preferentially degrades DNA at the internucleosomal sections, is a very characteristic (although not universal) event of apoptosis *(2,8,11,12)*. After DNA degradation the apoptotic cells have easily extractable low-mol-wt DNA of approx

180 bp and its multiples, and not degraded, high-mol-wt DNA associated with the nuclear matrix ***(2,8,10,40)***. The products of DNA degradation are visible in agarose gel electrophoresis as a characteristic "ladder." Necrotic cells when submitted to gel electrophoresis will yield a "smear" resulting from the presence of DNA fragments of highly heterogeneous sizes.

1.3. Flow Cytometry

The characteristic morphological, functional, and biochemical features of cells undergoing apoptosis or necrosis described in **Subheading 1.** provided the basis for development of many flow cytometric techniques to differentiate between the two modes of cell death ***(3,4,41–56)***. Changes in size and gross cell structure, in plasma membrane transport function or physical integrity, as well as in chromatin and DNA structure that occur during cell death all provided markers for the design of these methods.

1.4. Laser Scanning Cytometry (LSC)

The LSC is a microscope-based cytofluorometer ***(57,58)*** that can quantify fluorescence of individual cells in histologic sections or cytological preparations on slides with sensitivity, accuracy, and speed approaching that obtained by flow cytometry. It has several additional features, however, that can extend applicability in situations beyond the capabilities of flow cytometry. Thus, because the cells (or tissue sections) are attached to glass slides and no cell loss occurs during specimen preparation and staining, small samples (10^3–10^4 cells) are adequate for LSC. Other advantages include a record of the spatial position (coordinates) of every cell measured on the slide. The cell then can be relocated (after restaining, if desired) and classified visually or by other features subsequently measured. The time of each measurement also is recorded, permitting kinetic studies.

All the methods to identify apoptotic cells by flow cytometry presented in this chapter can be applied to LSC. The cells attached to the microscopic slides, preferentially by cytocentrifugation, are fixed with the appropriate fixative, and then subjected to the same sequence of treatments as described for the cells stained in suspension for flow cytometry. The staining procedures for LSC are simpler since there is no need for cell centrifugation and resuspension of the cell pellet. Cells on the slides are either stained in Coplin jars, or when the cost of reagents is an issue, stained by adding (and removing) small volumes of the reagents to horizontally placed slides located in a humid chamber. Incubations are carried on after covering the drop of the reagent with a small (i.e., 2 × 4 cm) polyethylene foil to prevent drying. **Figure 2** presents DNA strand break labeling in apoptotic cells by TdT assay (*see* **Subheading 1.6.**) measured by LSC.

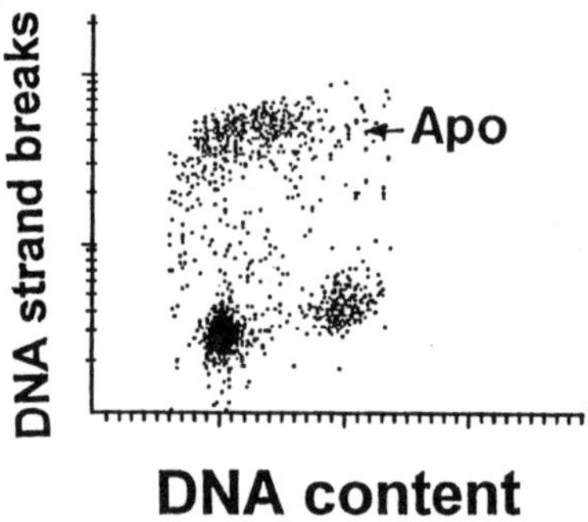

Fig. 2. Detection of apoptosis-associated DNA strand-break apoptotic cells. To induce apoptosis, HL-60 cells were treated with camptothecin as described in Fig. 1. DNA strand breaks in apoptotic cells were labeled with biotin-conjugated dUTP (b-dUTP) in the reaction catalyzed by TdT. The incorporated b-dUTP was detected by avidin-FITC; cellular DNA was stained with PI. Cellular fluorescence was measured by laser scanning cytometer (LSC), the instrument that combines the advantages of flow and image cytometry ***(57)***. Note the presence of DNA strand breaks in cells characterized by DNA content equivalent to that of S-phase cells (Apo).

1.5. Cellular DNA Content Measurements

As mentioned, apoptotic cells are characterized by activation of an endonuclease(s) that preferentially cleaves DNA at the internucleosomal (linker) sections ***(2,8)***. A fraction of DNA resists endonuclease action and remains associated with the nuclear matrix of the fragmented nuclei ***(40)***. Analysis of DNA content in the nuclei isolated by detergent-mediated cytolysis has become one of the most common methods for single-parameter flow cytometric measurement of apoptosis: the objects with fractional DNA content located on DNA frequency histograms to the left of the G_1 peak ("sub-G_1-cells," "sub-diploid cells") are then identified as apoptotic cells ***(3,9,59–63)***. The lysis of unfixed cells with detergent to analyze apoptosis, however, results in isolation of nuclei. Because apoptotic cells have fragmented nuclei, the frequency of the "sub-G_1 cells" of so-treated samples represents the number of individual nuclear fragments and not the number of apoptotic cells. Unless special care is used to lyse the cells, the detergent-based methods, cannot provide a quantitative estimate of the proportion of apoptotic cells. Fixation of cells in precipitating fixatives, such as alcohol or acetone, therefore has to be used to quantify apoptotic cells based on their fractional DNA content. After fixation, the degraded low-mol-wt DNA is extracted from the cell during subsequent rinsing and staining. As a result, apoptotic cells contain less DNA and stain less intensely with any DNA fluorochrome (**Fig. 3**) ***(3,40)***. It should be stressed

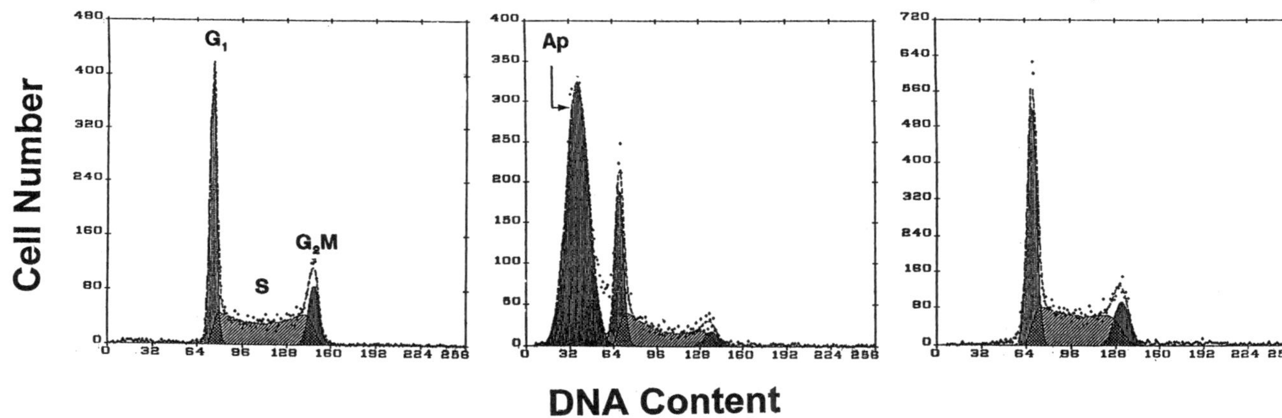

Fig. 3. DNA content frequency distribution histograms representing apoptotic and necrotic cells. To induce apoptosis, HL-60 cells were treated with 5 μ*M* DNA topoisomerase II inhibitor fostriecin for 4 h (**middle**). Necrosis of HL-60 cells was induced by their exposure to 1 m*M* fostriecin for 4 h (**right**). Cellular DNA was stained with DAPI as described in **Subheading 3.1.2.** (**Left**) Untreated cells. Note the presence of a population of apoptotic cells characterized by fractional DNA content (Ap) at 5 μ*M* fostriecin, and no change in DNA histogram (compared to the untreated cells) representing necrotic cells, in cultures containing 1 m*M* fostriecin (right).

that fixation with the crosslinking fixatives formaldehyde or glutaraldehyde prevents extraction of DNA from apoptotic cells and therefore can not be used for discrimination of apoptosis by analysis of DNA content.

The degree of DNA degradation varies depending on the stage of apoptosis, cell type, and often the nature of the apoptosis-inducing agent. High molarity phosphate-citrate buffer enhances extraction of the degraded DNA from apoptotic cells, allowing optimal separation of apoptotic "peak" from G_1 cells ***(40)***.

Application of a fixative like ethanol instead of detergent allows one to combine DNA staining with cell immunophenotyping. It is possible, therefore, to detect apoptosis in cell subpopulations within heterogeneous specimens ***(61,63,64)***. Similar analysis (discrimination of apoptosis and immunophenotyping) is also possible with application of 7-amino-actinomycin D staining.

1.6. DNA Strand-Break Labeling Assay

Endonucleolytic DNA cleavage results in the presence of extensive DNA breaks. Those breaks (3'OH termini) can be detected by labeling with biotin–digoxygenin or directly with fluorochrome-conjugated triphosphodeoxynucleotides in the reaction catalyzed by terminal deoxynucleotidyl transferase (TdT) ***(46,66,67)*** or DNA polymerase ***(45–47,68–70)***. DNA strand-break labeling assay appears to be the most specific in terms of positive identification of apoptotic cells. The assay based on the use of TdT, also called "nick end-labeling" or "TUNEL" assay, offers a better discrimination between apoptotic and nonapoptotic cell populations than the nick translation assay ***(47,69)***.

A new approach utilizing TdT has recently been introduced, in which DNA strand breaks are labeled BrdUTP. The incorporated BrdUrd is subsequently detected by FITC-conjugated BrdUrd antibody ***(19,48)***. This method, presented below, offers the greatest sensitivity and uses less expensive reagents, compared to the methods based on DNA strand-break labeling with digoxygenin–biotin or directly fluorochrome-conjugated nucleotides (**Fig. 4**).

1.7. Detection of Apoptotic Cells by Light Scatter Analysis

Intersection of a cell with the light of the laser beam in a flow cytometer results in light scatter. Analysis of the scattered light provides information about cell size and structure ***(50,73)***. The forward light scatter correlates with cell size and the scattered light measured at a right angle to the laser beam ("side scatter") correlates with granularity, refractiveness, and the ability of the intracellular structures to reflect the light. The cell's ability to scatter light is altered during cell death, reflecting the morphological changes, such as cell swelling or shrinkage, rupture of the plasma membrane, chromatin condensation, nuclear fragmentation, and shedding of apoptotic bodies.

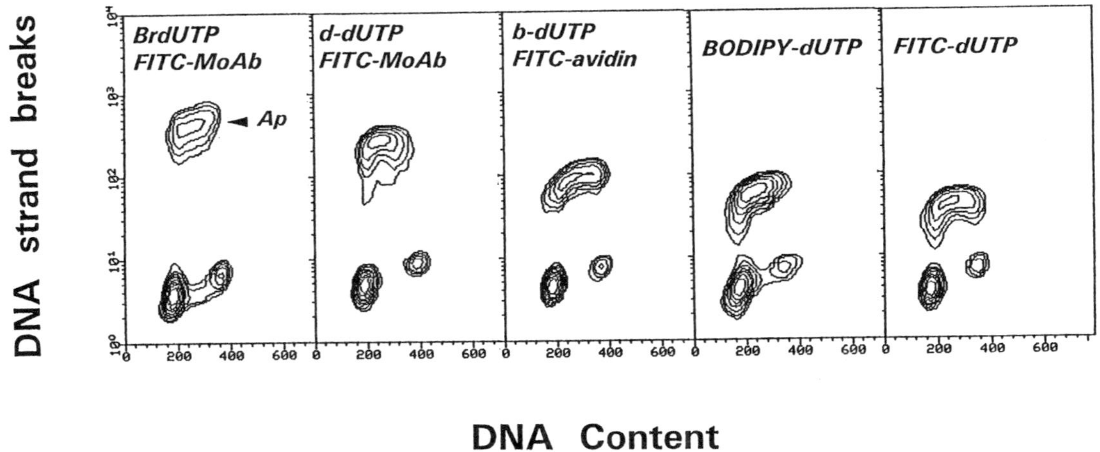

Fig. 4. Differences in sensitivity of detection of DNA strand breaks by various methods utilizing TdT and different nucleotides. Apoptosis of HL-60 cells was induced by their exposure to 0.15 μ*M* camptothecin for 3 h. The first three panels to left represent indirect labeling of DNA strand breaks, utilizing BrdUTP, digoxygenin conjugated dUTP (d-dUTP), or biotinylated dUTP (b-dUTP). The two right panels show cell distributions following a single-step DNA strand break labeling, either with BODIPY or FITC-conjugated dUTP. Note the highest discrimination between apoptotic (S-phase cells, Ap) and nonapoptotic cells (G_1 and G_2/M) after DNA strand-break labeling with BrdUTP, by the method presented in **Subheading 3.4.** (From **ref.** *77*, with permission).

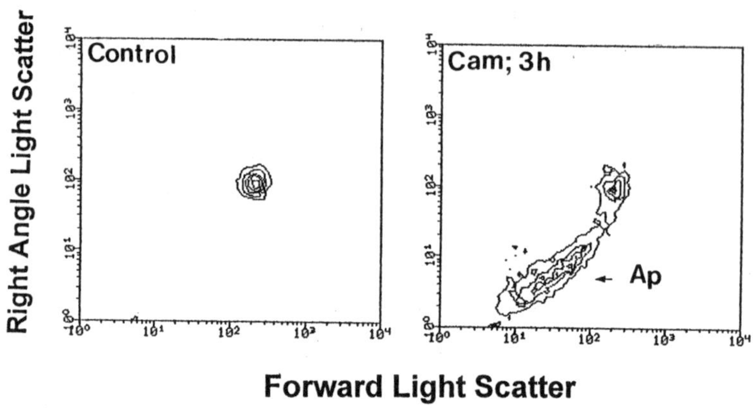

Fig. 5. Changes in forward and right-angle light scatter of cells during apoptosis. To induce apoptosis, HL-60 cells were treated with 0.15 μM camptothecin for 3 h **(right)**. Light scatter of the unfixed cells suspended in PBS was measured using standard settings of the FACScan flow cytometer (Becton Dickinson, San Jose, CA). Note the decrease in both right-angle and forward light scattering properties of apoptotic cells (Ap) from the camptothecin-treated culture, compared to the untreaded cells **(left)**.

The necrotic mode of cell death is characterized by a rapid initial increase in the cell's ability to scatter light simultaneously in the forward and right angle direction, which, in all probability, is a reflection of cell swelling. During apoptosis, the decrease in forward light scatter is often paralleled by a decrease in side scatter, although an increase in side scatter simultaneous with a decrease in forward scatter has been observed in some cell systems. In later stages of apoptosis the intensity of light scattered at both forward and right angles decreases (**Fig. 5**) ***(3,50)***.

1.8. Exclusion of PI Combined with Uptake of Mitochondrial Probe Rhodamine 123 (Rh 123)

Rh 123 is a cationic green fluorochrome that, as a consequence of the transmembrane potential of mitochondria (active proton pump), is taken up by these organelles ***(74)***. Live nonapoptotic cells with an intact plasma membrane and active (charged) mitochondria concentrate Rh 123 and exhibit strong green fluorescence when incubated with 0.5–1 µg/mL of Rh 123 and 5–20 µg/mL of PI. Because live cells exclude PI they have minimal red fluorescence. As cells undergo apoptosis their plasma membrane integrity and the ability to exclude PI remain preserved for some period of time. However, they lose the mitochondrial transmembrane potential and the ability to stain with Rh 123. At early stages of apoptosis, therefore, the cells incubated with Rh 123 and PI show a diminished Rh 123 (green) fluorescence and, similar to live cells, minimal PI (red) fluorescence. At later stages of apoptosis the ability of cells to exclude PI is compromised, and they stain intensively red with this fluorochrome while showing no significant Rh 123 fluorescence.

During necrosis an early transient moment occurs when cells appear to have hyperpolarized mitochondria and stain strongly with Rh 123 ***(75)***. Shortly thereafter, however, the loss of the plasma membrane integrity leads to intense staining of cellular DNA and ds RNA with PI. This step is paralleled by a loss of cells' ability to retain Rh 123 fluorescence. Thus, cell then stain weakly with rhodamine but strongly with propidium ***(75)***. Dual cell staining with Rh 123 and PI, therefore, provides an assay of mitochondrial transmembrane potential which can also serve as the assay of cell viability ***(74, 75)***.

1.9. Detection of Phosphatidylserine with Annexin V-FITC Conjugate

Plasma membrane phospholipids are asymmetrically distributed between inner and outer leaflets of the plasma membrane. In live cells phosphatidylserine (PS) is almost exclusively observed on the inner surface of the membrane ***(56, 76)***. Recently, it has been shown that loss of phospholipid asymmetry leading to exposure of PS on the outside of the plasma membrane is an early event in apoptosis ***(56, 76)***. Annexin V, a Ca^{2+}-dependent phospholipid-binding protein, has high affinity for negatively charged phospholipids like PS ***(56, 76)***. Hence this protein, when conjugated with a fluorochrome, can be used as a sensitive probe for the presence of PS on the outside of plasma membrane. Translocation of PS to the external cell surface is not unique to apoptosis, but occurs also during cell necrosis. However, by using a combination of Annexin V-FITC conjugate and PI, it has been possible to distinguish apoptotic from necrotic cells with flow cytometry. During apoptosis, the cells become reactive with Annexin V after the onset of chromatin condensation but prior to the loss of the plasma membrane's ability to exclude cationic dyes, such as PI ***(56, 76)***. Thus, nonapoptotic cells are Annexin V- and PI-negative ($FITC^-/PI^-$), early apoptotic cells are Annexin V-positive but PI-negative ($FITC^+/PI^-$), and late apoptotic cells as well as necrotic cells are stained intensely with PI. The Annexin V assay offers the possibility of detecting early phases of apoptosis before the loss of cell membrane integrity and permits measurements of the kinetics of apoptotic death in relation to the cell cycle.

2. Materials

2.1. Cellular DNA Content Measurements

1. Na_2HPO_4.
2. Citric acid.
3. Propidium iodide (PI, Sigma, St. Louis, MO).

4. DNase-free RNase (Sigma).
5. PBS.
6. Ethanol.
7. Triton X-100.
8. $MgCl_2$.
9. NaCl.
10. PIPES (Sigma).
11. DAPI (Molecular Probes, Inc., Eugene, OR).

2.2. DNA Strand-Break Labeling Assay

1. Methanol-free formaldehyde (Polysciences Inc., Warrington, PA).
2. Potassium (or sodium) cacodylate.
3. Tris-HCl.
4. Bovine serum albumin (BSA).
5. Cobalt chloride.
6. BrdUTP (Sigma).
7. BrdUrd-FITC (Becton Dickinson).
8. Triton X-100.
9. Propidium iodide (PI, Sigma).
10. DNase-free RNase (Sigma).

The TdT reaction buffer, $CoCl_2$, and TdT are available from Boehringer Mannheim (Indianapolis, IN).

2.3. Detection of Apoptotic Cells by Light Scatter Analysis

Light scatter properties of apoptotic cells can be analyzed routinely either alone or in combination with the measurement of cell fluorescence (as described in other methods).

2.4. Exclusion of PI Combined with Uptake of Mitochondrial Probe Rh 123

1. Rh 123 (Molecular Probes, Inc.).
2. PI (Sigma).
3. Tissue culture medium (or HBSS).

2.5. Detection of Phosphatidylserine with Annexin V-FITC Conjugate

1. FITC-labeled Annexin V (B.R.A.N.D Applications, Maastricht, The Netherlands; cat. no. A-700, 25 μg or A-800, 125 μg).
2. HEPES/NaOH.
3. NaCl.
4. $CaCl_2$.
5. PI (Sigma).

3. Methods

3.1. Cellular DNA Content Measurements

3.1.1. Cells Staining with PI

1. Prepare DNA extraction buffer by mixing 192 mL of 0.2 *M* Na_2HPO_4 with 8 mL of 0.1 *M* citric acid, pH 7.8.
2. Prepare DNA staining solution by dissolving 200 µg of PI in 10 mL of PBS. Add 2 mg of DNase-free RNase (solution should be prepared fresh before each use),
3. Fix the cells in suspension by adding 1 mL of cells suspended in PBS (10^6–5 × 10^6 cells) to 9 mL of ice-cold 70% ethanol (*see* **Note 1**). (Cells can be stored in the fixative at –20°C for several weeks.)
4. Centrifuge the cells, decant ethanol, suspend the cell pellet in 10 mL of PBS, and centrifuge.
5. Suspend the cell pellet in 0.5 mL of PBS, into which 0.2–1.0 mL of the DNA extraction buffer may be added.
6. Incubate the cells at room temperature for 5 min and centrifuge.
7. Suspend the cells in 1 mL of DNA staining solution.
8. Incubate for 30 min at room temperature.
9. Analyze cells by flow cytometry. Use 488 nm laser line or blue light (BG12 filter) for excitation and long-pass (>600 nm) fluorescence emission filter. Measure red fluorescence and forward light scatter (*see* **Notes 2–5**).

3.1.2. Cells Staining with DAPI

1. Prepare staining solution containing: 0.1% (v/v) Triton X-100, 2 m*M* $MgCl_2$, 0.1 *M* NaCl, 10 m*M* PIPES buffer (final pH 6.8; Sigma) and 1 µL/mL DAPI (Molecular Probes, Inc., Eugene, OR).
2. Fix the cells by admixing 1 mL of cells suspended in PBS (10^6–5 × 10^6 cells) into 9 mL of ice-cold 70% ethanol. (Cells can be stored in fixative at –20°C for several weeks.)
3. Centrifuge the cells, decant ethanol, suspend the cell pellet in 10 mL of PBS and centrifuge.
4. Suspend the cell pellet in 1 mL of the staining solution.
5. Analyze cells by flow cytometry. Use excitation with UV light (e.g., 351 argon ion line or UG1 filter for mercury lamp illumination. Measure blue fluorescence of DAPI in a band from 460 to 500 nm (*see* **Notes 3–5**).

3.2. DNA Strand Break Labeling Assay

1. Prepare the following solutions:
   a. Fixative: 1% methanol-free formaldehyde (Polysciences Inc., Warrington, PA) in PBS, pH 7.0.
   b. TdT reaction buffer (5X concentrated) containing 1 *M* potassium or sodium) cacodylate, 125 m*M* Tris-HCl (pH 6.6), 1.25 mg/mL bovine serum albumin (BSA).

c. 10 m*M* Cobalt chloride.
d. TdT in storage buffer, 25 U in 1 µL.
e. BrdUTP stock solution: 2 m*M* BrdUTP (100 nmol in 50-µL Sigma) in 50 m*M* Tris-HCl, pH 7.5.
f. FITC-conjugated anti-BrdUrd MAb solution (per 100 µL of PBS): 0.7 µg of anti-BrdUrd FITC-conjugated MAb (Becton Dickinson), 0.3% Triton X-100, 1% BSA.
g. Rinsing buffer: 0.1% Triton X-100 and 5 mg/mL BSA.
h. PI staining buffer: 5 µg/mL PI and 200 µg/mL DNase-free RNase A in PBS.

2. Fix the cells in suspension in 1% formaldehyde in PBS (pH 7.4) on ice for 15 min (*see* **Note 6**).
3. Centrifuge, resuspend the cell pellet in 5 mL of PBS, repeat centrifugation, and resuspend the cells (approx 10^6 cells) in 0.5 mL of PBS.
4. Add the above 0.5 mL aliquot of cell suspension into 5 mL of ice-cold 70% ethanol. (The cells can be stored in ethanol at –20°C for several weeks.)
5. Centrifuge, remove ethanol, resuspend cells in 5 mL of PBS and centrifuge.
6. Resuspend cell pellet in 50 µL of a solution containing 10 µL of the reaction buffer, 2.0 µL of BrdUTP stock solution, 0.5 µL (12.5 U) of TdT in storage buffer, 5 µL of $CoCl_2$ solution, and 33.5 µL distilled H_2O.
7. Incubate cells in this solution for 40 min at 37°C (or overnight at 22°C). There should be no more than 10^6 cells per 50 µL of the reaction solution.
8. Add 1.5 mL of the rinsing buffer and centrifuge.
9. Add 100 µL of FITC-conjugated anti-BrdUrd MAb solution.
10. Incubate at room temperature for 30 min (in the dark).
11. Rinse the cells in 1.5 mL of the rinsing buffer and centrifuge.
12. Resuspend the pellet in 1 mL of PI staining solution (with RNase).
13. Incubate 30 min at room temperature in the dark.
14. Analyze cells by flow cytometry using blue light (488 nm laser line or BG12 excitation filter). Measure green fluorescence of FITC at 530 ± 20 nm and red fluorescence of PI at >600 nm (*see* **Notes 7** and **8**).

Commercial kits are available for detection of apoptosis (*see* **Note 9**).

3.3. Detection of Apoptotic Cells by Light Scatter Analysis

The possibility of analysis of light scatter in the forward and right angle (90°) direction is a built-in feature of nearly every commercially available flow cytometer and is a routine measurement, either alone or in combination with the measurement of cell fluorescence (*see* **Note 10**). Because cell populations are generally characterized by great heterogeneity in light scatter of individual cells, an exponential scale (logarithmic amplifiers) should be used during analysis. It is important to select the correct angle for best discrimination of apoptosis in different cell systems ***(50)***.

3.4. Exclusion of PI Combined with Uptake of Mitochondrial Probe Rh 123

1. Stock solution of Rh 123: Dissolve 1 mg of Rh 123 (Molecular Probes, Inc.) in 1 mL of distilled H_2O.
2. Stock solution of PI: Dissolve 1 mg of PI in 1 mL of distilled H_2O. (Both stock solutions can be stored at 0–4°C in the dark for several weeks.)
3. Add 0.5 µL of Rh 123 stock solution to approx 10^6 cells suspended in 1 mL of tissue culture medium (or HBSS) and incubate 5 min at 37°C.
4. Add 20 µL of the PI stock solution and keep 5 min at room temperature.
5. Analyze cells by flow cytometry using excitation in blue light (e.g., 488 nm line of the argon ion laser) and light scatter to trigger cell measurement.
6. Measure green fluorescence of Rh 123 at 530 ± 20 nm and red fluorescence of PI at >600 nm (*see* **Notes 11** and **12**).

3.5. Detection of Phosphatidylserine with Annexin V-FITC Conjugate

1. Dissolve FITC-labeled Annexin V conjugate (1:1 stoichiometric complex) in binding buffer (10 m*M* HEPES/NaOH, pH 7.4, 140 m*M* NaCl, 2.5 m*M* $CaCl_2$) at the concentration of 1 µg/mL. This solution should be prepared fresh each time. (FITC-annexin V is available from B.R.A.N.D Applications, PO Box 864, 6200 AW Maastricht, The Netherlands; cat. no. A-700 for 25 µg or A-800 for 125 µg.)
2. Prepare stock solution of PI by dissolving 1 mg of PI (Sigma) in 1 mL of distilled H_2O.
3. Suspend cells in 1–2 mL of FITC-Annexin V solution. Various concentrations of FITC-Annexin V (0.1–1 µg/mL) can be explored for optimal results.
4. Add PI stock solution to the cell suspension 5 min prior to analysis (final PI concentration 1 µg/mL).
5. Analyze by flow cytometry using blue light excitation (e.g., 488 nm line of the argon ion laser) (*see* **Note 13**).
6. Measure green fluorescence of FITC at 530 ± 20 nm and red fluorescence of PI at >600 nm.

4. Notes

4.1. Cellular DNA Content Measurements

1. Degraded DNA extracted from the ethanol-prefixed apoptotic cells with phosphate-citrate buffer can be directly analyzed by gel electrophoresis ***(40)***.
2. Simplicity and low cost are the main attributes of the analysis of apoptosis by measurement of DNA content. Another advantage is a possibility of simultaneous analysis of DNA ploidy and cell cycle distribution of the nonapoptotic cell population. The method can be applied to a wide range of DNA fluorochromes and instrument types ***(62,63)***.
3. The disadvantage of the method is its low specificity. The sub-G_1 peak, in addition to apoptotic cells, can represent mechanically damaged cells, cells with lower

DNA content (e.g., hypodiploid cells in DNA multiploid tumors), or cells with different chromatin structure (which have lower stainability with DNA fluorochromes). Other factors, such as the presence of certain cations ($CaCl_2/MgCl_2$) or specific RNase in the lysing solution, may influence the extractability of DNA from the nuclei and affect flow cytometric detection of apoptosis by DNA staining ***(65)***.

4. Apoptotic cells have diminished PI or DAPI fluorescence when compared to the cells in the main G1 peak. The problem of overlapping of "apoptotic sub-G_1 peak" with G_1 peak can be solved by experimenting with the concentration and duration of incubation of the cells with extraction buffer. Advanced (late) apoptosis would require, obviously, less or no treatment at all with this buffer, whereas in the situation where DNA is not markedly degraded (early apoptosis), more concentrated buffer (up to 1 mL) would be needed to separate apoptotic cells from the nonapoptotic cell population. The method cannot detect apoptosis if DNA cleavage does not progress to the internucleosomal segments but stops at 50–300 kb, as it happens in some cell systems.
5. Because objects with very small DNA content (e.g., below 10% of DNA content of G_1 cells) may represent minute nuclear fragments, free individual chromosomes from lysed cells, individual apoptotic bodies, cell debris, and so forth, it is recommended to exclude them from the analysis by setting a minimum threshold at 10% of the G_1 peak.

4.2. DNA Strand Break Labeling Assay

6. The procedure of DNA strand break labeling is relatively complex and involves several steps, which may lead to false-negative or false-positive results. It is important, therefore, to include positive and negative controls while running the assay. Such control is provided by, for example, HL-60 cells incubated with 0.2 μ*M* DNA topoisomerase inhibitor camptothecin for 3–4 h. Because only S-phase cells undergo apoptosis under these conditions, the control sample consists of both nonapoptotic (G_1 and G_2/M) and apoptotic (S-phase) cell populations.
7. The method offers the possibility to analyze the cell cycle position and DNA ploidy of apoptotic cells ***(46,47,67,71)***. The number of DNA strand breaks in apoptotic cells appears to be of such large magnitude that the intensity of DNA strand-break labeling can be a specific marker of these cells ***(46)***. The assay is well suited to be applied in analyzing apoptosis in clinical material ***(26,47,67)***.
8. Necrotic cells, or cells with primary DNA breaks caused by X-ray irradiation (up to the dose of 25 Gy) or DNA damaging drugs, have, by order of magnitude, fewer DNA strand breaks compared to apoptotic cells and can be discriminated from apoptotic cells based on intensity of DNA strand-break labeling. It is unclear, however, whether the DNA strand-break labeling assay detects apoptotic cells when apoptosis is atypical, namely characterized by the lack of internucleosomal cleavage. Some data suggest apoptotic cells can be distinguished by such an assay even in the absence of internucleosomal DNA cleavage ***(72)***.

9. Commercial kits are available for detection of apoptosis by DNA strand-break labeling assay: Phoenix Flow Systems (San Diego, CA) provides a kit (Apo-Direct) to identify apoptotic cells based on a single-step procedure, utilizing TdT and FITC-conjugated dUTP, as well as the kit utilizing BrdUTP and BrdUrd MAb. Another kit (ApopTag), based on two-step DNA strand-break labeling with digoxygenin-16-dUTP by TdT, is provided by ONCOR, Inc. (Gaithersburg, MD).

4.3. Detection of Apoptotic Cells by Light Scatter Analysis

10. The advantages of this method are its simplicity and the possibility of combining light scatter with other types of cell analysis by flow cytometry, in particular with surface immunofluorescence (to identify the phenotype of the dying cell). The light scatter changes, however, are specific to neither apoptosis nor necrosis. Mechanically broken cells, isolated nuclei, cell debris, and individual apoptotic bodies also have low light scatter properties. Furthermore, the distinction between necrotic and apoptotic cells is not always apparent; therefore, the light scatter analysis requires several controls and should be accompanied by another, more specific cytofluorometric assay and by light microscopic inspection.

4.4. Exclusion of PI Combined with Uptake of Mitochondrial Probe Rh 123

11. Live nonapoptotic cells stain strongly green with Rh 123 and have minimal red PI fluorescence. Early apoptotic cells have diminished green fluorescence and slightly increased red fluorescence. Late apoptotic cells, similar to late necrotic cells, have strong red and minimal green fluorescence. Very early necrotic cells may have increased green fluorescence and somewhat increased red fluorescence compared to live, nonapoptotic cells. Isolated nuclei, or necrotic or mechanically broken cells that have damaged mitochondrial and/or plasma membranes stain intensely red with PI.
12. It should be stressed that exclusion of PI during apoptosis is not as efficient as in the case of live nonapoptotic cells and after prolonged incubation in the presence of this fluorochrome even early apoptotic cells show significantly higher PI stainability compared to nonapoptotic cells.

4.5. Detection of Phosphatidylserine with Annexin V-FITC Conjugate

13. This assay is primarily applicable to the cells growing in suspension or existing naturally as single cells in the host (e.g., blood or lymph node cells), since attempts to disaggregate solid tissue or to remove adherent cells from vessel walls may lead to damage of the plasma membrane. Any procedure that affects the integrity of the plasma membrane of what would otherwise be intact cells will allow access of Annexin V to the inner plasma membrane, resulting in a cells that will be scored as positive for apoptosis. The use of enzymes to disrupt cell clumps or to remove adherent cells from culture walls as well as the mechanical manipulations (e.g., cells detachment by the "rubber policeman") may affect the detection of apoptotic cells by this assay. Ethidium bromide may be substituted for PI ***(76)***.

References

1. Arends, M. J., McGregor, A. H., and Wyllie, A. H. (1994) Apoptosis is inversely related to necrosis and determines net growth in tumors bearing constitutively expressed myc, ras, and HPV oncogenes. *Am. J. Pathol.* **144,** 1045–1057.
2. Compton, M. M. (1992) A biochemical hallmark of apoptosis: internucleosomal degradation of the genome. *Cancer Metast. Rev.* **11,** 105–119.
3. Darzynkiewicz, Z., Bruno, S., Del Bino, G., et al. (1992) Features of apoptotic cells measured by flow cytometry. *Cytometry* **13,** 795–808.
4. Dive, C., Gregory, C. D., Phipps, D. J., Evans, D. L., Milner, A. E., and Wyllie, A. H. (1992) Analysis and discrimination of necrosis and apoptosis (programmed cell death) by multiparameter flow cytometry. *Biochim. Biophys. Acta* **1133,** 275–285.
5. Wyllie, A. H. (1992) Apoptosis and the regulation of cell numbers in normal and neoplastic tissues: an overview. *Cancer Metast. Rev.* **11,** 95–103.
6. Kerr, J. F. R., Wyllie, A. H., and Curie, A. R. (1994) Apoptosis: a basic biological phenomenon with wide-ranging implications in tissue kinetics. *Br. J. Cancer* **26,** 239–257.
7. Vaux, D. L. (1993) Toward an understanding of the molecular mechanisms of physiological cell death. *Proc. Natl. Acad. Sci. USA* **90,** 786–789.
8. Arends, M. J., Moris, R. G., and Wyllie, A. H. (1990) Apoptosis: the role of endonuclease. *Am. J. Pathol.* **136,** 593–608.
9. Machaca, K. and Compton, M. M. (1993) Analysis of thymic lymphocyte apoptosis using in vitro techniques. *Dev. Comp. Immunol.* **17,** 263–276.
10. Wyllie, A. H., Arends, M. J., Moris, R. G., Walker, S. W., and Evan, G. (1992) The apoptosis endonuclease and its regulation. *Semin. Immunol.* **4,** 389–398.
11. Cohen, G. M., Su, X. M., Snowden, R. T., Dinsdale, D., and Skilleter, D. N. (1992) Key morphological features of apoptosis may occur in the absence of internucleosomal DNA fragmentation. *Biochem. J.* **286,** 331–334.
12. Collins, R. J., Harmon, B. V., Gobe, G. C., and Kerr, J. F. R. (1992) Internucleosomal DNA cleavage should not be the sole criterion for identifying apoptosis. *Int. J. Radiat. Biol.* **61,** 451–453.
13. Hockenbery, D. (1995) Defining apoptosis. *Am. J. Pathol.* **146,** 16–19.
14. Majno, G. and Joris, I. (1995) Apoptosis, oncosis, and necrosis. An overview of cell death. *Am. J. Pathol.* **146,** 3–16.
15. Cory, S. (1995) Regulation of lymphocyte survival by the bcl-2 gene family. *Ann. Rev. Immunol.* **13,** 513–543.
16. Hockenberry, D. M. (1995) bcl-2 in cancer, development and apoptosis. *J. Cell Sci.* **18,** 51–55.
17. Raffo, A. J., Perlman, H., Chen, M. W., Day, M. L., Streitman, J. S., and Buttyan, R. (1995) Overexpression of bcl-2 protects prostate cancer cells from apoptosis in vitro and confers resistance to androgen depletion in vivo. *Cancer Res.* **55,** 4438–4445.
18. Reed, J. C. (1994) bcl-2 and the regulation of programmed cell death. *J. Cell Biol.* **124,** 1–6.
19. Lomo, J., Smeland, E. B., Krajewski, S., Reed, J. C., and Blomhoff, H. K. (1996) Expression of bcl-2 homologue mcl-1 correlates with survival of peripheral blood B lymphocytes. *Cancer Res.* **56,** 40–43.

20. Ryan, J. J., Prochownik, E., Gottlieb, C. A., et al. (1994) c-myc and bcl-2 modulate p53 function by altering p53 subcellular trafficking during the cell cycle. *Proc. Natl. Acad. Sci. USA* **91,** 5878–5882.
21. Schott, A. F., Apel, I. J., Nunez, G., and Clarke, M. F. (1995) Bcl-XL protects cancer cells from p53-mediated apoptosis. *Oncogene* **11,** 1389–1394.
22. Wang, H. G., Miyashita, T., Takayama, S., et al. (1994) Apoptosis regulation by interaction of Bcl-2 protein and Raf-1 kinase. *Oncogene* **9,** 2751–2756.
23. Oltvai, Z. N. and Korsmeyer, S. J. (1994) Checkpoints of dueling dimers foil death wishes. *Cell* **79,** 189–192.
24. Cohen, J. J. (1993) Apoptosis. *Immunol. Today* **14,** 126–130.
25. Darzynkiewicz, Z. (1995) Apoptosis in antitumor strategies: modulation of cell cycle or differentiation. *J. Cell Biochem.* **58,** 151–159.
26. Gorczyca, W., Bigman, K., Mittelman, A., et al. (1993) Induction of DNA strand breaks associated with apoptosis during treatment of leukemias. *Leukemia* **7,** 659–670.
27. Green, D. R. and Cotter, T. G. (1992) Introduction: apoptosis in the immune system. *Semin. Immunol.* **4,** 355–362.
28. Martin, S. J. and Green, D. R. (1994) Apoptosis as a goal of cancer therapy. *Curr. Opin. Oncol.* **6,** 616–621.
29. Fisher, D. E. (1994) Apoptosis in cancer therapy: crossing the threshold. *Cell* **78,** 539–542.
30. Kerr, J. F., Winterford, C. M., and Harmon, B. V. (1994) Apoptosis. Its significance in cancer and cancer therapy (published erratum appears in *Cancer* [1994] **73[12],** 3108). *Cancer* **73,** 2013–2026.
31. Laidlaw, I. J. and Potten, C. S. (1992) Reduction of apoptosis relative to mitosis in histologically normal epithelium accompanies fibrocystic change and carcinoma of the premenopausal human breast. *J. Pathol.* **167,** 25–32.
32. Cotter, T. G. (1992) Induction of apoptosis in cells of the immune system by cytotoxic stimuli. *Semin. Immunol.* **4,** 399–406.
33. Kornbluth, R. S. (1994) The immunological potential of apoptotic debris produced by tumor cells and during HIV infection. *Immunol. Lett.* **43,** 125–132.
34. Oyaizu, N., McCloskey, T. W., Than, S., Hu, R., and Pahwa, S. (1995) Mechanism of apoptosis in peripheral blood mononuclear cells of HIV-infected patients. *Adv. Exp. Med. Biol.* **374,** 101–114.
35. Schuler, D., Szende, B., Borsi, J. D., et al. (1994) Apoptosis as a possible way of destruction of lymphoblasts after glucocorticoid treatment of children with acute lymphoblastic leukemia. *Pediatr. Hematol. Oncol.* **11,** 641–649.
36. Echaniz, P., de Juan, M. D., and Cuadrado, E. (1995) DNA staining changes associated with apoptosis and necrosis in blood lymphocytes of individuals with HIV infection. *Cytometry* **19,** 164–170.
37. Estaquier, J., Idziorek, T., de Bels, F., et al. (1994) Programmed cell death and AIDS: significance of T-cell apoptosis in pathogenic and nonpathogenic primate lentiviral infections. *Proc. Natl. Acad. Sci. USA* **91,** 9431–9435.

38. Razvi, E. S. and Welsh, R. M. (1993) Programmed cell death of T lymphocytes during acute viral infection: a mechanism for virus-induced immune deficiency. *J. Virol.* **67,** 5754–5765.
39. Darzynkiewicz, Z., Juan, G., Li, X., Gorczyca, W., Murakami, T., and Traganos, F. (1997) Cytometry in cell necrobiology: analysis of apoptosis and accidental cell death (necrosis). *Cytometry* **27,** 1–20.
40. Gong, J., Traganos, F., and Darzynkiewicz, Z. (1994) A selective procedure for DNA extraction from apoptotic cells applicable for gel electrophoresis and flow cytometry. *Anal. Biochem.* **218,** 314–319.
41. Crompton, T., Peitsch, M. C., MacDonald, H. R., and Tschopp, J. (1992) Propidium iodide staining correlates with the extent of DNA degradation in isolated nuclei. *Biochem. Biophys. Res. Commun.* **183,** 532–537.
42. Darzynkiewicz, Z., Li, X., Gong, J., Hara, S., and Traganos, F. (1995) Analysis of cell death by flow cytometry, in *Cell Growth and Apoptosis. A Practical Approach* (Studzinski, G. P., ed.), IRL, Oxford, UK, pp. 143–168.
43. Dolzhanskiy, A. and Basch, R. S. (1995) Flow cytometric determination of apoptosis in heterogeneous cell populations. *J. Immunol. Methods* **180,** 131–140.
44. Endresen, P. C., Prytz, P. S., and Arbakke, J. (1995) A new flow cytometric method for discrimination of apoptotic cells and detection of their cell cycle specificity through staining of F-actin and DNA. *Cytometry* **20,** 162–171.
45. Gold, R., Schmied, M., Rothe, G., et al. (1993) Detection of DNA fragmentation in apoptosis: application of in situ nick translation to cell culture systems and tissue sections. *J. Histochem. Cytochem.* **41,** 1023–1030.
46. Gorczyca, W., Bruno, S., Darzynkiewicz, R. J., Gong, J., and Darzynkiewicz, Z. (1992) DNA strand breaks occurring during apoptosis: their early *in situ* detection by the terminal deoxynucleotidyl transferase and nick translation assays and prevention by serine protease inhibitors. *Int. J. Oncol.* **1,** 639–648.
47. Gorczyca, W., Gong, J., and Darzynkiewicz, Z. (1993) Detection of DNA strand breaks in individual apoptotic cells by the *in situ* terminal deoxynucleotidyl transferase and nick translation assays. *Cancer Res.* **52,** 1945–1951.
48. Li, X., Melamed, M. R., and Darzynkiewicz, Z. (1996) Detection of apoptosis and DNA replication by differential labeling of DNA strand breaks with fluorochromes of different color. *Exp. Cell. Res.* **222,** 28–37.
49. Nicoletti, I., Migliorati, G., Pagliacci, M. C., Grignani, F., and Riccardi, C. (1991) A rapid and simple method for measuring thymocyte apoptosis by propidium iodide staining and flow cytometry. *J. Immunol. Methods* **139,** 271–279.
50. Ormerod, M. G., Cheetham, F. P. M., and Sun, X. M. (1995) Discrimination of apoptotic thymocytes by forward light scatter. *Cytometry* **21,** 300–304.
51. Ormerod, M. G., Sun, X. M., Brown, D., Snowden, R. T., and Cohen, G. M. (1993) Quantification of apoptosis and necrosis by flow cytometry. *Acta Oncol.* **32,** 417–424.
52. Philpott, M. G., Turner, A. J. C., Scopes, J., et al. (1996) The use of 7-amino actinomycin D in identifying apoptosis: simplicity of use and broad spectrum of application compared with other techniques. *Blood* **87,** 2244–2251.

53. Schmid, I., Uittenbogaart, C. H., Keld, B., and Giorgi, J. V. (1994) A rapid method for measuring apoptosis and dual-color immunofluorescence by single laser flow cytometry. *J. Immunol. Methods* **170,** 145–157.
54. Sun, X. M., Snowden, R. T., Skilleter, D. N., Dinsdale, D., Ormerod, M. G., and Cohen, G. M. (1992) A flow-cytometric method for the separation and quantitation of normal and apoptotic thymocytes. *Anal. Biochem.* **204,** 351–356.
55. Swat, W., Ignatowicz, L., and Kisielow, P. (1991) Detection of apoptosis of immature CD4+8+ thymocytes by flow cytometry. *J. Immunol. Methods* **137,** 79–87.
56. Vermes, I., Haanen, C., Steffens Nakken, H., and Reutelingsperger, C. (1995) A novel assay for apoptosis. Flow cytometric detection of phosphatidylserine expression on early apoptotic cells using fluorescein labelled Annexin V. *J. Immunol. Methods* **184,** 39–51.
57. Kamentsky, L. A. and Kamentsky, L. D. (1991) Microscope-based multiparameter laser scanning cytometer yielding data comparable to flow cytometry. *Cytometry* **12,** 381–387.
58. Martin Reay, D. G., Kamentsky, L. A., Weinberg, D. S., Hollister, K. A., and Cibas, E. S. (1994) Evaluation of a new slide-based laser scanning cytometer for DNA analysis of tumors. Comparison with flow cytometry and image analysis. *Am. J. Clin. Pathol.* **102,** 432–438.
59. Darzynkiewicz, Z., Li, X., and Gong, J. (1994) Assays of cell viability: discrimination of cells dying by apoptosis. *Methods Cell Biol.* **41,** 15–38.
60. Elstein, K. H. and Zucker, R. M. (1994) Comparison of cellular and nuclear flow cytometric techniques for discriminating apoptotic subpopulations. *Exp. Cell. Res.* **211,** 322–331.
61. Fraker, P. J., King, L. E., Lill Elghanian, D., and Telford, W. G. (1995) Quantification of apoptotic events in pure and heterogeneous populations of cells using the flow cytometer. *Methods Cell Biol.* **46,** 57–76.
62. Telford, W. G., King, L. E., and Fraker, P. J. (1991) Evaluation of glucocorticoid-induced DNA fragmentation in mouse thymocytes by flow cytometry. *Cell Prolif.* **24,** 447–459.
63. Telford, W. G., King, L. E., and Fraker, P. J. (1992) Comparative evaluation of several DNA binding dyes in the detection of apoptosis-associated chromatin degradation by flow cytometry. *Cytometry* **13,** 137–143.
64. Telford, W. G., King, L. E., and Fraker, P. J. (1994) Rapid quantitation of apoptosis in pure and heterogeneous cell populations using flow cytometry. *J. Immunol. Methods* **172,** 1–16.
65. Elstein, K. H., Thomas, D. J., and Zucker, R. M. (1995) Factors affecting flow cytometric detection of apoptotic nuclei by DNA analysis. *Cytometry* **21,** 170–176.
66. Gorczyca, W., Traganos, F., Jesionowska, H., and Darzynkiewicz, Z. (1993) Presence of DNA strand breaks and increased sensitivity of DNA *in situ* to denaturation in abnormal human sperm cells. Analogy to apoptosis of somatic cells. *Exp. Cell Res.* **207,** 202–205.

67. Gorczyca, W., Tuziak, T., Kram, A., Melamed, M. R., and Darzynkiewicz, Z. (1994) Detection of apoptosis-associated DNA strand breaks in fine-needle aspiration biopsies by in situ end labeling of fragmented DNA. *Cytometry* **15,** 169–175.
68. Gold, R., Schmied, M., Giegerich, G., et al. (1994) Differentiation between cellular apoptosis and necrosis by the combined use of in situ tailing and nick translation techniques [see comments]. *Lab. Invest.* **71,** 219–225.
69. Gorczyca, W., Melamed, M. R., and Darzynkiewicz, Z. (1993) Apoptosis of S-phase HL-60 cells induced by DNA topoisomerase inhibitors: detection of DNA strand breaks by flow cytometry using the in situ nick translation assay. *Toxicol. Lett.* **67,** 249–258.
70. Toyoshima, H. and Hunter, T. (1994) p27, a novel inhibitor of G1 cyclin-Cdk protein kinase activity, is related to p21. *Cell* **78,** 67–74.
71. Gorczyca, W., Gong, J., Ardelt, B., Traganos, F., and Darzynkiewicz, Z. (1993) The cell cycle related differences in susceptibility of HL-60 cells to apoptosis induced by various antitumor agents. *Cancer Res.* **53,** 3186–3192.
72. Chapman, R. S., Chresta, C. M., Herberg, A. A., et al. (1995) Further characterisation of the in situ terminal deoxynucleotidyl transferase (TdT) assay for the flow cytometric analysis of apoptosis in drug resistant and drug sensitive leukaemic cells. *Cytometry* **20,** 245–256.
73. Melamed, M. R., Lindmo, T., Mendelson, M. L. (1990) *Flow Cytometry and Sorting,* 2nd ed. Wiley-Liss, New York.
74. Johnson, L. V., Walsh, M. L., and Chen, L. B. (1980) Localization of mitochondria in living cells with rhodamine 123. *Proc. Natl. Acad. Sci. USA* **77,** 990–994.
75. Darzynkiewicz, Z., Traganos, F., Staiano-Coico, L., Kapuscinski, J., and Melamed, M. R. (1982) Interactions of rhodamine 123 with living cells studied by flow cytometry. *Cancer Res.* **42,** 799–806.
76. Koopman, G., Reutelingsperger, C., Kuiijten, G. A. M., Keehnen, R. M. J., Pals, S. T., and van Oers, M. H. J. (1994) Annexin V for flow cytometric detection of phosphatidylserine expression of B cells undergoing apoptosis. *Blood* **84,** 1415–1420.
77. Li, X. and Darzynkiewicz, Z. (1995) Labeling DNA strand breaks with BrdUTP detection of apoptosis and cell proliferation. *Cell Prolif.* **28,** 571–579.

21

Chromosome Sorting and Analysis by FACS

Simon P. Monard

1. Introduction

Many types of cancer and genetic diseases are characterized by chromosomal aberrations. Conventional cytogenetics, the analysis of banded metaphase chromosomes, can be very time consuming and, in many cases, marker chromosomes cannot be identified by their banding pattern alone. For some years there has been interest in a more rapid and objective method of analyzing chromosomes. Flow cytometry offers an alternative machine-based approach in which a suspension of metaphase chromosomes is prepared, stained with one or two DNA-binding fluorochromes, and passed through a flow cytometer. The first human flow karyotypes were demonstrated in the mid-1970s using chromosomes isolated from fibroblast cells *(1,2)*. High resolution flow karyotypes were subsequently obtained from PHA-stimulated peripheral blood lymphocytes *(3)* and lymphoblastoid cell lines *(4)*. In practice, chromosomes can be prepared from almost any culture of growing cells. In the case of the single-laser system in which one nonbase-specific dye, such as ethidium bromide, is used, the intensity of fluorescence from each chromosome is proportional to its DNA content. The signal intensity from each chromosome is measured and recorded as it passes through the focused laser beam. Accumulating data from 10,000–50,000 chromosomes and presenting it as a histogram of fluorescence intensity against frequency produces a distinctive species specific pattern of peaks. Such a histogram is known as a univariate flow karyotype. Not all human chromosomes appear as separate peaks. An improvement of this system involves staining the chromosome preparation with two dyes that have a base pair preference, such as Hoechst 33258 and chromomycin A3. The intensity of the signal from each chromosome is influenced not only by their DNA content but also by their base- pair make up. The intensity of both fluorescent signals

From: *Methods in Molecular Biology, Vol. 91: Flow Cytometry Protocols*
Edited by: M. J. Jaroszeski and R. Heller © Humana Press Inc., Totowa, NJ

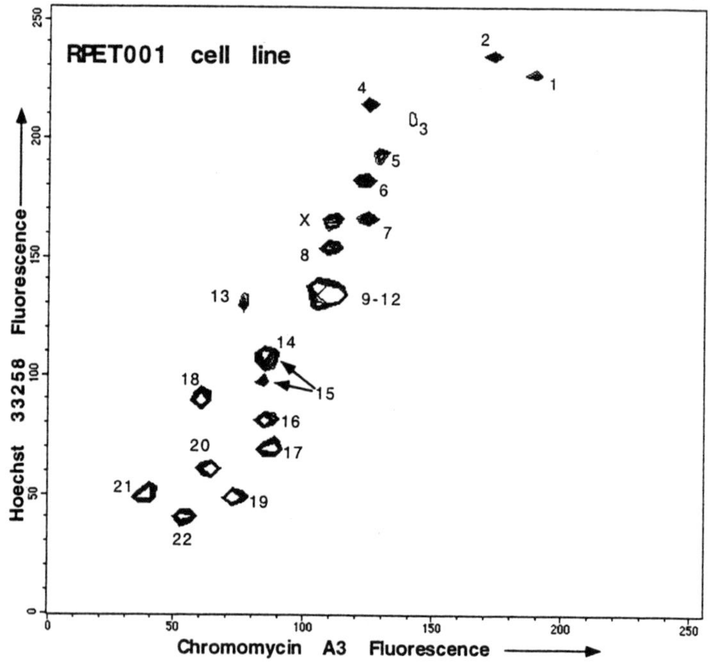

Fig. 1. Bivariate flow karyotype of EBV transformed lymphoblastoid cell line RPET001. One homolog of chromosome 15 appears in the same peak as chromosome 14.

are recorded for each chromosome and the data accumulated can be presented as a dot-plot or contour map see **Fig. 1**. All human chromosomes can be resolved as separate peaks except for chromosomes 9–12. The flow karyotype provides no information about the individual cell but can provide an accurate measurement of the frequency of the different chromosome types, for instance trisomy 21 would appear as a 50% increase in the frequency of chromosome 21 as compared with the other chromosomes. Translocations resulting in two derivative chromosomes that differ in either DNA content or base pair ratio from the chromosomes from which they are derived will appear as two separate peaks in positions where there are normally none (*see* **Fig. 2**). Small marker chromosomes and deletions can also usually be detected (*see* **Fig. 3**). Why then has this rapid, highly reproducible machine-based approach not completely displaced conventional cytogenetics? There are several reasons: Sometimes the derivative chromosomes appear in the same position as other chromosomes, making them difficult to detect, also a reciprocal translocation resulting in two derivative chromosomes that have the same DNA content and base pair ratio as the parent chromosomes would remain undetected. Another reason is the

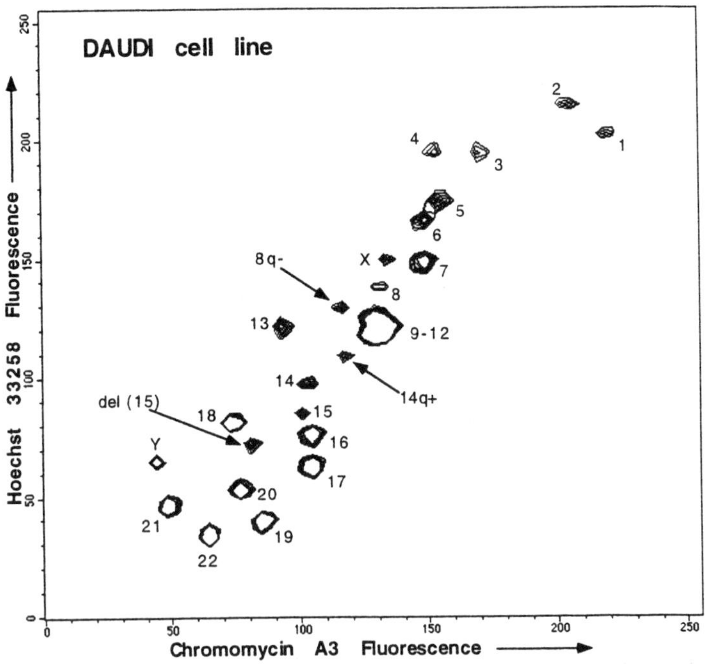

Fig. 2. Bivariate flow karyotype of Daudi cell line, which carries the t(8;14) translocation found in Burkitts lymphoma. The two translocation products are identified. One copy of chromosome 15 has a deletion and appears as a separate peak.

expense of buying a dual laser flow cytometer and the high level of expertise needed to operate it. However, the major obstacle is the polymorphic nature of the population. The two homologs of the same chromosome type in a normal individual often appear as separate peaks (*see* **Fig. 3**). Therefore, in practice it would be difficult to ascertain whether the two peaks seen in a flow karyotype were the two normal homologs of a chromosome or whether one was abnormal. Certain chromosomes have regions of centric heterochromatin that can vary considerably in size, resulting in microscopically visible differences ***(5)***. Variations in flow karyotypes have been correlated with specific C or Q band heteromorphisms ***(3,6)***. As chromosome heteromorphisms appear to be inherited unaltered in size ***(7)***, any feature that cannot be seen in either of the parental flow karyotypes can be assumed to be of *de novo* occurrence ***(8)***. Such family studies would prove difficult in cancer cytogenetics in which the age of presentation is often in late middle age and are hardly a step towards a rapid machine-based approach to chromosome analysis.

The real power of the flow system is the ability of the flow cytometer to physically separate the different chromosome types. Any chromosome that can

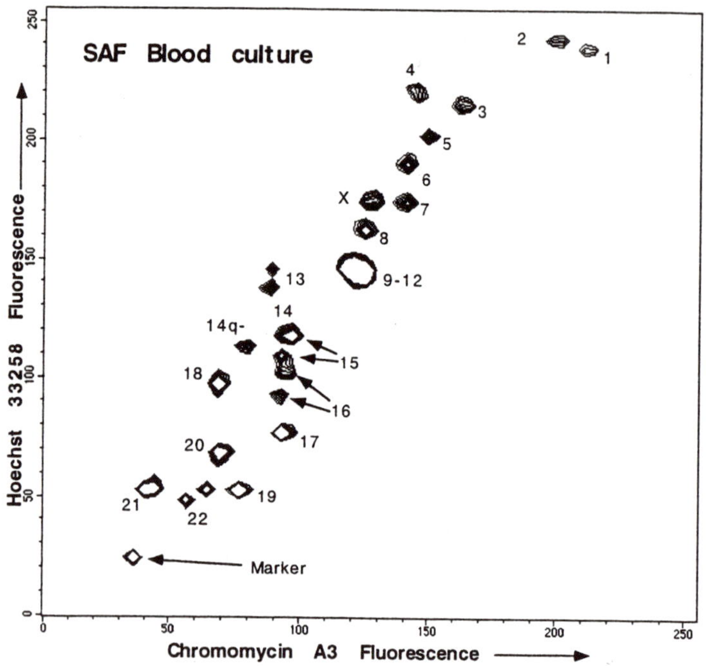

Fig. 3. Bivariate flow karyotype of phytohemagglutinin (PHA) stimulated peripheral blood culture from a female with a psychodevelopmental disorder. The picture is normal except for a deletion of one chromosome 14 and a small marker that is about half the size of a G group chromosome. The two homologs of chromosome 22, although normal, appear as separate peaks. Reverse chromosome painting was used to investigate the composition of this unusual marker chromosome ***(24)***.

be resolved as a separate peak can be sorted with a high degree of purity, a purity of 95% is typical, and up to 99% purity is possible. Phage libraries ***(9)*** and more recently cosmic libraries ***(10,11)*** have been constructed from flow-sorted chromosomes. These libraries have proved to be an essential resource for the study of the human genome. Suspension cell lines are the most convenient type of cells when very large numbers of chromosomes are required, such as for sorting for chromosome library construction, and monolayer-type cells can produce chromosome preparations with less debris, because dead cells can be removed by washing. It should be remembered that transformed cell lines especially rodent cell lines often develop chromosome aberrations that were not present in the primary tissue from which they were derived.

Sorting chromosomes for library construction is time consuming, it may take several days of sorting before enough chromosomes are separated. More appealing to the flow cytometrist is the combination of flow sorting and polymerase chain reaction (PCR) ***(12)***. Enough chromosomes for a PCR reaction can be sorted in only a few seconds. If Alu-PCR ***(13)*** or degenerate oligonucleotide primed PCR (DOP-PCR) ***(14)*** are used to amplify chromsomal DNA, a library of PCR products will be produced ranging from approx 500 bp to 3 kb in length and all having sequence identity to the chromosome from which they were generated. These PCR products can be used as chromosome-specific probes for the purpose of isolating cosmids or YACS of interest ***(15)*** or they can be labeled with biotin-dUTP or fluorescent dUTPs and used as *in situ* hybridization probes ***(13,16)***. Using such fluorescent probes directed against the whole chromosome is termed chromosome painting and the probes are referred to as chromosome paints. The resultant signal identifies the chromosome type from which the paint was prepared. Using chromosome paints prepared from normal chromosomes and applying these paints to metaphases containing abnormal chromosomes can reveal the identity of aberrant chromosomes, this is termed forward chromosome painting. Chromosome paints can be prepared from any of the chromosomes that resolve as separate peaks. The positions of the different human chromosomes in the bivariate flow karyotype are well-established, so paints can readily be prepared from all the human chromosomes except chromosomes 9–12. A more rapid way of identifying the make-up of an unidentified marker chromosome is to flow sort that chromosome, prepare a chromosome paint from it, and apply this paint to a metaphase spread from a normal individual. The signal will be seen to be restricted to those chromosomes from which the marker is composed ***(16)***. This technique has been termed reverse chromosome painting. Chromosome paints can also be generated from other species. Paints have been produced from porcine chromosomes ***(17)***, mouse chromosomes ***(18)***, and rat chromosomes ***(20)*** (*see* **Fig. 4**). Chromosome painting can be used for the detection of structural and numerical aberrations in tissue preparations and cell suspensions fixed on slides ***(18,19)***. Such techniques can be used for the purpose of evaluating the toxological properties of chemical mutagens and radiation on animal cells.

Chromosome paints generated by Alu-PCR do not paint the chromosome evenly, a series of bright and dark bands are seen that correspond to the it-banding pattern. This is because of the Giemsa dark bands having few Mu-repeats. More even painting can be achieved by using DOP-PCR. The primer used in this PCR reaction, unlike Alu primers is not species-dependent and produces a signal that is more evenly distributed over the chromosome, although, as with the Alu-PCR-generated paints, the centromere is not usually painted. There are two main ways the PCR products can be labeled: During the PCR reaction by addition of labeled dNTPs or after the PCR reaction using one of the commercially available nick translation kits or random prime labeling kits.

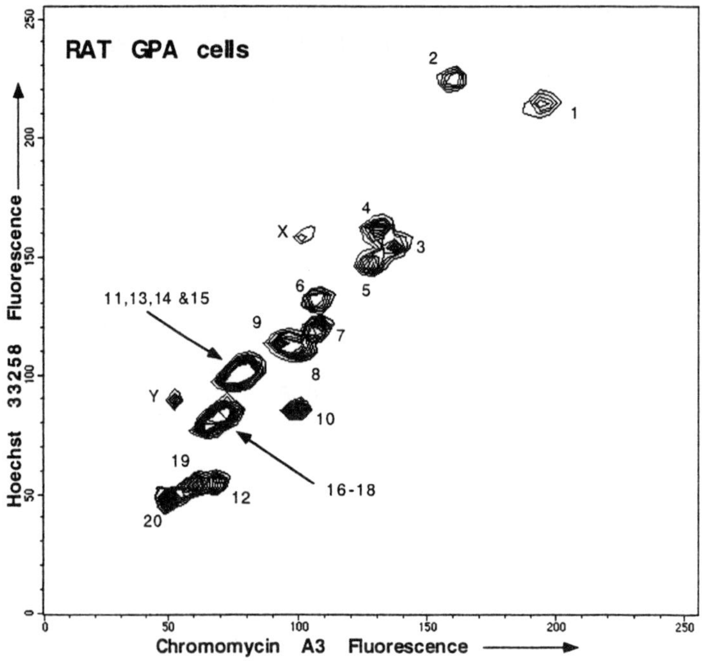

Fig. 4. Bivariate flow karyotype of the rat *Rattus norvegicus*. The granuloma pouch assay (GPA) was exploited to avoid the aberrent chromosomes that can appear in immortal rodent cell lines.

2. Materials

2.1. Chromosome Preparation and Sorting

1. Sheath buffer: 100 m*M* NaCl, 10 m*M* Tris-HCl, 1 m*M* EDTA, pH 8.0, in distilled water.
2. Colcemid solution (Life Technologies).
3. Hypotonic solution: 75 m*M* KCl.
4. 10X Chromosome isolation buffer 1 (10X CIB1): 200 m*M* NaCl, 800 m*M* KCl, 150 m*M* Tris-HCl, 2 m*M* spermine (free base), 5 m*M* spermidine (free base) pH 7.2 in autoclaved distilled water. This buffer may be stored at 4°C for several months.
5. Hoechst 33258 (Sigma, St. Louis, MO) 100 µg/mL in distilled water.
6. Chromomycin A_3 (Sigma) 2 mg/mL in ethanol.
7. Digitonin (Sigma).
8. Propidium iodide (PI) 50 µg/mL in phosphate-buffered saline (PBS).
9. Chromosome isolation buffer 2 (CIB2): 40 m*M* KCl, 5 m*M* HEPES, 10 m*M* $MgSO_4$, 3 m*M* dithiothreitol, pH 8.0, in autoclaved distilled water.
10. Triton X-100.

2.2. Library Preparation

1. t-RNA (Life Technologies).
2. Proteinase K (BDH).
3. Stock solution 500 m*M* EDTA, pH 8.0.
4. Stock solution 20% w/v n-lauroylsarcosine (sodium salt).

2.3. Polymerase Chain Reaction

1. 10X DOP-Buffer 20 m*M* $MgCl_2$, 500 m*M* KCl, 100 m*M* Tris-HCl, 1 mg/mL gelatin, pH 8.4.
2. dNTPs (Pharmacia Ultrapure, supplied as 100 m*M* solutions) working solution 10 m*M* each dNTP.
3. 6 MW DOP-PCR primer (5' CCG ACT CGA GNN NNN NAT GTG G 3'), where N=A,C,G, or T in approximately equal proportions) working solution at 1 μg/μL.
4. Taq polymerase (AmpliTaq, Perkin Elmer Cetus).
5. Distilled water (autoclaved distilled water is acceptable).
6. Mineral oil (Sigma).
7. 10X ALU-buffer: 15 m*M* $MgCl_2$, 500 m*M* KCl, 100 m*M* Tris-HCl, 1.0% Triton X-100, 1 mg/mL gelatin, pH 8.9.
8. BK33 primer (5' CTG GGA TTA C2AG GCG TGA GC 3'), working solution at 1 μg/μL.

2.4. Instrumentation

Most commercially available flow sorters can be used for univariate chromosome sorting and analysis. The computer system should allow the histogram to have at least 256 channels. Chromosomes are usually stained with ethidium bromide that can be excited with the 514 nm line, or more usually with the 488 nm line of an argon ion laser. The emission signal can be collected using a 580 nm long pass filter. Other fluorochromes can be used for univariate analysis and sorting if they give a clearer separation of the chromosome peak of interest, the choice of fluorochrome often depends on the light source available. Bench-top analyzers with small air-cooled lasers, such as the FACScan (Becton Dickinson, San Jose, CA) cannot resolve the separate peaks sufficiently well to be useful for chromosome analysis. In our experience the best univariate flow karyotypes are obtained with Hoechst 33258.

Bivariate chromosome analysis and sorting using Hoechst 33258 and chromomycin A3 requires a flow cytometer equipped with two argon ion lasers. The primary laser, i.e., the laser that intersects the sample stream nearest the nozzle, is usually tuned to the UV lines from 351.1–363.8 nm and is used to excite the Hoechst dye. The secondary laser is usually tuned to 457.9 nm and is used to excite the chromomycin. The signal from the Hoechst dye is collected using a 390-nm long pass and a 480-nm short-pass filter. The chromomycin signal is collected using a 490-nm long-pass filter only. All filters are of the colored glass variety. The signal from the primary laser is usually designated fluorescence 1 (FL1) and that from the secondary laser fluorescence 2 (FL2). Bivariate analysis is preferable as more of the different chromosome types are represented as separate peaks and these peaks are more clearly separated. With Becton Dickinson dual-laser instruments the two signals from the different dyes are separated in three ways. First there is temporal separation, a particle is illuminated by the secondary laser about 20 μs after it is illuminated by the primary laser, so the signal from the chromomycin is collected 20 μs after the instrument has been triggered by the Hoechst signal. Second, there is spatial separation, the primary laser strikes the stream above the secondary laser, the fluorescent signals are inverted by the collection optics, and the secondary signal is reflected by a half mirror into the FL2 channel. The third mode of signal separation is by optical filters, colored glass filters are used in front of the photomultiplier tubes (PMTs), the filters in front of the FL1 PMT allow only blue light through from 390–480 nm. The FL2 signal is collected above 490 nm. These three methods of separation ensure that the signals from the two dyes are measured completely independently from one another.

The alignment of the laser beams is of critical importance. Special care should be taken to ensure that they do not pass too close to the edge of any of the prisms because this can cause diffraction and subsequent loss of resolution. It is important that the lasers are functioning in the TEM00 mode. The laser focusing lens is usually an achromatic doublet, i.e., two lenses of different materials positioned close to each other or cemented together. If the lens is of the cemented type (such as that used in a FACS 440 (Becton Dickinson) the cement may discolor after prolonged use in the UV and so should be inspected periodically. The usual nozzle orifice size used is 50 μm, although a 70 μm nozzle can be used. A dirty nozzle can cause poor resolution, increased noise, and irregular deflection streams (*see* **Note 1**). When sorting chromosomes, particularly for library construction, keeping the sample and sorted fraction cool with ice or circulating cold water reduces the chance of DNA degradation.

The instrument should be triggered on the Hoechst signal and the minimum possible threshold should be used as a high threshold would allow small particles, such as chromosome fragments, to pass undetected through the instrument and into the sorted droplets. Although separation of the different chromosome types is optimal with the primary laser tuned to the UV and secondary laser tuned to 457.9 nm if necessary, these can be reversed.

Sorting chromosomes for library construction is a time consuming process. With a well-setup conventional cytometer and good preparation it should be possible to pass 1500–2500 chromosomes/s through the instrument without much deterioration of resolution. Thus it should be possible to sort a single copy chromosome at a rate of 30–50/s. High-speed sorters should easily exceed these rates. When running chromosomes for analysis only, collect 20,000–50,000 events, the best discrimination between chromosome peaks will be obtained using a low sample pressure.

Care should be exercised when aligning the lasers, especially in the UV. The lowest laser powers should be used and protective goggles must be worn.

3. Methods

3.1. Chromosome Preparation

Chromosome suspensions are prepared by adding an agent, such as colcemid or vinblastine, to a culture of growing cells and incubating at 37°C for several hours or overnight until sufficient cells have accumulated in mitosis. Leaving for too long will result in death and necrosis of some cells, which will produce DNA debris indistinguishable from chromosomes. Incubating for too short a period results in too few cells in mitosis. The ideal length of time depends on the rate at which the cells are growing. The key to making a good chromosome preparation is to start with a cell culture of healthy cells that are growing optimally. If the cells grow as an attached layer, mitotic shake-off can be used to obtain an enriched population of metaphase cells. The electronics of conventional flow sorters can only deal with flow rates of up to approx 5000 events per second, if only a small proportion of these events are chromosomes, then the actual number of chromosomes sorted will be low. For instance, ignoring abort and coincidence rates, at a sample rate of 5000 events per second assuming all events are chromosomes, a single copy chromosome will be isolated at a rate of approx 100/s. If only 10% of the events are chromosomes for the same sample rate, only 10 chromosomes per second will be isolated. A new generation of commercially available high-speed sorters is now emerging, but the author is to-date unable to comment on their effectiveness as chromosome sorters.

Sufficient chromosomes for flow karyotyping can be obtained from as little as 2 mL of human peripheral blood. The use of short-term cultures has the advantage that there is little opportunity for karyotype alterations that can occur in established cell lines. Also, the flow karyotype can be compared directly with the conventionally banded karyotype normally performed for routine analysis. Recently established EBV transformed cell lines, on the other hand, represent an ideal source when large numbers of chromosomes are required, such as for sorting for library construction. Some carry known aberrant chromosomes that may be sorted to facilitate analysis of the genotype of the aberration or to simplify subchromosome gene mapping with known karyotype abnormalities.

There are two main methods used for the preparation of chromosomes for flow sorting, one uses polyamines to stabilize chromosomal DNA, the other uses magnesium sulfate. The polyamine method ***(22)*** offers good discrimination between the chromosome types and the DNA after sorting is of very high molecular weight, This method is described in detail below in **Subheading 3.1.1.** The magnesium-sulfate method *(23)* offers excellent discrimination between the chromosomes and is rapid and simple to perform, the DNA however may not be of such good quality, this method is described in **Subheading 3.1.2.** Some additional comments on chromosome preparation will be found in **Note 2**.

3.1.1. Chromosome Preparation: Polyamine Method

This is the method used in our laboratory, it was first described in 1981 and represented a breakthrough as it enabled commercially available instruments with 5-W lasers to be used for sorting chromosomes. The preparations can be stored for several weeks or even months, the DNA is of high quality and the resolution is good. We use the same protocol for preparing chromosomes from human and rodent cells.

1. Monolayer or suspension cell lines may be used, as may phytohemaglutinin-stimulated peripheral blood cells. Whichever type of cell is used, the best preparations will be made from healthy cells growing optimally. Subculture cells 24 h before blocking with colcemid.
2. Block cells with 0.05 μg/mL colcemid for 5–16 h depending on the rate of growth. Usually blocking overnight gives good results.
3. The proportion of monolayer cells in mitosis can be estimated on an inverted microscope. Mitotic cells are round and can usually be shaken off into the medium by giving the flask a sharp rap (this is known as mitotic shakeoff). Some monolayer cell lines may require the use of trypsin. Once in suspension centrifuge all types of cells at 100*g* for 10 min in 50-mL plastic tubes, discard the supernatant, and resuspend the cells in fresh medium before a further 10 min centrifugation at 100*g*.
4. Discard the supernatant by inverting the tube. Remove the last few drops from inside the tube with a tissue. Disaggregate the cell pellet by vortexing gently or by flicking the tube. Add 5 mL hypotonic solution, mix gently, and leave for 10–30 min at room temperature (lymphoblastoid cell lines usually require 20 min, fibroblastoid cell lines usually require 30 min). This is a convenient time to pool the contents of several tubes.
5. While the cells are in the swelling solution, dissolve 12 mg of digitonin in 5 mL distilled water by heating on a hot plate or in a microwave oven. Allow the digitonin solution to cool then add l mL 10X CIB1 and make the volume up to 10 mL with distilled water. Adjust to pH 7.2 if necessary and place on ice.
6. Centrifuge the tubes for 10 min at 100*g*. Following centrifugation, carefully remove the supernatant with a Pasteur pipet and agitate the tube gently to disaggregate the cells. Add about 10 times the volume of the cell pellet in cold CIB1 (1X with digitonin as prepared in **step 5**), aspirate gently a few times with a Pasteur pipet. Mix a small amount of the preparation with an equal volume of PI

(50 μg/mL in PBS) and view with a fluorescence microscope. If the chromosomes are not monodispersed, aspirate the preparation more vigorously or vortex gently. Avoid vortexing too vigorously as it can result in chromosome damage.

7. The chromosome suspension may be stored at 4°C for several weeks with little deterioration of flow karyotype.
8. To stain, transfer 1 mL of the chromosome suspension into a 12 × 75 mm plastic test-tube, add 30 μL Hoechst 33258 (100 μg/mL in distilled water), mix immediately. Add 40 μL of 15 m*M* $MgCL_2$ and 50 μL Chromomycin A_3 (2 mg/mL in ethanol), mix, and leave the sample at 4°C for 2 h in the dark.
9. The chromosome profile can be improved if 100 μl sodium citrate (100 m*M*) and 100 μL sodium sulfite (250 m*M*) are added at least 15 min prior to running on cytometer. Aggregates and intact nuclei in the sample can be removed by centrifuging at 200*g* for 1 min, then transferring the supernatant to a new tube. Centrifugation selectively depletes the larger chromosomes so should be avoided if the flow karyotype is to be analyzed.

3.1.2. Chromosome Preparation: Magnesium Sulfate Method

1. Prepare colcemid-blocked cells as with the polyamine method.
2. Centrifuge cells at 300*g* for 10 min at room temperature and decant supernatant, draining tubes on an absorbent paper towel.
3. Add 1 mL of CIB2 to 6 × 10^5 cells, resuspend gently, and incubate at room temperature for 10 min.
4. Add 0.1 mL of Triton X-100 solution (2.5% in distilled water) and incubate on ice for 10 min.
5. Vortex for 10–20 s to disrupt the cells and incubate at room temperature for 10 min (monitor using phase contrast microscopy).
6. Stain for bivariate analysis as in **Subheading 3.1.1., steps 8** and **9**.

3.2. Flow Sorting Chromosomes for Library Construction

Cosmid libraries constructed from flow sorted chromosomes ***(10,11)*** have proved to be a vital resource for the analysis of the human genome. Flow sorting for such a purpose can be time-consuming, so it is essential to sort at the highest rate possible without compromising purity. Keeping both the sample and sorted fractions cool reduces the chance of DNA degradation. A concentrated chromosome suspension with few interphase nuclei allows a higher sample rate for the same sample pressure.

1. Prepare sterile sheath buffer containing 500 μg/mL t-RNA. Dispense 50 μL of this solution into sterile 1.5-mL conical tubes and vortex vigorously to coat the inside of the tubes. They can be stored by freezing quickly on dry ice and stored at –20°C.
2. Sort 5 × 10^5 chromosomes into each tube.
3. Prepare a working solution of 250 m*M* EDTA with 10% w/v n-lauroylsarcosine. This solution is added to the sorted chromosome suspension to make a final con-

centration of 25 m*M* EDTA and 1% n-lauroylsarcosine. When using a 50 µm nozzle and a three-drop deflection, 5×10^5 chromosomes should occupy a volume of about 700–800 µL, thus 70–80 µL working solution should be added.
4. Add 180 µg of proteinase K to each tube, vortex, and incubate at 42°C overnight.
5. The resulting DNA preparation can be stored at 4°C for many months.

3.3. Generation of Chromosome Paints

3.3.1. Degenerate Oligonucleotide-Primed Polymerase Chain Reaction (DOP-PCR)

DOP-PCR using the primer 6–MW can be used to generate chromosome paints from flow-sorted chromosomes from any species. The paints generated using this primer paint the chromosome evenly along its length, although the centromere does not usually label. The method described below is a modified protocol of Telinius et al. ***(14)***.

1. Combine the reagents in one tube to produce PCR mix with final concentrations of: 1X DOP Buffer, 200 µ*M* each dNTP, 15 µg/mL 6-MW, and 25 U/mL Taq polymerase (*see* **Note 3**).
2. Aliquot either 50 µl or 100 µL into PCR tubes.
3. Chromosomes can be sorted directly into the tubes; 500–1000 chromosomes for a 100 µL reaction seems optimal. A brief spin in a microcentrifuge ensures the sorted droplets and hence chromosomes are in the bottom of the tube.
4. Overlay with 50 µL mineral oil.
5. Place tubes in thermal cycler and execute a program with the following characteristics: Initial denaturation 5 min at 94°C followed by seven cycles of 30 s at 95°C, 1.5 min at 30°C, 3 min at 72°C with the transition from 30–72°C taking 3 min. This is followed by 35 cycles of 30 s at 95°C, 1 min at 56°C, and 3 min 72°C with an addition of 1 s per cycle to the extension step. The last cycle is followed by an additional 10 min at 72°C.
6. Combine the reagents in one tube to produce a final concentration of: 1X DOP buffer, 200 µ*M* each dNTP, 20 µg/mL 6-MW, and 25 U/mL Taq polymerase. Proceed exactly as in **step 1**, except add 10 µL/mL of a previously amplified sample. Labeled dNTPs can be included.
7. Dispense 50 or 100 µL into PCR tubes and overlay with 50 µL mineral oil.
8. Place tubes in thermal cycler and execute a program with the following characteristics: Initial denaturation 5 min at 95°C followed by 10–25 cycles (if the yield of the primary amplification is high 10–15 cycles is enough) of 1 min at 95°C, 1.5 min at 56°C, and 4 min 72°C with an addition of 1 s per cycle to the extension step. The last cycle is followed by an additional 10 min at 72°C.
9. The PCR products can be visualized by running 5 µL on a 1.5% agarose gel. The products should be in the size range of 300 bp to 3 kb. The *Phi*X174 *Hae*III markers are a suitable DNA standard. Some additional comments concerning PCR problems will be found in **Note 3**.

3.3.2. Alu-Polymerase Chain Reaction

It may sometimes be desirable to generate paints using primers directed against repeat sequences, such as Alu repeats. Mu-primers can be used to generate paints from human chromosomes in somatic cell hybrids in which only the human material is amplified. Alu repeats occur on average approx every 4 kb in the genome but they are not evenly spaced. Painting using Alu paints results in uneven painting along the length of the chromosome, giving a pattern corresponding to it-banding. There is a risk that the region of interest will not be amplified when Alu paints are used. On the other hand, the DOP-PCR paints do not paint entirely evenly. As with Alu paints, the centromere is rarely painted. Alu paints give low background and the pattern along the length of the chromosome can aid chromosome identification.

1. Combine the reagents in one tube to form a final concentration of: 200 μ*M*, each dNTP, 1X ALU buffer, 10 μg/mL BK33, and 25 U/mL Taq polymerase. Aliquot into PCR tubes either 100 μL or 50 μL *see* **Note 3**.
2. Sort 500–1000 chromosomes directly into each PCR tube leaving one tube free for the control. Ideally the control tube should have the appropriate amount of sheath buffer.
3. Execute a PCR program with the following characteristics: 2 min initial denaturation at 95°C then 35 cycles of 30 s 95°C, 1 min 55°C, and 4 min at 68°C. Finally extend for 10 min at 68°C.
4. The PCR products can either be labeled by reamplifying 5 μL of the PCR products with a labeled dNTP, using 10 more cycles of the same protocol, or by using a nick translation kit.

4. Notes

1. It is not always easy to determine if the poor discrimination between chromosomes seen on the cytometer because of poor preparation, or poor cytometer or laser performance. The instrument should be monitored regularly with fluorescent microspheres. The coefficients of variation (CV) of the fluorescent peaks and the intensity of the signal give an indication of how the instrument is performing. A dirty nozzle is a common cause of high CVs. The nozzle orifice can be viewed by placing the nozzle on a microscope slide and viewing with an inverted microscope. If several nozzles of the same size are available, then it is worth while to try each of them and *see* which one gives the lowest CV. A small sonic water bath is a useful device for cleaning nozzles. Place the nozzle in a small tube such as a 5 mL bijou, add 2 mL of water, then sonicate for a few seconds. Do not sonicate the nozzle holder as it can disintegrate.
2. Two main problems can occur when preparing chromosomes. The first and most common problem is with cell culture, if there are very few chromosomes in the preparation and few intact cells in metaphase, then almost certainly the cells are not growing well. There can be many reasons for this ranging from media prob-

lems to mycoplasma infection. The proportion of cells in mitosis can be estimated by pelleting the cells from 1 mL of the blocked cell culture, discarding the supernatant and resuspending in PBS containing 50 μg/mL PI and 0.1% Triton X-100. The estimation can be made either using a fluorescence microscope or a bench-top flow cytometer. It should be possible to get 40–60% of the cells in mitosis with suspension cell lines. The other main problem is that the preparation has many cells in metaphase but few of the chromosomes are released into suspension. Some cell types are very resistant to lysis. To overcome this problem, one can increase the swelling time in KCl, use more detergent, and vortex more vigorously. Some types of cells seem to die when left in the presence of colcemid for long periods, resulting in a mucus-like pellet after swelling. One solution to this problem is to block the cells for a shorter time, for instance, two hours. With very slow growing cells, one strategy that could be employed is to synchronize the cells, monitor the stage of the cell cycle, and block the cells as they approach metaphase. It is also possible to remove dead cells by spinning the cells though a density gradient after blocking, removing the cells at the interface, then washing in PBS before swelling them in KCl. To view the chromosome suspension microscopically, mix 10 μL of the chromosome preparation and 10 μL of propidium iodine solution (50 μg/mL in PBS) together, transfer onto a glass microscope slide, and drop on a cover slip then view with a fluorescence microscope using the 40X objective. Chromosomes will appear as brightly stained "X" shapes and should be abundant. Finally, it is invaluable to exchange preparations with another laboratory experienced in chromosome sorting and analysis.

3. Thermal cyclers can be tested for accuracy using a thermocouple inside a PCR tube. Assuming the thermal cycler is functioning correctly problems with PCR usually fall into three categories. The first is low yield. This can be due to reagent problems, insufficient template DNA, or inappropriate reaction conditions inside the PCR tube. If one or more of the reagents are suspected of denaturing or being inactive, rather than spend many hours substituting new reagents for old one at a time it is often more economical to replace all of the reagents (with the possible exception of the primer). Another reason for low yield can be incomplete denaturation of the template DNA caused by having too low a denaturation temperature. The second common problem encountered particularly when using Alu and DOP primers is the template-independent so called primer-dimer. This presents itself as either a smear or a ladder on the gel. It is not always easy to distinguish primer-dimer from genuine contamination. The only certain way is to sequence some of the PCR products. Primer-dimer is undesirable as these products compete for the other reagents in the reaction and so reduce yield of the chromosome PCR products. This artifact can be reduced or eliminated by reducing the primer concentration or reducing the number of cycles. Using "hot-start" may also help to reduce primer-dimer. Third, contamination of the reaction by DNA can be a problem when doing PCR from small numbers of chromosomes. Certain precautions can be taken such as keeping a set of DNA 'clean' pipet and performing gel electrophoresis in a different area to that used for preparing the PCR reagents. A useful device for destroying DNA in PCR tubes is an Eprom Eraser available

from RS Components, UK. This relatively inexpensive device is used in the electronics industry and uses a UV light source to erase the memory from EPROMS (programmable silicon chips). In our hands, the PCR buffers used in DOP-PCR and Alu-PCR are **not** interchangeable. It is recommended that a large batch of 10X buffer be made, aliquoted, and frozen.

References

1. Gray, J. W., Carrero, A. V., Moore, I. D. H., Steinmetz, L. L., Minkler, J., Mayall, B. H., Mendelsohn, M. L., and Van Dilla, M. A. (1975) High speed quantitative karyotyping by flow microfluorometry. *Clin. Chem.* **21,** 1258–1262
2. Gray, J. W., Carrano, A. V. , Steinmetz, L. L., Van Dilla, M. A., Moore II, D. H., Mayall, B. H., and Mendelsohn, M. L. (1975) Chromosome measurement and sorting by flow systems. *Proc. Natl. Acad. Sci. USA* **72,** 1231–1234.
3. Young, B. D., Ferguson-Smith, M. A., Sillar, R., and Boyd, E. (1981) High resolution analysis of human peripheral lymphocyte chromosomes by flow cytometry. *Proc. Natl. Acad. Sci. USA* **78,** 7727–7731.
4. Krumlauf, R., Jeanpierre, M., and Young, B. D. (1982) Construction and characterization of genomic libraries from specific human chromosomes. *Proc. Natl. Acad. Sci. USA* **79,** 2971–2975.
5. Trask, B., van den Engh, G., Mayall, B., and Gray, J. W. (1989) Chromosome heteromorphism quantified by high-resolution bivariate flow karyotyping. *Am. J. Hum. Genet.* **45,** 739–752.
6. Langois, R. G., Yu, L. C., Gray, L. W., and Carrano, A. V. (1982) Quantitative karyotyping of human chromosomes by dual beam flow cytometry. *Proc. Natl. Acad. Sci USA* **79,** 7876–7880.
7. Robinson, J. A., Buckton, K. E., Spowart, G., Newton, M., Jacobs, P. A., Evans, H. J., and Hill, R. (1976) The segregation of human chromosome polymorphisms. *Am. J. Hum. Genet.* **40,** 113–121.
8. Trask, B., van den Engh, G., and Gray, J. W (1989) Inheritance of chromosome heteromorphisms analized by high-resolution, bivariate flow karyotyping. *Am. J. Hum. Genet.* **45,** 753–760.
9. Krumlauf, R., Jeanpierre, M., and Young, B. D. (1982) Construction and characterization of genomic libraries from specific human chromosomes. *Proc. Natl. Acad. Sci. USA* **22,** 2971–2975.
10. Nizetic, D., Zehetner, G., Monaco, A. P., Young, B. D., and Lehrach, H. (1991) Construction, arraying and high density display of large insert libraries of the human chromosomes X and 21: their potential use as reference libraries of these chromosomes. *Proc. Natl. Acad. Sci. USA* **88,** 3233–3238.
11. Nizetic, D., Monard, S., Young, B., Cotter, F., Zehetner, G., and Lehrach, H. (1994) Construction of cosmic libraries from flow-sorted human chromosomes 1, 6, 7, 11, 13, and 18 for reference library resources. *Mamm. Genome* **5,** 801,802.
12. Saiki, R. K., Gelfand, D. H., Stoffel, S., Scharf, S. J., Higuchi, R., Horn, G. T., Mullis, K. B., and Erlich, H. A. (1988) Primer-directed enzymatic amplification of DNA with a thermostable DNA polymerase. *Science* **239,** 487–491.

13. Nelson, D. L., Ledbetter, S. A., Corbo, L., Victoria, M. F., Ramirez-Solis, R., Webster, T. D., Ledbetter, D. H., and Caskey, C. T . (1989) Alu polmerase chain reaction: a method for rapid isolation of human-specific sequences from complex DNA sources. *Proc. Natl. Acad. Sci. USA* **86,** 6686–6690.
14. Telinius, H., Pelmear, A., Tunnacliffe, A., Carter, N., Behmet, A., Ferguson-Smith, M. E., Nordenskjold, M., Pfragner, R., and Ponder, B. (1992) Cytogenetic analysis by chromosome painting using DOP-PCR amplifies flow sorted chromosomes. *Genes, Chromosomes Cancer* **4,** 257–263.
15. Cotter, F. E., Das, S., Douek, E., Carter, N., and Young, B. D. (1991) The generation of DNA probes to chromosome 11q23 by Alu PCR on small numbers of flow sorted 22q-derivative chromosomes. *Genomics* **2,** 473–480.
16. Suijkerbuijk, R., Matthopoulos, D., Kearney, L., Monard, S., Dhut, S., Cotter, F., Herbergs, J., Gerts van Kessel, A., and Young, B. (1992) Fluorescent *in situ* identification of human marker chromosomes using flow sorting and Alu element-mediated PCR. *Genomics* **13,** 355–362.
17. Langford, C. F., Telinius, H., Carter, N. P., Miller, N. G. A., and Tucker, E. M. (1992) Chromosome painting using chromosome specific probes from flow sorted pig chromosomes. *Cytogenet. Cell Genet.* **61,** 221–223.
18. Rabbitts, P., Impey, H., Heppel-Parton, A., Langford, C., Tease, C., Lowe, N., Bailey, D., Ferguson-Smith, M., and Carter, N. (1995) Chromosome specific paints from a high resolution flow karyotype of the mouse. *Nat. Genet.* **9,** 369–375.
19. Hoebee, B., de Stoppelaar, J. M., Suijkerbuik, R. F., and Monard, S. (1993) Isolation of rat chromosome-specific paint probes by bivariate flow sorting followed by degenerate oligonucleotide primed-PCR. Cytogenet. *Cell Genet.* **66,** 277–282.
20. Hopman, A. H. N., Ramaekers, F. C. S., and Vooijs, G. P. (1990) Interphase cytogenetics of solid tumours, in *In Situ Hybridisation: Principles and Practice* (Polak, J. M. eds.), Oxford University Press, Oxford, UK.
21. Cremer, T., Lichter, P., Borden, J., Ward, D. C., and Manuelidis, L. (1988) The detection of chromosome aberrations in metaphase and interphase tumour cells by *in situ* hybridisation using chromosome specific library probes. *Hum. Genetics* **80,** 235–246.
22. Sillar, R. and Young, B. (1981) A new method for the preparation of metaphase chromosomes for flow analysis. *J. Histochem. Cytochem.* **29,** 74–78.
23. van den Engh, G. J., Trask, B., Gray, J. W., Langlois, R. G., and Yu. L. C. (1985) Preparation and bivariate analysis of suspensions of human chromosomes. *Cytometry* **6,** 92–100.
24. Sacchi, N., Magnani, I., Fuhrman-Conti, A. M., Monard, S. P., and Darfler, M. (1996) A stable marker chromosome with a cryptic centromere: evidence for centromeric sequences associated with an inverted duplication. *Cytogenet. Cell Genet.* **73,** 123–129.

22

Simultaneous Five-Six Color Multiparameter Analysis

Kenneth J. Pennline

1. Introduction

The development of simultaneous multiparameter analysis protocols has facilitated the use of flow cytometry in many diverse and complex research and clinical programs. This effort has been advanced by the availability of a wide range of fluorochrome-conjugated antibody reagents and the use of fluorescent particles and probes that can be employed for the analysis of numerous antigens that define cell type and function in many diverse applications. These include clinical diagnoses and monitoring of disease states ***(1,2)***, bone marrow progenitor cell isolation ***(3)***, ion flux measurements ***(4)***, nucleic acid quantitation ***(5,6)***, cell migration ***(7)***, metabolic activation ***(8)*** and routine phenotyping ***(9)***. In addition, technical advances in flow cytometric instrumentation have allowed investigators to perform more complex experiments requiring the acquisition of an increasing number of parameters. Correlation of the multiparametric data obtained from these studies significantly increases the accuracy in identifying and defining selected subpopulations in a hetergeneous cell suspension. On the other hand, multicolor immunofluorescence may also be advantageous when cell numbers are limited. Typically, peripheral blood or lymphoid tissues provide a high cell number with which multiparametric determinations can be obtained easily when stained for two-, three-, four-, or five-color analysis. However, extensive flow cytometric analysis of samples with relatively few cells (i.e., fine needle aspirates, lung lavages) would necessitate simultaneous multiparametric determinations consisting of five- and possibly six-color immunofluorescence.

The development of new and improved fluorochromes has facilitated many of the advances in multicolor fluorescent analyses ***(10–14)***. These fluorescent

From: *Methods in Molecular Biology, Vol. 91: Flow Cytometry Protocols*
Edited by: M. J. Jaroszeski and R. Heller © Humana Press Inc., Totowa, NJ

reagents exhibit a variety of excitation and emission maxima that allows the selection of a combination of fluorochromes to complement the optical configuration of individual flow cytometers for the purpose of multicolor flow cytometric analysis. A new tandem reagent, ALLO-7 (APC-CY7), was recently described for use in multicolor flow cytometry ***(15)*** and will be discussed in this chapter. In addition, reports have shown that fluorescent spheres or particles offer a versatile alternative to the more conventional fluorochrome–antibody or fluorochrome–ligand/receptor systems for use in flow cytometric analysis ***(16)***. These particles, which display a unique range of excitation and emission wavelengths, can spontaneously adsorb specific monoclonal antibodies (MAbs) and subsequently are used in combination with other fluorescent reagents such as direct antibody conjugates and biotin/streptavidin systems. Finally, the selection of specific fluorescent reagents that can be used in simultaneous multicolor flow cytometry is limited by several factors, including the excitation wavelengths available, the number and type of lasers, optical filters, and the number and configuration of optical detectors.

Recently, my laboratory described a new tandem dye ***(15)***, ALLO-7 (excitation: 647 nm; emission: 780 nm), for use in multicolor flow cytometry. Although this new fluorochrome is not presently commercially available, it is appropriate to describe its use and versatility in a five-color protocol. Materials and methods are described below for five-color analysis using standard reagents on mouse cells ***(1)*** or using standard reagents and ALLO-7 on human leukocytes ***(2)***. We also describe a protocol that combines the use of standard flow cytometry reagents with fluorescent spheres (fluorospheres) possessing unique emission wavelengths. All of these analyses were performed on a FACS Vantage. The latter protocol required minor instrument modifications that would not be necessary on flow cytometers equipped with cross-laser compensation.

2. Materials

2.1. Five-Color Analysis

1. The MAbs L3T4–PE (CD4) and THY1.2 (CD90)-biotin are available from Becton Dickinson Immunocytometry Systems (San Jose, CA); I-Ad (class II)-FITC, LYT2 (CD8)-APC, and Fc-block can be purchased from Pharmingen (San Diego, CA), and the B220 (CD45R)-RED613 can be obtained from Gibco-BRL (Grand Island, NY). Streptavidin-Cascade Blue (CB) is available from Molecular Probes (Eugene, OR).
2. The MAbs CD4, CD8-biotin, CD3-FITC, and CD14-PE can be obtained from Becton Dickinson Immunocytometry Systems (BDIS, San Jose, CA); CD19-ECD is available from Coulter (Hialeah, FL); and goat-antimouse-APC (GaM-APC) can be purchased from Caltag (San Francisco, CA). The CY7 can be supplied by Amersham Life Sciences (Pittsburgh, PA), and the streptavidin-APC-CY7

(SA-ALLO-7) tandem reagent described here was conjugated by Chromaprobe (Mountain View, CA).

2.2. Six-Color Analysis: Fluorospheres and Fluorescent Reagents

1. Sky Blue fluorospheres (2.08-μm diameter) can be purchased from Spherotech (Libertyville, IL).
2. The MAbs L3T4(CD4)-PE and THY1.2(CD90)-Biotin can be obtained from Becton Dickinson Immunocytometry Systems.
3. The IAd-FITC, LYT2(CD8a)-APC and Fc-Block can be purchased from Pharmingen.
4. A source for B220(CD45R)-RED613 is Gibco-BRL.
5. The purified CD11b can be purchased from Devaron (Dayton, NJ).
6. Streptavidin-CB can be obtained from Molecular Probes.

3. Methods

3.1. Five-Color Analysis

3.1.1. Cell Preparations

1. Sacrifice normal male Balb/C mice (Taconic Farms, Germantown, NY) with CO_2 and remove their spleens. Prepare single-cell suspensions by teasing the tissue apart with forceps in 2 mL phosphate-buffered saline (PBS) and filtering the cell mixture through gauze. Wash the cells three times in cold PBS, enumerate using a Coulter counter, and then resuspend in PBS containing 2% fetal calf serum and 0.1% sodium azide (PBS-FCS) at a concentration of 1×10^7 cells/mL.
2. Obtain peripheral blood from healthy donors by venupuncture and collect into vacutainer tubes containing EDTA. Aliquot the blood into polypropylene tubes (100 μL/tube) prior to staining with appropriate MAbs for five-color immunofluorescence.

3.1.2. Immunofluorescence Staining

1. Aliquot cells (1×10^6/100 μL) into wells of a 96-well polypropylene plate, centrifuge (200*g*, 5 min, 4°C) and discard the supernatant.
2. Resuspend the cell pellets in 50 μL PBS-FCS and incubate with 5 μL Fc-block (0.1 mg/mL) per well for 2 min at room temperature.
3. Add the following MAbs to appropriate wells as single color samples or in combination for a five-color sample: I-Ad-FITC (10 μL of 1:10 dilution), L3T4(CD4)-PE (4 μL), THY1.2(CD90)-biotin (4 μL), LYT2(CD8a)-APC (10 μL) and B220(CD45R)-RED613 (10 μL of 1:25 dilution). Include isotypic control antibodies or streptavidin-cascade blue as nonspecific binding controls. Equalize all volumes with PBS-FCS and incubate for 30 min at 4°C. Wash cells three times with PBS-FCS (200 μL/well) and resuspend in streptavidin-Cascade blue (10 μL of 1:10 dilution) or PBS-FCS (10 μL). Following an incubation for 30 min at 4°C, wash the cells three times with PBS-FCS, lyse RBCs with FACS

lysing solution (150 µL/well, 2 min at room temperature), wash twice in PBS-FCS and fix in 1% paraformaldehyde (200 µL).

4. Add the following antibodies as single-color controls or five-color samples: CD4 (20 µL)/GaM-APC (10 µL), CD3-FITC (20 µL), CD14-PE (20 µL), CD19-ECD (10 µL), or CD8–biotin (20 µL) or HLAABC-biotin (5 µL). For multicolor analysis, incubate the cells with CD4 for 20 min at 4°C, wash twice in 2 mL PBS-FCS and then stain with GaM-APC for an additional 20 min. After this time, wash the cells and incubate with CD3-FITC, CD14-PE, CD19-ECD, and CD8-biotin.
5. Stain the cells with SA-ALLO-7 for 20 min and then treat with FACS lysing solution to remove RBC prior to fixing in 1% paraformaldehyde (500 µL/tube). Include isotypic control antibodies and SA-ALLO-7 alone nonspecific binding controls.

3.1.3. Instrument Configuration

1. Complex multiparameter analysis can only be performed on a multilaser flow cytometer equipped with the proper optics. In data shown here cells were analyzed for fluorescence using a dual laser, triple-beam FACS Vantage flow cytometer (Becton Dickinson) equipped with a Coherent dual-beam Enterprise laser (300 mW at 488 nm, 60 mW at 351–364 nm), a Coherent Spectrum laser (tuned to 30 mW at 647 nm), and a seventh detector option for five-color fluorescence measurements. For this configuration, regulate the Enterprise laser in 488 nm light output mode at 300 mW, attenuate with a 30% transmittance neutral density filter, to give 90 mW output with a simultaneous UV emission at 50 mW. The 488 nm line from the Enterprise laser was the primary beam (set threshold trigger set on forward scatter) and configues the 351–364-nm beam (from the Enterprise) and the 647-nm beam (from the Spectrum) coincident as secondary beams with a 15 µs time delay. **Figure 1** shows the optical configuration that can be used for five-color analysis employing the antibodies and fluorochromes listed above. As shown in the diagram, the optics are configured so that the primary laser fluorescence signals are separated using a 610 short-pass dichroic (SP Di) and a 560 SP Di and then collected with a 630/22 bandpass (BP) (FL3 RED613), a 575/26 BP (FL2 PE), and a 530/30 BP (FL1 FITC). The secondary laser fluorescence signals are separated with a 640 long-pass dichroic and can be collected with a 424/44 BP (FL4 CB) and a 670/14 BP (FL5 APC). With this configuration, and employing appropriate photomultiplier tube (PMT) and fluorescence compensation settings, simultaneous five-color analysis can be performed to generate data shown in **Fig. 2.** All fluorescent signals are resolved from each other with virtually no spectral overlap. **Table 1** compares the percentages of cells obtained with single or five-color analysis and illustrates the consistency of measurements in this multiparameter protocol.
2. The laser configuration is the same as above. Separate the primary laser fluorescence signals with a 610 SP Di and a 560 SP Di and collected with a 630/22 BP (FL3 ECD), a 575/26 BP (FL2 PE) and a 530/30 BP (FL1 FITC). Separate the fluorescence signals generated from the second laser with a 730 SP Di and collect with a 790/50 BP (FL4 ALLO-7) and a 670/14 BP (FL5 APC). **Figure 3**

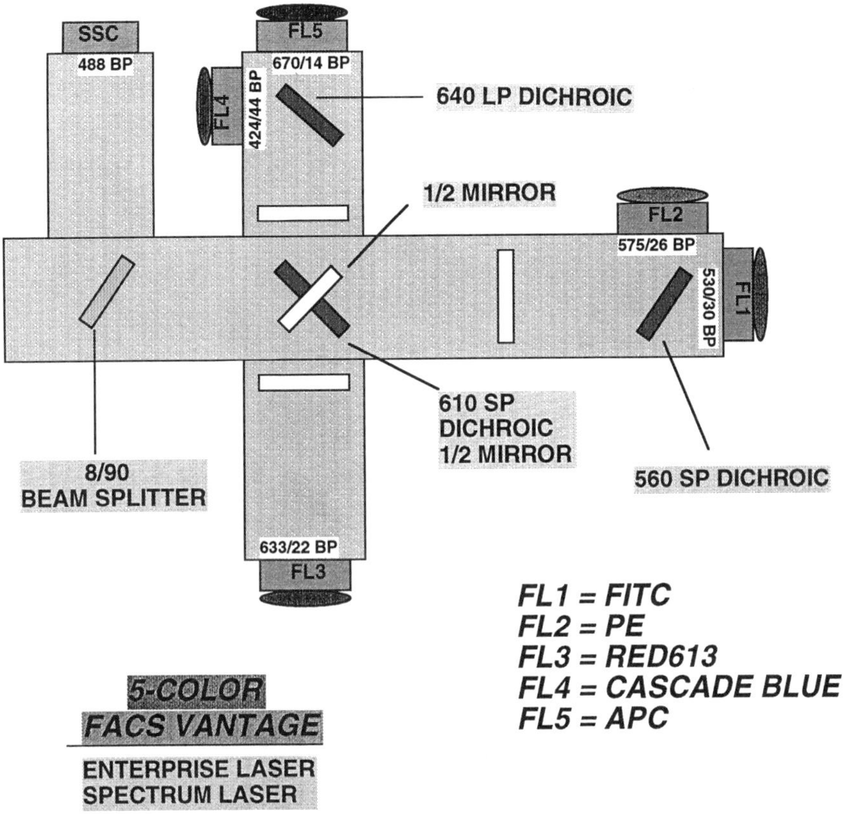

Fig. 1. Optical configuration of the FACS Vantage equipped with a Spectrum and Enterprise Laser. This figure gives a diagramatic representation of the position and type of band pass (BP), short pass (SP), 1/2 mirrors, and beam splitter optics that were used to perform simultaneous five-color analysis with the following fluorochromes: fluoroscein isothiocyanate (FITC), FL1; phycoerythrin (PE), FL2; RED613, FL3; cascade blue (CB), FL4; allophycocyanin (APC), FL5. (From Beavis and Pennline[1996] *Biotechniques* **21,** 498–503.)

clearly illustrates that this optical configuration can be successfully used to resolve the fluorescence of ALLO-7 from four other fluorescent signals.

3.2. Six-Color Analysis

3.2.1. Conjugation of the Fluorospheres with MAbs

Monoclonal antibodies can be conjugated to the fluorospheres by passive adsorption. Briefly, add 300 μL of 2.08 μm Sky Blue fluorospheres (1% w/v) to 100 μL of CD11b (1 mg/mL) and 1.6 mL phosphate buffer (0.1 *M*, pH 7.0).

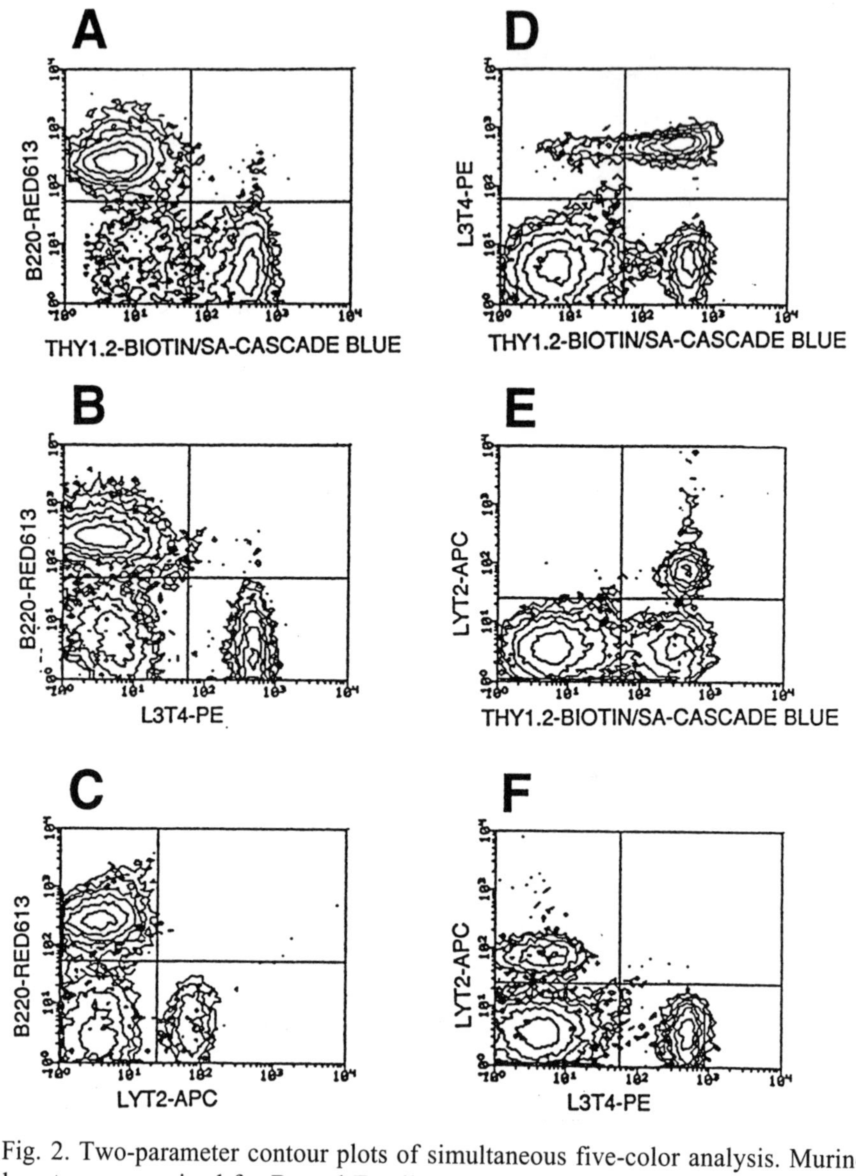

Fig. 2. Two-parameter contour plots of simultaneous five-color analysis. Murine leukocytes were stained for B- and T-cell markers as well as the activation antigen IAd. Cells were gated on forward and side scatter (not shown) and fluorescence was collected on a four-decade log scale. **(A)** Thy1.2-biotin/streptavidin-CB and B220-RED613; **(B)** L3T4-PE and B220-RED613; **(C)** LYT2-APC and B220-RED613; **(D)** ThY1.2-biotin/streptavidin-CB; **(E)** ThY1.2 biotin/streptavidin-CB; and **(F)** L3T4-PE and LYT2-APC. Quadrants were set according to fluorescent profiles of isotypic controls or CB alone (data not shown) (from ref. *9*).

Table 1
Comparison of Cells Positive for the Expression of Cell Surface Markers Determined from Single- and Five-Color Immunofluorescence Samples

		% Positive	
MAb	CD	Single-color	Five-color
THY1.2	CD90	33.2	32.7
L3T4	CD4	24.5	23.3
LYT2	CD8a	11.9	11.6
B220	CD45R	56.6	58.0
I-Ad	—	56.9	58.4

Vortex the fluorosphere-antibody mixtures and incubate for 60 min at room temperature. After centrifugation at 3000*g* for 15 min, wash the ACF twice in 4 mL of PBS and incubate as above in 2 mL PBS containing 10% FBS. After this time, wash the AFC twice in PBS, resuspend in 4 mL PBS (0.25% w/v) solution, and store at 4°C in the dark until use.

3.2.2. Six-Color Immunofluorescence

The Sky Blue ACF can be used for the simultaneous detection of six cell surface antigens by six-color immunofluorescence. Stain cells with the following MAbs as single color controls and as a six-color sample: IAd-FITC (10 μL of 1:10 dilution), L3T4(CD4)-PE (4 μL), THY1.2(CD90)-Biotin (4 μL), LYT2(CD8)-APC (10 μL), and B220(CD45R)-RED613 (10 μL of a 1:25 dilution). After all volumes are equalized with PBS-FCS, incubate the cells for 30 min at 4°C. After this time, wash the cells, incubate with streptavidin-CB (10 μL of a 1:10 dilution) and CD11b-Sky Blue (50 μL) for 30 min, wash again, and fix in 1% paraformaldehyde. Include isotypic control antibodies, streptavidin-CB or serum-adsorped Sky Blue particles as nonspecific binding controls.

3.2.3. Instrument Settings

For six-color immunofluorescence analysis the Sky Blue fluorescence signal (FL6) is collected instead of side scatter (*see* **Note 5**). Separate the primary laser fluorescence signals using a 610 SPDi and a 560 SPDi and collect with a 630/22 BP (FL3 RED613), a 575/26 BP (FL2 PE), and a 530/30 BP (FL1 FITC). Separate the secondary laser fluorescence signals with a 640 LPDi and collect with a 424/44 BP (FL4 CB) and a 670/14 BP (FL5 APC). A diagram of this optical configuration on a FACS Vantage is shown in **Fig. 4**. All

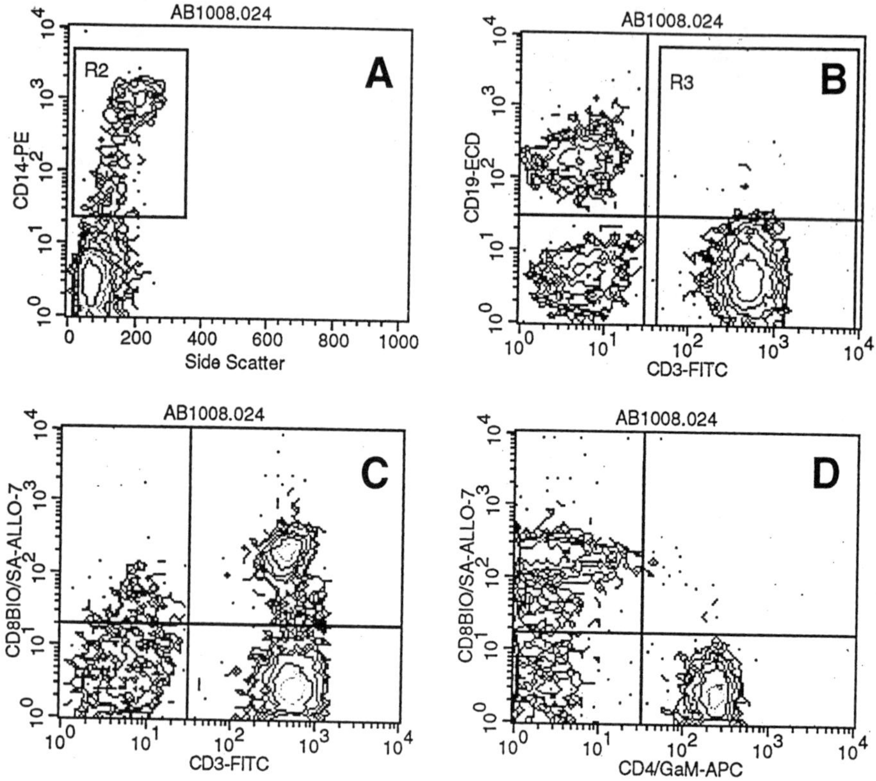

Fig. 3. Two-parameter contour plots of simultaneous five-color analysis utilizing ALL0–7 with standard fluorochromes. Human peripheral blood leukoctes were stained with monoclonal antibodies to identify T-cells, B-cells, and macrophages with the following reagents: CD3-FITC; CD14-PE; CD19-ECD; CD4/GaM-APC, and CD8-biotin/streptavidin-ALL0-7. Monocytes (CD14+) were identified in the R2 gate **(A)** and excluded from the analysis of B- and T-cells shown in **(B–D)**. Analysis shown in **(D)** was restricted to CD3+ T-cells in the R3 gate that was defined in **(B)**. All data was contored in log density starting At 50% of the peak height (from ref. *15*).

fluorescence signals should be collected in logarithmic mode with pulse processing on. Using this configuration, with the proper PMT and compensation settings, six-color analysis can be performed with sufficient resolution of fluorescent signals to quantitate all six parameters (**Table 2**).

4. Notes

1. The predominant excitation wavelength on standard flow cytometers is the 488-nm beam. It is primarily used for FITC (fluorescein isothiocyanate) and PE (phycoerythrin), fluorochromes that emit at wavelengths that are clearly resolved spec-

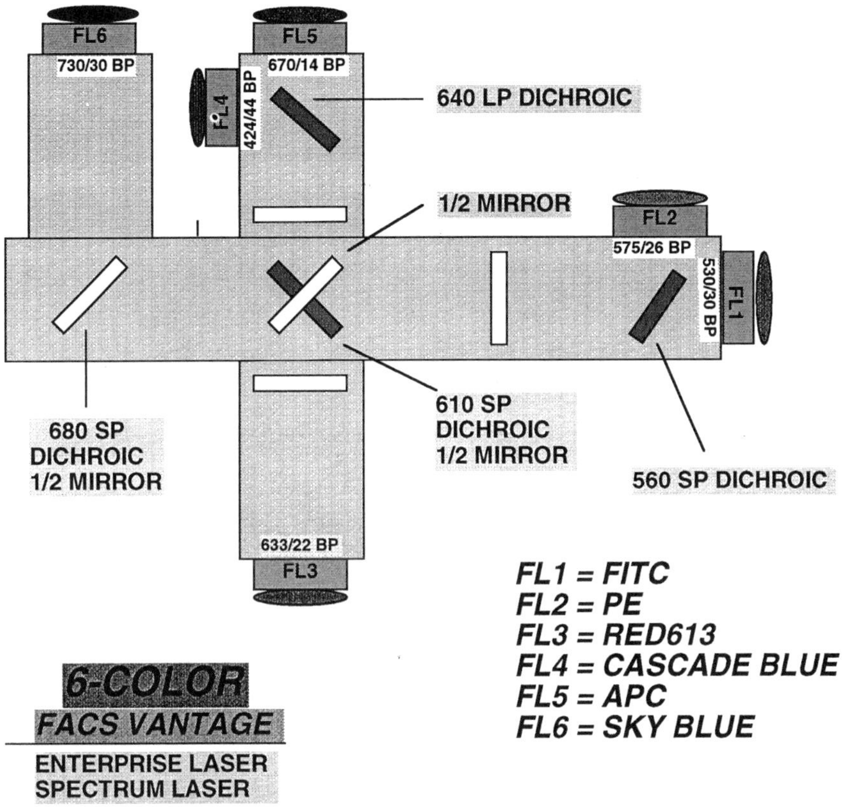

Fig. 4. Customized optical configuration on a FACS Vantage equipped with a Spectrum and Enterprise laser. This figure gives a diagramatic representation of the position and type of band pass (BP), short pass (SP), 1/2 mirrors, and beam splitter optics to perform simultaneous six-color analysis with the following fluorochromes: FITC (FL1), PE (FL2), RED613 (FL3), CB (FL4), APC (FL5), and SKY BLUE (FL6). Side-scatter optics (8/90 beam splitter; and 488 BP) were replaced with a 680 SPDi (1/2 mirror) and 730/30 BP in order to acquire the FL6 fluorescence.

trally and which are available in a variety of configurations, easily facilitating their use in two-color immunofluorescence. New fluorochrome tandem dyes (Duochrome, Tricolor, Cychrome, PERCP, and so on) developed for single-laser systems in order to facilitate three-color analysis utilize the excitation of one dye at 488 nm and the red emission of another to produce a large Stokes shift and energy transfer to emit at a higher wavelength. Care must be taken with these dyes in multicolor fluorescence on instruments with more than one light source since they will also be excited by UV and 647-nm laser lines resulting in spectral overlap in the FL5 channel. Even slight spectral overlap can be problematic in

Table 2
Comparison of Cells Positive for the Expression of Cell Surface Antigens as Determined by Single-, Five-, and Six-Color Immunofluorescence

	% Cells positive		
Phenotype	Single-color	Five-color	Six-color
IAd (Class II)	50.0	53.6	51.9
L3T4 (CD4)	23.2	24.0	23.7
B220 (CD45R)	47.1	51.6	51.6
THY1.2 (CD90)	36.7	33.8	34.1
LYT2 (CD8a)	14.3	14.8	15.2
MAC1 (CD11b)	6.3	—	6.6

cases in which dim fluorescence, associated with low antigen density, is being measured on cells.

2. In the first five-color protocol (on murine cells) described above, FITC, PE, and RED613 were selected as fluorochomes to be excited by the primary 488-nm laser line. RED613 is a tandem dye of PE and Texas red ***(17)*** with excitation at 488 nm and red emission at 613 nm. This dye was chosen as the third primary laser fluorochrome because it was not excited by the UV or 647 nm laser lines. The only caveat here was the need for 50% compensation (FL2-%FL3) in order to resolve the considerable emission of fluorescence in the PE channel. This could also be problematic when measuring dim fluorescence and should be taken into account when selecting the fluorochromes to detect specific markers. Allophycocyanin (APC) produced the best fluorescent signal generated by the 647-nm laser line in this protocol. This dye can also be excited by the 633-nm line on instruments configured with HeNe laser yielding a consistent emission at 670 nm. There are numerous direct APC-conjugated reagents available that provide flexibility in choosing the right combination of antibodies and dyes. Cy5, a low molecular weight carboxymethylindocyanine derivative, is also excited by the 647- and 633-nm laser lines and could be used as a possible alternative to APC, although it has less emission intensity ***(18)***. For the last laser line, the ultraviolet (UV) light line (excitation <400 nm), CB was the dye chosen and displayed emission at 410–430 nm ***(19)***. This fluorochrome was only available in the form of a streptavidin (SA) conjugate and configured well with the four direct antibody-fluorochrome conjugates. Another UV dye, SA-7-amino-4-methyl coumarin-3-acetic acid (AMCA), was evaluated, and in our hands it did not produce the spectral resolution obtained with CB, but does provide a possible alternative. In general, background fluorescent signals are higher for UV-excited dyes, therefore, it is advantageous to employ an SA conjugate to amplify the signal and provide better resolution between negative and positive fluorescence.

3. In the second five-color protocol (on human cells) described here, the fluorochromes excited by the primary 488-nm laser line were the same and included FITC, PE, and ECD (energy coupled dye, Coulter Red613). The fluorochromes selected for the secondary 647-nm laser line were APC and ALLO-7 (APC-CY7 tandem dye). The ALLO-7 dye was configured on the basis of a sufficient overlap of APC emission and CY7 absorbance ***(15)***. CY7 (indotricarbocyanine) fluoresces in the deep red region between 760- and 850 nm and is unsuitable for use on standard flow cytometers because it requires a xenon or tungsten lamp for excitation at 744 nm. However, this tandem dye exploits the APC excitation/emission and through Stokes shift and energy transfer produces a fluorescent signal >760 nm. This unique dye has the potential to increase the versatility of standard flow cytometers by providing the capability for dual-color immunofluorescence with a 647-nm laser line (or HeNe at 633 nm) or three-color analysis with a dye-head laser, neither of which was possible before the development of ALLO-7. Here the ALLO-7 was excited by the 647-nm line from the Spectrum laser resulting in bright fluorescence emission that was collected at approx 780 nm. The ALLO-7 emission was separated from that of APC alone with a 730SP Di. There are advantages associated with the use of red-excited fluorochromes. Flavoproteins exhibit high levels of autofluorescence ***(20,21)*** because of absorbance at lower wavelengths that result in reduced signal-to-noise ratios and potential loss of positive fluorescence resolution above background. These problems are circumvented when red excitation beams from krypton (647 nm) or HeNe (633 nm) lasers are used with APC or CY5. In addition, red-excited dyes exhibit minimal spectral overlap with standard fluorochromes, thereby facilitating their use in multicolor, multiparametric analyses.
4. This five-color protocol illustrated the use of ALLO-7 in combination with FITC, PE, ECD, and APC. As shown in **Fig. 3**, the fluorescence emission of ALLO-7 (FL4) was clearly resolved from that of APC (FL5) and from those fluorescent signals generated by the others (FL1, FL2, FL3). The minimal spectral overlap between APC and ALLO-7 was easily corrected with electronic compensation (FL4-%FL 5 = 8% and FL5-%FL4 = 25%). In the analysis shown in **Fig. 3**, a lymphocyte region was defined from forward and slide light-scatter signals (not shown) by positive exclusion of the CD14(+) cells (PE fluorescence). The major lymphocyte populations, B- and T-cells, were identified by positive fluorescence for CD19 (ECD fluorescence) and CD3 (FITC florescence), respectively. A gate was established for CD3(+) cells from which the anlayses for the T-cell subsets CD8 (ALLO-7 fluorescence) and CD4 (APC fluorescence) were obtained.
5. The six-color protocol described here on murine cells utilizes standard fluorochrome dyes as well as antibody-conjugated Sky Blue fluorescent particles. In order to perform this protocol on a standard dual (or triple) laser beam instrument it is necessary to modify the optical configuration. In this case, the SKY BLUE signal was collected instead of side scatter by replacing the 8/90 beam splitter, 488/10 BP and the SSC PMT IP28A with a 680 SPDi (1/2 mirror), 730/30 BP (FL6), and a red-sensitive PMT, respectively. These modifications were

necessary since on standard instrumentation, where selection of reagents is limited, spectral overlap of fluorescent signals from different laser lines can not be compensated (*see* **Notes 6** and **7**). However, the availability of cross-laser (cross-beam) compensation (Becton Dickinson) negates the need for instrument modification and increases the versatility of the cytometer with respect to number and type of fluorochromes that can be spectrally resolved in multicolor analysis.

6. Fluorospheres offer a versatile alternative to conventional fluorochrome reagents. In general, these particles are uniform in size and display a wide range of excitation/emission wavelengths that result in bright, homogeneous, fluorescent signals. Fluorospheres such as these can be used for instrument calibration *(22)*, surface antigen detection *(23)* cell migration *(24)*, and quantitation of phagocytic cell function *(25,26)*. The application illustrated here takes advantage of the ability of these fluorospheres to spontaneously adsorb MAbs to their surface. These ACF are suitable for use in a variety of flow cytometric analyses including multicolor determinations. There were two important considerations for selecting the Sky Blue fluorospheres as the sixth reagent in our multiparametric protocol. First, in the absence of cross-laser compensation, the use of any other commercially available fluorochrome conjugates would have resulted in significant spectral overlap with dyes already comprising the five-color scheme. The range of wavelength emissions for the first five fluorochromes was somewhat congested and the chance of a sixth signal from one of the commonly used reagents fitting in without considerable problems in resolution was unlikely. Second, it was required that the configuration of the sixth fluorochrome be in the form of a direct conjugate or a unique ligand/ligand indirect system since the protocol already included one two-step biotin/strepavidin system (cascade blue). Also, there was reservation for the use of a fluorochrome-conjugated secondary antibody resulting from the high probability of nonspecific binding or crossreactivity.
7. The antibody CD11b (macrophage specific) was conjugated to the Sky Blue fluorospheres in this multicolor protocol. The fluorescent signal obtained with the Sky Blue ACF was resolved from those of FITC, PE and APC. Most important was the resolution between Sky Blue and APC since both dyes are excited by the same laser line. As in the case of ALL0–7, this reagent represents an important finding since it demonstrates that the 647-nm line from the spectrum (or 633-nm line from an HeNe laser) can now be used to collect two fluorescent signals simultaneously. In this analysis, macrophages were identified by the Sky Blue ACF and then excluded by gating to allow for the analysis of T- and B-cell surface antigens on CD11b (-) negative cells alone (data shown in **Table 2**). Analysis of this data is most accurately measured on Attractors software (Becton Dickinson) in order to show the resolution of all signals in this mulitparametric scheme. The clear resolution for fluorochromes identifying the T-cell markers, including CB (Thy1.2–CD90), PE (L3T4–CD4), and APC (LYT2–CD8), as well as the B-cell markers including RED613 (B220–CD45R) and FITC (IAd-class II antigen), is supported by the similarity in percent positive cells for single-, five- and six-color analysis (**Table 2**). Using the ACF in this manner has provided a accurate

method to eliminate the macrophage cell population (and problems associated with autofluorescence) using a phenotypic gate rather than light-scatter properties alone. Equally important, the number of parameters that could be used for the analysis of leukocytes was maximized. The one disadvantage to using the Sky Blue fluorospheres is the moderate spectral overlap with FITC and CB caused by minimal excitation by 488 nm and UV lines. This problem can be minimized by using smaller microspheres (0.45 μm) that emit at a lower fluorescent intensity (Beavis and Pennline, unpublished data). This procedure shows that using a combination of conventional fluorochrome-conjugated antibodies and ACF six different cell surface antigens on murine leukocytes could be analyzed simultaneously without the benefit of cross-laser compensation. The use of fluorescent microspheres in flow cytometric analysis is attractive in multiparametric protocols because of the uniqueness of the dyes, their emission wavelengths, and the fact that they add versatility to MAbs that are only available in purified form. Instrumentation that has cross-laser compensation may be able to take full advantage of these unique reagents and in the process conserve the side scatter parameter of the analysis.

Acknowledgment

I would like to acknowledge the excellent work of Andrew J. Beavis. He was responsible for planning and performing all of the experimentation that is discussed in this chapter, in particular the ALL0–7 work. It was his initial idea to construct the tandem dye and was solely responsible for contacting the companies involved, having the tandem made and testing the utility of the tandem in multiparameter analysis. His contributions in the area of multicolor immunofluorescence are invaluable to the field of flow cytometry.

References

1. Gale, H. B. and Henry, K. (1992) Measuring percent lymphocytes by flow cytometry to calculate absolute lymphocyte subset counts for HIV+ specimens. *Cytometry* **13,** 175–181.
2. Ichikawa, Y., Shimizu, H., Yoshida, M., and Arimori, S. (1990) Activation of T-cell subsets in the peripheral blood of patients with Sjogren's syndrome. Multicolor flow cytometric analysis. *Arthritis and Rheumatism* **33,** 1674–1681.
3. Huang, S. and Terstappen, L. W. (1992) Formation of haematopoietic microenvironment and haematopoietic stem cells from single human bone marrow stem cells. *Nature* **360,** 745–749.
4. Rabinovitch, P. S., June, C. H., Grossman, A., and Ledbetter, J. A. (1986) Heterogeneity among T-cells in intracellular free calcium responses after mitogen stimulation with PHA or anti-CD3. Simultaneous use of INDO-1 and immunofluorescence with flow cytometry. *J. Immunol.* **137,** 952–961.
5. Traganos, F., Darzynkiewicz, Z., Sharpless, T., and Melamed, M. R. (1977) Simultaneous staining of ribonucleic and deoxyribonucleic acids in unfixed cells

using acridine orange in a flow cytofluorometric system. *J. Histochem. Cytochem.* **25,** 46–56.
6. Pennline, K. J., Pellerito-Bessette, F., Umland, S. P., Siegel, M. I., and Smith, S. R. (1992) Detection of in vivo-induced IL-1 mRNA in murine cells by flow cytrometry and fluorescent *in situ* hybridization (FISH). *Cytokine Res.* **11,** 65–71.
7. Beavis, A. J. and Pennline, K. J. (1994) Tracking of murine spleen cells in vivo: detection of PKH26–labeled cells in the pancreas of non-obese diabetic (NOD) mice. *J. Immunol. Methods* **170,** 57–65.
8. Bass, D. A., Parce, J. W., DeChatelet, L. R., Szejda, P., Seeds, M. C., and Thomas, M. (1993) Flow cytometric studies of oxidative product formation by neutrophils: a graded response to membrane stimulation. *J. Immunol.* **130,** 1910–1917.
9. Beavis, A. J. and Pennline, K. J. (1994) Simultaneous measurement of five cell surface antigens by five-colour immunofluorescence. *Cytometry* **15,** 371–376.
10. Ernst, L A., Gupta, R. K., Mujumdar, R. B., and Waggoner, A. S. (1989) Cyanine dye labeling reagents for sulfhydryl groups. *Cytometry* **10,** 3–10.
11. Minta, A., Kao, J. P. Y., and Tsien, R. Y. (1989) Fluorescent indicators for cytosolic calcium based on rhodamine and fluorescein chromophores. *J. Biol. Chem.* **264,** 8171–8178.
12. Horan, P. K., Melnicoff, M. J., Jensen, B. D., and Slezak, S. E. (1990) Fluorescent cell labeling for in vivo and in vitro cell tracking, in *Methods in Cell Biology, vol. 3, Flow Cytometry* (Darzynkiewicz, Z., and Crissman, H. A., eds.), Academic, New York, pp. 469–490.
13. Whitaker, J. E., Haugland, R. P., Ryan, D., Hewitt, P. C., Haugland, R. P., and Prendergast, F. G. (1992) Fluorescent rhodol derivatives: versatile, photostable labels and tracers. *Anal. Biochem.* **207,** 267–279.
14. Whitaker, J. E., Haugland, R. P., Moore, P. L., Hewitt, P. C., Reese, M., and Haugland, R. P. (1991) Cascade Blue derivatives: water soluble, reactive, blue emission dyes evaluated as fluorescent labels and tracers. *Anal. Biochem.* **198,** 119–130.
15. Beavis, A. J. and Pennline, K. J. (1996) ALLO-7: a new fluorescent tandem dye for use in flow cytometry. *Cytometry* **24,** 390–394.
16. Beavis, A. J. and Pennline, K. J. (1996) Detection of cell surface antigens using antibody-conjugated fluorospheres (ACF): application for six-color immunofluorescence. *Biotechniques* **21,** 498–503.
17. Glazer, A. and Stryer, L. (1983) Fluorescent tandem phycobiliprotein conjugates emission wavelength shifting by energy transfer. *Biochem. J.* **43,** 383–386.
18. Southwick, P. L., Ernst. L. A., Tauriello, E. W., Parker, S. R., Mujumdar, R. B., Mujumdar, S. R., Clever, H. A., and Waggoner, A. S. (1990) Cyanine dye labeling reagents-carboxymethylindocyanine succinimidyl esters. *Cytometry* **11,** 418–430.
19. Whitaker, J. E., Haugland, R. P., Moore, P. L., Hewitt, P. C., Reese, M., and Haugland, R.P. (1991) Cascade Blue derivatives: water soluble, reactive, blue emission dyes evaluated as fluorescent labels and tracers. *Anal. Biochem.* **198,** 119–130.
20. Aubin, J. (1979) Autofluorescence of viable cultured mammalian cells. *J. Histochem. Cytochem.* **27,** 36.

21. Benson, H. C., Meyer, R. A., and Zaruba, M. E. (1979) Cellular autofluorescence- Is it due to flavins? *J. Histochem. Cytochem.* **27,** 44.
22. Paulis, M., Robins, A., and Powell, R. (1993) Quantitative analysis of lymphocyte CD11a using standardized flow cytometry. *Scand. J. Immunol.* **38,** 559–564.
23. Cosio, F. G., Xiao-Ping, S., Birmingham, D. J., Van Aman, M., and Herbert, L. A. (1990) Evaluation of the mechanisms responsible for the reduction in erythrocyte complement receptors when immune complexes form in vivo in primates. *J. Immunol.* **145,** 4198–4206.
24. Technau, U. and Holstein, T. W. (1992) Cell sorting during the regeneration of Hydra from reaggregated cells. *Developmental Biol.* **151,** 117–127.
25. Bender, J. G., Unverzagt, K. L., Maples, P. B., Mehrotra, Y. J., Mellon, Y. J., Van Epps, D. E., and Stewart, C. C. (1992) Functional characterization of mouse granulocytes and macrophages produced in vitro from bone marrow progenitors stimulated with interleukin 3 (IL-3) or granulocyte-macrophage colony stimulating factor (GM-CSF). *Exp. Hematol.* 1135–1140.
26. Blair, O. C., Carbone, R., and Sartorelli, A. C. (1986) Differentiation of HL-60 promyelocytic leukemia cells: simultaneous determination of phagocytic activity and cell cycle distribution by flow cytometry. *Cytometry* **7,** 171–177.

Index

P

R

S

T